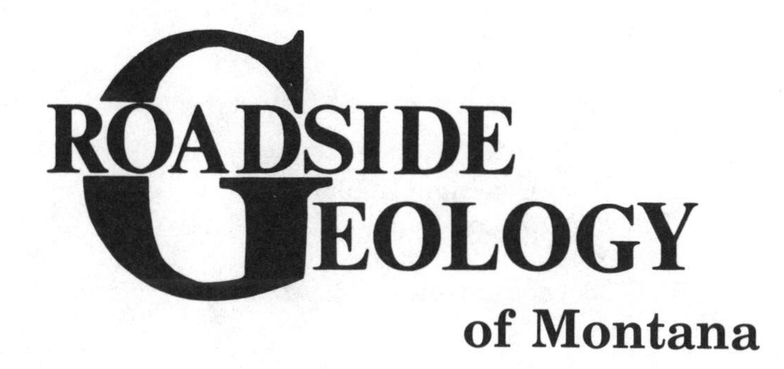

# of Montana

David Alt
Donald W. Hyndman

MOUNTAIN PRESS PUBLISHING COMPANY
MISSOULA

Seventh Printing, May 1997

Library of Congress Cataloging in Publication Data

Alt, David D.
Roadside geology of Montana.

Bibliography: p.
Includes index.
1. Geology — Montana — Guide-books. I. Hyndman, Donald W. II. Title.
QE133.A674 1986 557.86 86-17954
ISBN 0-87842-202-1 (pbk.)

Mountain Press Publishing Company
P.O. Box 2399
Missoula, MT 59806
1-800-234-5308

Preface

We never dreamed when we wrote the *Roadside Geology of the Northern Rockies* about 15 years ago that ideas about rocks would change so much so soon. But they have. Nothing looks quite the same anymore now that the results of another 15 years of research are in. So this is a completely new book that uses nothing from the old one, not so much as a paragraph or map, and hardly a photograph. We hope that people who enjoyed our earlier book will use and enjoy this one even more. And we hope the next 15 years of research will change the looks of the rocks enough to require another new book.

We wrote this book for people who are not geologists but would like to know something about the rocks and landscapes of Montana. We hope our professional colleagues will enjoy it, too.

The book consists of five large chapters: a general introductory narrative followed by chapters dealing with the four major geologic provinces in Montana. Each starts with its own more specific narrative that introduces a series of roadguides, each complete with a local geologic map. We hope that people who read the roadguide and use the geologic map will be able to identify most of the rocks they see, and know something about how they formed and what they mean.

Nearly every geologist who has ever worked in Montana contributed something to this book. We read almost everything our colleagues wrote, and packed as much of their information as we could between the covers of this book. Each of those people will recognize his or her own contribution, and we want them all to know that we recognize it, too. We thank them collectively, regretting that it isn't possible to acknowledge individual sources in a book of this type.

Several people made a special effort to help us with photographs. Hal James, Larry French, Jack Horner, and Rick Hull all supplied us with pictures, many of them taken specifically for this book. We needed their help, and thank them. Kathy Spitler and Dave Flaccus of Mountain Press devoted a great deal of time and effort to the forbidding task of artfully compressing a large manuscript into an attractive book. We are grateful for their efforts.

We compiled the geologic maps from the best available and most recent sources. As we worked on those maps and read the literature to get the information for this book, we arrived at a few new ideas of our own. Those, along with the results of some of our other research, are published here for the first time.

Missoula, Montana
July, 1986

Contents

GEOLOGIC TIME SCALE

Age	Eon	Era	Period	Epoch	Important Events in Montana
		CENOZOIC	QUATERNARY	Holocene (Recent) Pleistocene (glacial)	Pinedale ice age ended about 10,000 years ago Bull Lake ice age, 70,000-130,000 years ago Yellowstone volcano starts, 1.8 million years ago modern streams begin to flow
			2.5 MILLION YEARS AGO		
			TERTIARY	Pliocene	Sixmile Creek and Flaxville gravels
				Miocene	time of wet tropical climate
				Oligocene	Renova formation
				Eocene	mountains form in central Montana Lowland Creek volcanics Fort Union formation
			65 MILLION YEARS AGO		extinction of the dinosaurs
		MESOZOIC	CRETACEOUS		Elkhorn Mountains volcanics Boulder batholith, Idaho batholith Rocky Mountains form conglomerate, sandstone, shale
			JURASSIC TRIASSIC		shale, limestone, and sandstone — Atlantic Ocean begins to open
			210 MILLION YEARS AGO		
		PALEOZOIC	PERMIAN		sandstone and limestone
			PENNSYLVANIAN		Madison limestone
			MISSISSIPPIAN		limestones
			DEVONIAN		
			SILURIAN		no rocks
			ORDOVICIAN		
					sandstone and limestone first animal fossils
			600 MILLION YEARS AGO		
PRECAMBRIAN	PROTEROZOIC				continental rifting Belt sedimentary formations
			2,500 MILLION YEARS AGO		
	ARCHEOZOIC				most basement rock in Montana, 2,700 million years ago Stillwater layered intrusion oldest basement rock in Montana, 3,200 million years ago

MAP SYMBOLS

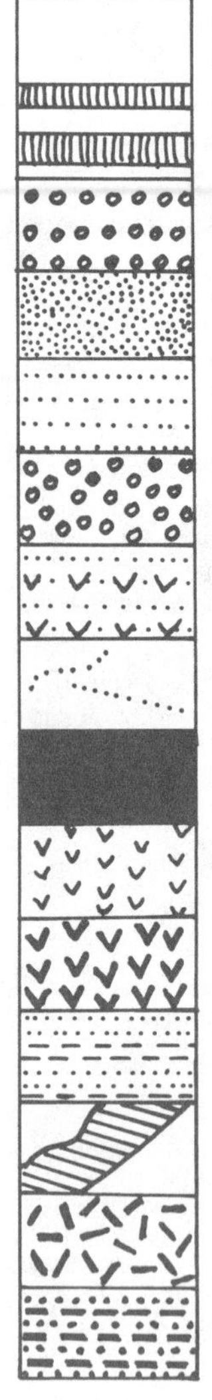

Pleistocene to Recent: sand, gravel

basalt: Tertiary, Pleistocene; Precambrian Purcell basalt

Miocene-Pliocene: Flaxville gravels; upper Tertiary gravels

Tertiary basin fill

lower Tertiary sediments: siltstone, sandstone, mudstone, mudflows; Fort Union fm.

Beaverhead conglomerate

Livingston volcanic sediments

Precambrian, Tertiary: diabase sills, dikes

Tertiary intrusions: granite, granodiorite, alkaline rocks in central Montana

volcanic rocks, mostly Eocene: Challis, Lowland Creek, Absaroka, central Montana alkaline

volcanic rocks, late Cretaceous: Elkhorn Mountains volcanics

Cretaceous sedimentary rocks: sandstone, shale

mylonite

Cretaceous granite, granodiorite

Triassic, Jurassic: sandstone, shale

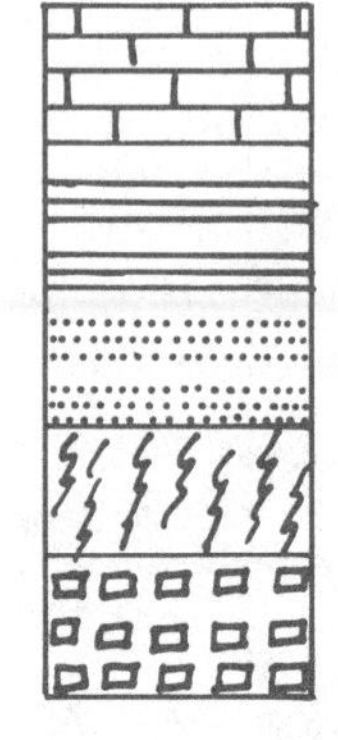

Paleozoic: Cambrian, Devonian, Mississippian, Pennsylvanian: limestone and dolomite Cambrian sandstone

Proterozoic: upper Belt sediments

Proterozoic: lower Belt sediments

Archean basement rocks: gneiss, schist, granite

Stillwater complex: peridotite, gabbro, anorthosite

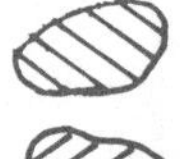

area of natural gas fields

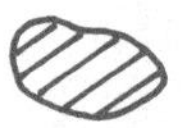

area of oil fields

glacial moraines

petrified wood area in Tertiary volcanic sediments

fault with sideways movement

fault with up and down movement: points on side that moved down

thrust fault: points on side that thrust over other rocks

I
The Big Picture

Geologists are still a long way from fitting all the rocks of Montana together into a coherent picture. Nevertheless, we can see enough of the broad outlines to be sure that when all the story of Montana's geology is finally told, it will be a tale of a part of a continent that remained low, flat and geologically quiet for a very long time, then got scrambled in a long and grinding collision with the Pacific Ocean.

The details of the story are hard to read because the geology of Montana is complex in the way that a single sheet of paper would be complex if it had several different pages printed on it, one on top of another. Each pass through the press leaves a new impression overprinted on those left from earlier passes. The problem of decipherment is a matter of sorting out the different impressions, then reading them in order. In Montana, as elsewhere in the Rocky Mountains, a long succession of geologic events created the rocks, then changed them, then changed many of them again. Much of the complicated geologic record remains to be read from the rocks.

Geologists read important parts of the geologic story of Montana primarily by applying the principles of plate tectonics, a powerful set of geologic theories developed during the 1960s and 1970s. So we will start with a few general words about our planet and the plate tectonics processes that create and change its rocky crust.

Like a golf ball, the earth has a small core in the center, a thick zone that surrounds the core, and a thin outer skin. We won't worry about the core because it seems to have little influence on events at the surface. The thick intermediate zone of the earth, the mantle, consists of heavy black rocks called peridotite that we rarely see. Although most of the mantle is

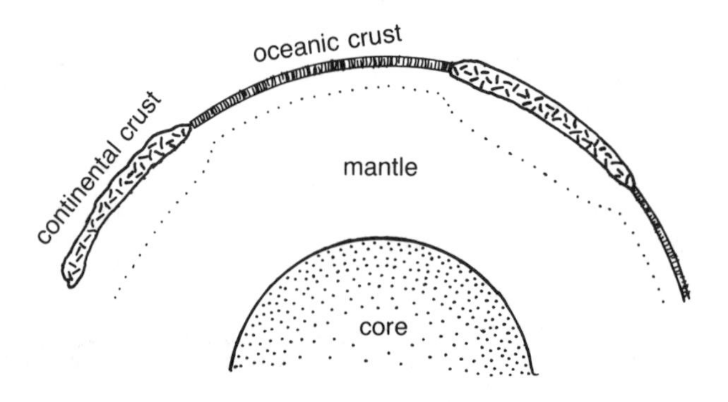

The earth as it would look if it were sawed in half.

solid, it is hot enough that the rocks within it flow fairly freely, as though they were modeling clay. The relatively cool outer 60 miles of the mantle forms a fairly rigid rind, the lithosphere. Everywhere, the lithosphere bears a thin outer skin, the crust.

Oceanic crust, which is several miles thick, consists almost entirely of basalt, the common black lava that forms when rocks in the mantle partially melt. The continental crust is about 25 miles thick in most areas. It is a raft of lighter rocks, granite mostly, embedded in the upper part of the lithosphere.

Plate tectonics begins with the idea that the earth's lithosphere consists of a mosaic of pieces called plates that fit together about like the bones in a skull. A dozen or so large plates and several small ones completely tile the earth's surface. They all seem to move in a more or less random pattern at rates in the vicnity of an inch or two per year. In some places, the moving plates pull away from each other; in other places they collide; and in still others they slide past each other. Those plate motions helped create many of the rocks and landscapes of western Montana.

Plates pull away from each other where masses of hot rock rising through the mantle break the lithosphere, and sweep the pieces aside. The broken pieces of lithosphere become different plates that move away from each other. Fissures open between them as they separate. That relieves pressure on the mantle rocks below, which are already so hot that they would melt if they were not under great pressure. The molten basalt magma that forms in the mantle rises and erupts to form lava flows. The process repeats itself as the plates continue to separate and

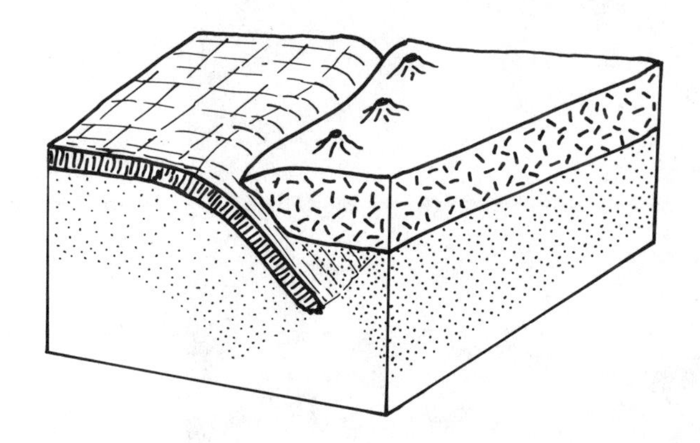

Oceanic crust sinking into an oceanic trench, causing melting, and supplying a chain of volcanoes at the surface with molten magma.

the lava flows form new oceanic crust, an outer skin of basalt on the lithosphere.

Where two plates collide, the heavier one sinks beneath the other into the mantle. The granitic rocks of the continental crust are too light to sink into the mantle, so it is the plate with oceanic crust on its surface that sinks. A deep oceanic trench forms on the surface of the sinking plate. As the sinking plate descends, it heats up, and finally blends into the rest of the mantle. Lithosphere is, after all, merely the cold outer rind on the mantle, so it loses its identity as it gets hot.

As the sinking plate reaches a depth of about 60 miles, it gets so hot that the rocks within it begin to lose enormous volumes of red hot steam at a temperature of about 1200 degrees centigrade. Steam so hot that it glows in the dark where we see it blowing out of a volcano causes melting in the mantle rocks of the overlying plate. In most cases, the now molten basalt erupts to form a chain of volcanoes parallel to the trench. In other cases, including the one that involves Montana, the molten basalt plays like a giant blowtorch on the overlying continental crust causing melting to form granite magma. This magma crystallizes before it reaches the surface to form large granite batholiths.

BASEMENT ROCK, THE CONTINENTAL CRUST

A continent is basically a thick slab of relatively light rock that floats on the much denser rock of the mantle, just as an air

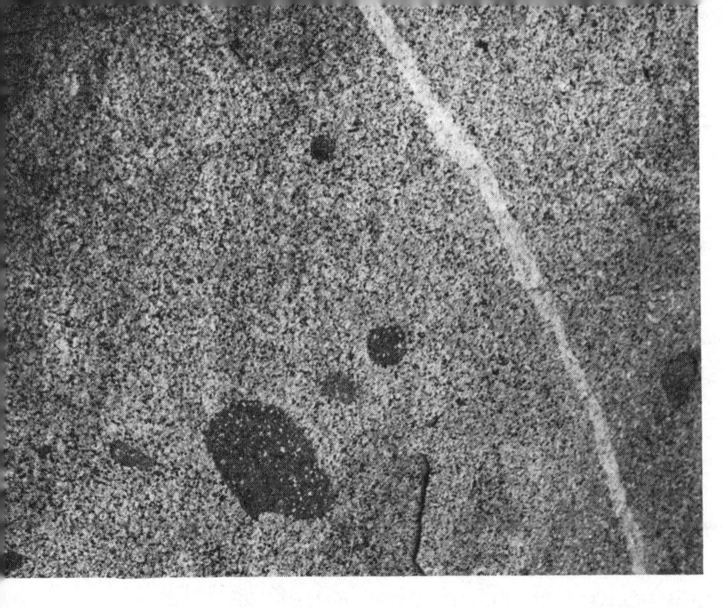

An outcrop of granite cut by a light-colored pegmatite dike and containing dark inclusions.

mattress floats on a lake. Its floating keeps continental crust permanently at the earth's surface, where it retains a lasting record of the geologic events that formed and later changed it. Continents are the keepers of the planetary archives. Part of those records exist in the crustal rocks themselves, others in the younger sedimentary rocks that cover them in most areas. So we will start our story with those rocks that lie beneath all the others, the ancient basement rocks of the continental crust.

The most abundant crustal rock is granite, an igneous rock formed through crystallization of molten magma. Other abundant crustal rocks include gneiss and schist, metamorphic rocks that form as older volcanic or sedimentary rocks recrystallize at very high temperature. All granites and most schists and gneisses consist largely of the same basic minerals: feldspar, quartz, mica, and hornblende, and differ mostly in the arrangement of their mineral grains. The chemical compositions of granite, gneiss, and schist fall within the same general range as the chemical compositions of ordinary mud—the same stuff in a different package.

Geologists often call the granites, gneisses, and schists of the continental crust "basement rock," because they seem to continue down to indefinite depth without any surface indication that anything may lie beneath them. If you drill deep enough almost anywhere on the continent, the hole will eventually penetrate basement rock. Echoes of reflected

Gneiss and schist, among the commonest rocks of the continental crust. Gneiss and schist have a streaky or flaky appearance because their mineral grains are lined up parallel to each other. The mineral grains in granite are not normally aligned.

earthquake waves show that the basement rocks probably do continue to the base of the continental crust, where they lie on the heavier black peridotite of the mantle. No one has ever drilled that deep.

Granites are pale gray or pink rocks; gneisses and schists come in many colors, most commonly gray, pale pink, and dark brown. Granite is typically a massive rock nearly devoid of internal structure, whereas gneisses and schists are banded and layered. A close view of basement rocks generally reveals shiny black needles of hornblende or glittering flakes of black or white mica scattered through a matrix composed mostly of pale feldspar and gray quartz. Many gneisses and schists also

This specimen of gneiss still retains clear traces of its original sedimentary layering. Original minerals are now recrystallized to feldspar, quartz, and black biotite mica.

contain bright red crystals of garnet or sky blue prisms of kyanite scattered through them. Truly, these are beautiful rocks.

Age dates on the basement rocks of Montana show that most of them crystallized into essentially their present form about 2.7 billion years ago. Obviously, the original sedimentary and volcanic rocks that became gneiss and schist must have existed before they were metamorphosed, perhaps long before. Those figures place formation of the continental crust of Montana in the inconceivably remote period of Precambrian time that geologists call Archean.

Unless they happened on a few bare spots where they could see wood, a swarm of ants crawling about on a board might conclude that it consists of paint. We are in a somewhat similar situation as we wander about the continents because we see mostly sedimentary and volcanic rocks that cover the continental crust in most areas. Only in the bare places do we see the basement rocks of the continental crust. Although the younger sedimentary and volcanic rocks are typically some thousands of feet thick, that is no more than the paint on the board if compared to the 25-mile thickness of the continental crust. The continental crust beneath Montana has been accumulating its veneer of younger rocks off and on for at least 1500 million years.

The Precambrian Belt Formations

About a billion years after the continental crust of Montana formed, sometime around a billion and a half years ago, thick deposits of sandy and muddy sediments began to accumulate in most of what is now the western third of our state, as well as in nearby parts of Idaho and British Columbia. Those early sediments continued to accumulate for another 600 million or so years, at least until about 800 million years ago. They survive as the Belt formations of western and west-central Montana, so named because they were first studied in the Belt Mountains.

Geologists describe the age of the Belt formations as Precambrian, meaning merely that they are older than Cambrian rocks. That distinction is important because the earliest animal fossils exist in Cambrian sedimentary rocks laid down about 600 million years ago. Precambrian sedimentary rocks,

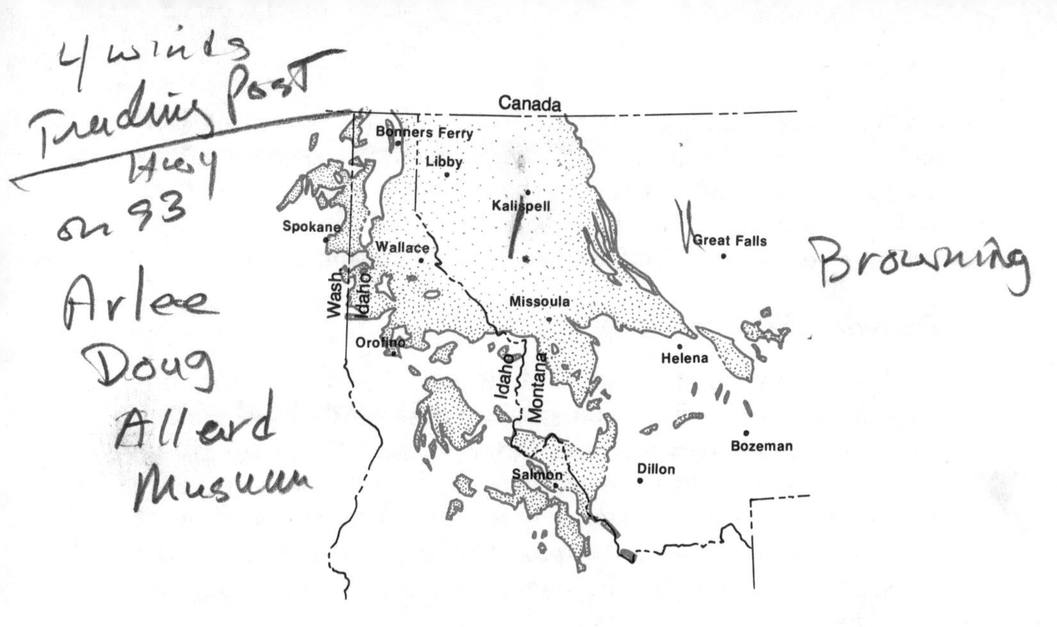

Distribution of the Belt formations.

including the Belt formations of Montana, contain abundant fossils of extremely primitive plants such as algae, but not the slightest trace of animal life.

Belt rocks accumulated generally north of two lines: the Willow Creek and Horse Prairie faults, which trend almost exactly west to east. The high region south of those Precambrian faults shed sediment into lower country to the north. Several other faults, which also trended from west to east, divided the low northern region into zones in which the thickness and character of the sedimentary deposits varied considerably. Deep deposits of sand, mud, and lime mud that accumulated then are now thick formations of very hard sandstone, mudstone, and limestone. The stack of Belt formations is shaped like a wedge that is thin in the east and thickens westward to some tens of thousands of feet in parts of northwestern Montana. The exact thickness is unknown, largely because the rocks contain no animal fossils that might identify layers of different ages.

The wedge of Belt formations ends abruptly at its thickest part, along a line just west of the western border of Idaho. It doesn't seem likely that the original sediments were deposited that way, sharply chopped off just where they reached their greatest thickness. In fact, it seems that they really were chopped off. Evidently, hot rock rising through the mantle split the continent, along with its accumulating cover of Belt sedimentary rocks, and swept the pieces apart, opening a new ocean between them. Besides the abrupt western edge of the Belt sedimentary pile, the best evidence of that event lies in igneous rocks intruded into the Belt formations.

Diabase Dikes and Sills

Throughout northwestern Montana, the Belt formations contain numerous dikes and sills composed of diabase, a black igneous rock identical to basalt except that its mineral grains are large enough to see without a microscope. Wherever geologists find other evidence of a continental rift, swarms of diabase dikes and sills are part of the picture. Some of the basalt magma that rises in the rift squirts into the continental crust as it begins to break.

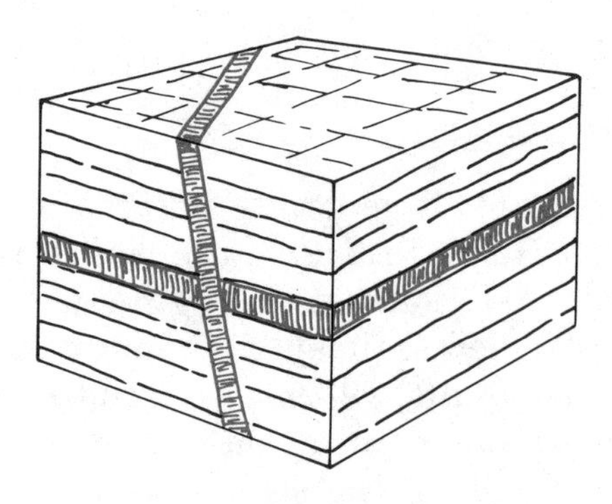

A dike fills a fracture; a sill forms a sandwich of igneous rock between layers of sedimentary rock.

Our best clue to when the continent split is in the age of the diabase dikes and sills. Age dates range from slightly less than 800 to a bit more than 1400 million years. Ordinarily, it only takes a few million years to split a continent, so that span of 600 million years is unreasonably long for a single such event. Either some of those dates are wrong, or they record more than the one event that geologists now recognize.

Several fairly convincing age dates cluster around 800 million years, and that is also the age of some sedimentary formations, the Windermere rocks, in Idaho and British Columbia

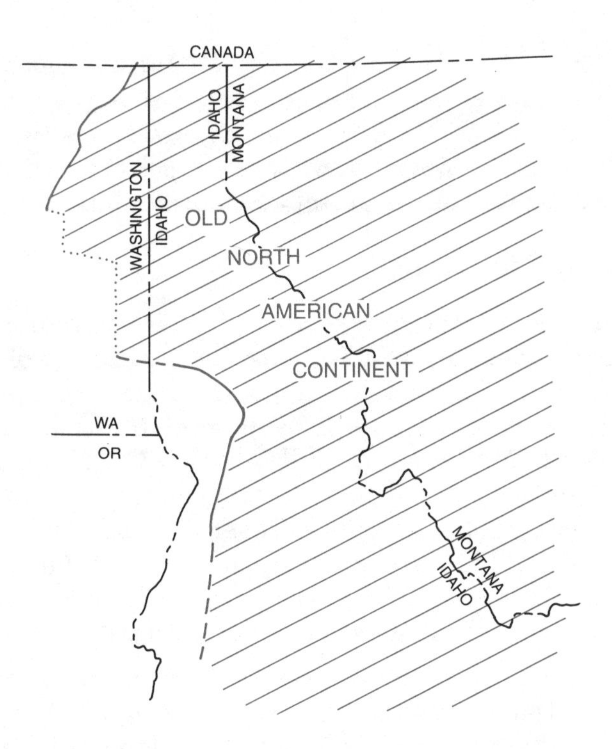

The western edge of North America as it was from about 800 million years ago until the Rocky Mountains began to form about 100 million years ago.

that appear to have accumulated along a coastline. So, about 800 million years ago seems a good guess for the time of the continental fracture that established a new western edge of the continent. That old coast, now long-vanished except for the patterns it left on geologic maps, followed a line that trends south from northeastern Washington through western Idaho and eastern Oregon into the Sierra Nevada of California. That line was to remain the west coast of North America until about 100 million years ago.

Where Is the Missing Piece?

Continental crust is too light to sink into the mantle, so the piece that rifted off North America about 800 million years ago must still exist somewhere. Where? Continental basement rocks in much of eastern Asia closely resemble those in the northern Rocky Mountains, and they are the same age. The Precambrian sedimentary rocks of eastern Asia also look like the Belt rocks of the northern Rocky Mountains, complete with diabase dikes and sills. Many geologists strongly suspect that the detached piece of North America now forms the large region that includes Korea, Manchuria, much of northern China, and nearby parts of the eastern Soviet Union.

Paleozoic Formations

Geologists call the long span from the beginning of Cambrian time, about 570 million years ago, to the end of Permian time, about 240 million years ago, the Paleozoic Era. During most of Paleozoic time, large parts of Montana were slightly below sea level, therefore shallowly flooded. Those areas accumulated deposits of sediment that cover the Belt formations and the basement to a depth of several thousand feet in most areas. Areas of the state that were above water during parts of Paleozoic time were almost certainly low islands or coastal plains only slightly above sea level, certainly not mountainous.

Most of the Paleozoic sedimentary formations contain animal fossils that geologists use to assign an age to the rocks that contain them. Most kinds of animals prefer to live in certain environments, so their fossils also tell of the conditions that existed when the rock formed, whether in sea water or a lake, for example. So fossils make the record of the rocks that contain them far more informative than that of the barren Precambrian formations.

Mesozoic Time

The Mesozoic Era was the time of the dinosaurs in all their abundance and lavish variety. That era lasted from about 240 million years ago, when the earliest dinosaurs began to appear, until 65 million years ago, when a sudden catastrophe exterminated them all, abruptly cut them off just as they reached the very height of their glory.

Most of Montana remained near sea level throughout Mesozoic time. Shallow inland seas, their shorelines shifting back and forth as the continental crust rose and sank, flooded all the eastern two thirds of the state, as well as much of the western part. Like the older Paleozoic formations, the Mesozoic sedimentary rocks accumulated to thicknesses of several thousand feet in most parts of the state, more in some places.

By the end of Mesozoic time, the Rocky Mountains were forming in western Montana, the continent was finally rising above sea level, and the inland sea retreated from the eastern part of Montana for the last time. Those were related events.

Shortly before the beginning of Mesozoic time, the restlessly shifting lithospheric plates happened to assemble most of the earth's inventory of continental crust into a single supercontinent. It was a chance aggregation not destined to last. A few tens of millions of years after all the pieces came together, a new continental rift split the supercontinent along a line that survives as the trace of the mid-Atlantic ridge, right along the middle of the Atlantic Ocean. Rocks along the east coasts of the Amerian continents and the established coasts of Europe and Africa contain diabase sills and dikes like those in the Precambrian Belt formations of western Montana. Age dates on those diabase intrusions show that the Atlantic Ocean began to open a little less than 200 million years ago.

Unlikely as it may seem, those remote events happening thousands of miles away started the long series of crustal movements in western North America that finally created the Rocky Mountains.

THE COLLISION BEGINS

As the plate bearing the North American continent began to move west away from the newly opening Atlantic Ocean, North America collided head on with the floor of the Pacific Ocean. Something had to give. The plate broke along the old western margin of the continent to form a trench along a line through eastern Washington and Oregon. Then the floor of the Pacific Ocean began to slide through the new trench and into the mantle beneath western North America. That began to happen sometime around 175 million years ago. Most of the events that formed the Rocky Mountains came much later, during the period between about 70 and 90 million years ago, almost at the end of Cretaceous time. That was a busy 20 million years.

Deep Thrusting

The plate collision at the trench off the old west coast compressed the western edge of the North American continent

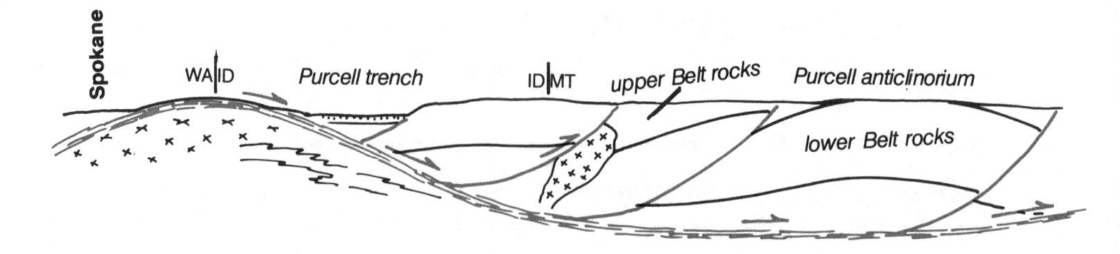

A cross section through northwestern Montana along a line approximately from Spokane to Eureka showing how slabs of rock moved from west to east along deep thrust faults.

as though it were an accordion. Thick slices of the sedimentary veneer, probably also of the continental crust, jammed eastward along deep fractures called thrust faults. The most important effect of that deep thrusting was to thicken the western margin of the continent, creating a broad highland. That helped set the stage for later formation of the overthrust belt.

Batholiths

Molten basalt and flaming hot steam rising above the oceanic crust that was sinking through the trench torched the lower part of the continent. The intense heat melted the ancient basement magma. The magma was granite because

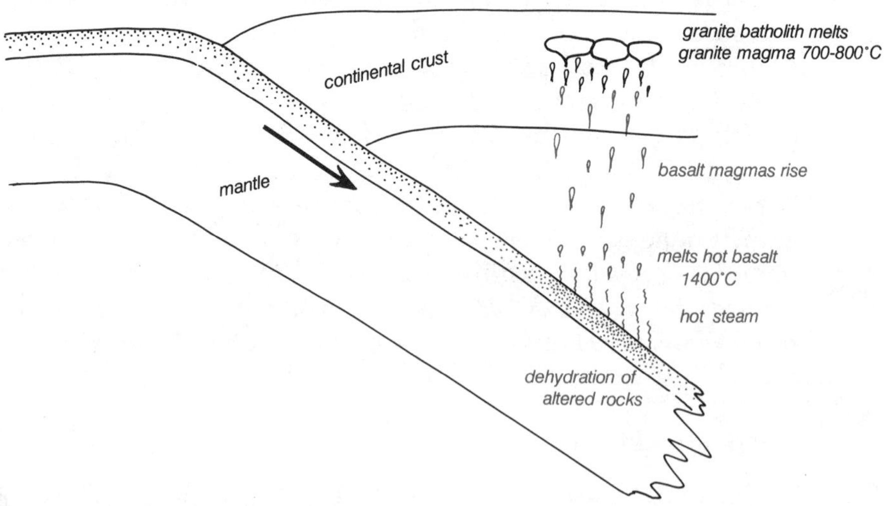

Steam and molten basalt magma rising from the sinking oceanic crust melted the lower part of the continent.

that is the continental crustal composition that melts at the lowest temperature. Once melted, the magma rose because it was lighter than the still solid rocks above.

The first fairly small masses of granite invaded the upper part of the continental crust in what is now western Idaho about 100 million years ago. Then, between 90 and 70 million years ago, enormous volumes of magma rose into the upper continental crust in a more or less continuous trend east of the old coastline from southern California north through Idaho and western Montana well into British Columbia. Those masses of magma crystallized to form a series of batholiths—any large mass of granite is a batholith. The eastern edge of the Idaho batholith is in the Bitterroot Range of southwestern Montana. Smaller satellite masses of that granite are widely scattered east of the Bitterroot Range, far into southwestern Montana.

Meanwhile, the Boulder batholith was forming in the area between Butte and Helena. That granite is the same age, 70 to 80 million years, as that in the batholiths farther west, but it is not the same granite. The main problem is that the Boulder

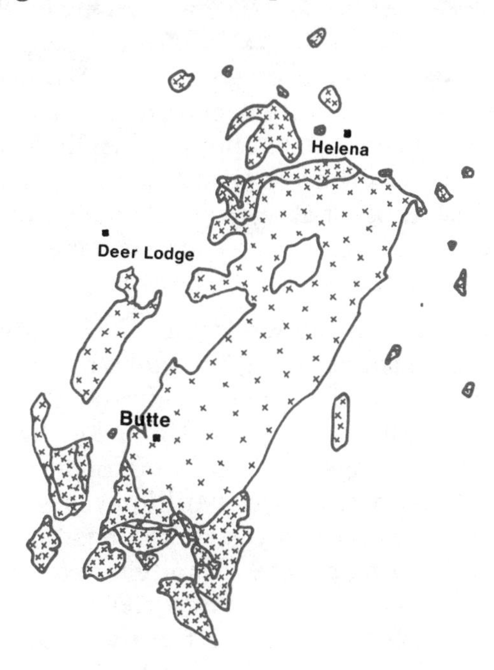

The Boulder batholith. Darker-colored granitic rocks are in darker pattern.

batholith is in the wrong place, an isolated major mass of granite about 100 miles east of the long trend that runs parallel to the old continental margin. And instead of crystallizing deep within the crust, some of the magma that became the Boulder batholith erupted voluminously to form the enormous Elkhorn Mountains volcanic pile.

THE ROOF SLIPS

Meanwhile, the earth's surface, already raised by the deep thrust faulting, probably bulged higher as heat expanded the rocks at depth. That bulge eventually became too high to remain stable. In some places, the sedimentary veneer came off piecemeal in great slabs. In other places, great blocks of the upper continental crust broke off all at once, and moved east.

The Overthrust Belt

In northwestern Montana, the sedimentary veneer appears to have peeled off the bulged crust in a series of slabs as much as several thousand feet thick. They came to rest stacked on each other, slab by slab, to become the overthrust belt that forms the east front of the Rocky Mountains from near Helena to the Canadian border. That movement involved slippage of the thin outer skin of the sedimentary veneer, a kind of movement quite unlike the deep thrusting that preceded it. Rocks in northwestern Montana west of the overthrust belt are mostly the oldest Belt formations, those that were deeply buried at the bottom of the pile. They are at the surface now because the younger rocks that once covered them moved east into the overthrust belt.

The Sapphire block, an enormous slab of the upper crust about ten miles thick, broke off the Idaho batholith and moved some 50 miles east into Montana. It includes the territory enclosed within the arc of the Garnet, Flint Creek, and Anaconda-Pintlar ranges. As the Sapphire block moved east, a trip that probably lasted several million years, it crumpled its leading edge and bulldozed the rocks ahead of it to form the tightly folded fringing ranges. Granite magma from the Idaho batholith evidently smeared along the base of the moving block, and then penetrated the tightly folded rocks of the Garnet, Flint Creek, and Anaconda-Pintlar ranges. The Bitterroot Valley is the gap behind the trailing edge of the Sapphire block.

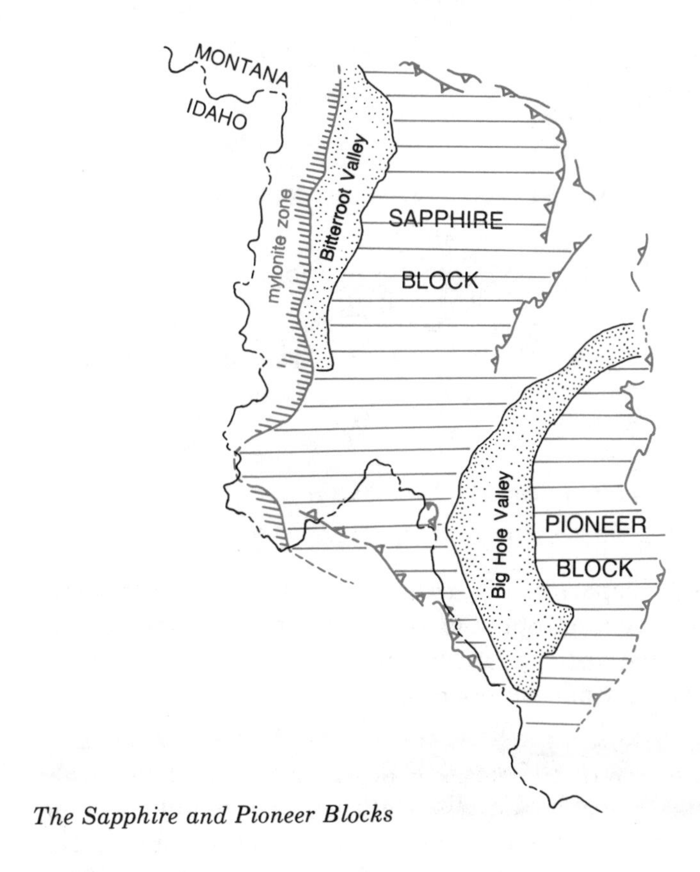

The Sapphire and Pioneer Blocks

The Pioneer Range is probably another big chunk that broke off the top of the Idaho batholith and glided east into Montana. Sedimentary rocks along its eastern edge are crumpled into extremely tight folds, as though they had been in the way of the moving block. The Big Hole, west of the Pioneer Range, appears to be the gap behind the Pioneer block, the counterpart of the Bitterroot Valley.

THE LEWIS AND CLARK FAULT ZONE

A squinting look at the geologic map of the northwest reveals a broad swarm of faults that trends from northwestern Washington southeast at least as far as Helena. We will call it the Lewis and Clark fault zone. The very feel of the landscape changes across that broad zone of crustal breakage. Northwestern Montana is mostly mountains with relatively

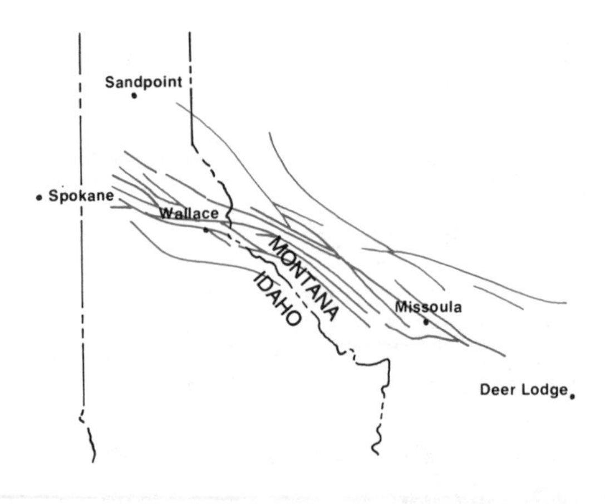

The Lewis and Clark fault zone

few valleys to separate the ranges, a landscape that some people find a bit claustrophobic. In southwestern Montana, the broad valleys between mountain ranges dominate the scene, giving the countryside an open and spacious feel.

All the faults that belong to the Lewis and Clark zone moved horizontally, one side sliding past the other as though you were sliding two books past each other on the top of a table. No one knows the cumulative displacement across all the faults in the entire zone, but it must be large, probably some tens of miles at the very least. Geologists disagree about when and in what direction those faults moved.

Imagine yourself looking in either direction along any of the faults and watching it move. Does the side on your left or right come toward you? We think the major movement on the faults in the Lewis and Clark zone brought the side on your left toward you—the side north of the fault west. The main reason we think that is because that is the direction the fault zone seems to displace the old continental margin. The faults break folds that appear to have formed while the big masses of granite were invading the crust, so they probably moved about 70 million years ago. But there is more to the story.

In northern Idaho, some of the faults of the Lewis and Clark zone cut igneous rocks that are only 50 million years old and move the south side west, so it seems that those faults reversed their direction of movement. One of the faults in the Missoula

Valley breaks rocks that are probably no more than about 25 million years old, so movement continued at least until then. There is no evidence, not the slightest sign of earthquake activity, to suggest that any of the faults in the Lewis and Clark zone are still moving.

THE TERTIARY SCENE

After the 20 million or so years of intense and varied activity during late Cretaceous time, Montana seems to have settled into about 20 million years of geologic quiet. Then, about 50 million years ago, another relatively brief period of intense geologic activity began with several kinds of events involving nearly the entire state. It is difficult to relate this later spasm of activity to the trench off the west coast because the west coast had by then moved farther west.

Sometime near 100 million years ago, the plate sinking into the trench off the old west coast swallowed enough of the Pacific Ocean floor to reel a small continent, now the Okanogan Highlands of northeastern Washington and central British Columbia, out of the vastness of the Pacific and onto the western margin of North America. Docking of that small continent jumped the trench to a new west coast along the present line of the Okanagon Valley of central Washington and British Columbia. Then, just 50 million years ago, another small continent, now the Northern Cascades of Washington and the Coastal Ranges of British Columbia, joined North America. Arrival of that second small continent forced the trench to jump again, this time to its present position off the modern west coast. The distance to either of those later trenches is so large that it is difficult to imagine any connection between them and events in Montana. Nevertheless, our part of the continent continued its mountain-forming dance.

Igneous Rocks of 50 Million Years Ago

Fifty million-year-old volcanic rocks, apparently the remains of a major chain of volcanoes, cover large parts of the Absaroka and Gallatin ranges, and extend well south into Wyoming. At the same time, widely scattered volcanoes erupted in central Montana, intense volcanic activity resumed in the area of the Boulder batholith, and other eruptions spread

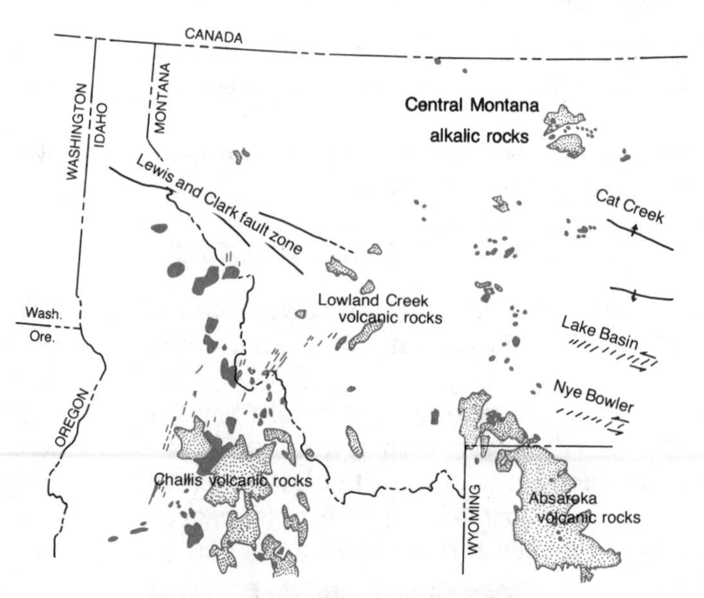

The 50 million-year-old igneous rocks. Granite in solid color; volcanic rocks shaded.

volcanic rocks over much of south-central Idaho. Masses of 50 million-year-old granitic rocks exist here and there throughout much of central and western Montana. An enormous swarm of large dikes, fractures filled with igneous rock, trends north east from south-central Idaho through southwestern Montana. Because those dikes fill fractures, we can be sure that the earth's crust was stretching in a direction perpendicular to their trend.

The Younger Mountains

Central and eastern Montana contain a number of faults and broad crustal folds, arches and troughs buckled in a complex pattern into the rocks beneath the plains. Some are big enough to form several of the large mountain ranges of central Montana, others appear only to people who study geologic maps. Several folds trend generally southeast, parallel to the trend of the Lewis and Clark fault zone. Another set of faults and folds trends slightly west of north. All those structures may have formed at about the same time the region rose several thousand feet above sea level, perhaps about 50 million years ago.

We can be quite sure that central and eastern Montana were

near sea level at least until the end of Cretaceous time, about 65 million years ago, because many of the late Cretaceous sedimentary formations accumulated in shallow sea water or along a coast. By about 55 million years ago, the shallow sea had retreated to North Dakota and Saskatchewan.

Remember that the continental crust is, in effect, a raft of light rocks floating on the heavier rocks of the earth's mantle. If small boys find their log raft beginning to sink, they shove more logs underneath to make it thicker, so it will float higher. Similarly, if a compressional force shoves the continent, that will thicken the crust by telescoping the rocks together, thus making the continental raft float higher. If the continental crust under Montana was under compression 50 million years ago, that could have thickened the crust and might also have wrinkled the mountain ranges into the plains of central Montana. If such a compressional force were directed from the southwest, it could also have opened the fractures that filled with magma to become the dike swarm of southwestern Montana.

If all the links in that long and rather circumstantial chain of reasoning actually do connect, then the continental crust wrinkled to form the mountains of central Montana at the same time that central and eastern Montana rose high above sea level and magma filled fractures to form the dikes of southwestern Montana. Numerous age dates show that the dikes are 50 million years old.

THE VALLEYS FILL

Approximately 40 million years ago, as the rush of crustal movements associated with the 50 million-year-old igneous rocks was winding down, the climate of Montana became dry enough to weaken the ability of the streams to carry sediment. Montana stayed fairly dry through Oligocene and early Miocene time, until the next drastic climatic change happened about 20 million years ago. Dry climatic periods leave their record in distinctive sedimentary rocks.

Soil erodes very rapidly in dry regions because the plant cover is too sparse to shelter the ground from splashing raindrops. That effect was probably more pronounced 40 million years ago than today because many of the plants that now thrive in dry regions had not yet evolved. Meanwhile, a dry climate makes stream flow too weak to carry all the eroded soil. Throughout Oligocene and early Miocene time, shriveled streams deposited large quantities of sediment in the broad valleys of the northern Rocky Mountains and all across the vast plains of eastern Montana into the Dakotas.

The layers of Oligocene and Miocene sediment that filled the valleys of western Montana are tilted nearly everywhere, steeply tilted in many places. Those tilted beds tell us as clearly as anything can that crustal movements were continuing in western Montana. Unfortunately, so little is known about the pattern of that fairly recent deformation that it is impossible to describe what happened.

The Renova Formation

The deep deposits of sediment that filled the mountain valleys of western Montana during Oligocene and early Miocene time includes an endlessly varied assortment of gravel, sand, mud, volcanic ash, limestone, and coal. In most of the large valleys, those sediments are more than 2000 feet thick. Geologists call that mess the Renova formation in western Montana, the White River Oligocene beds in the eastern part of the state. They recognize the Renova formation by its tendency to consist dominantly of pale gray and tan rocks that run mostly to sand and silt and are soft enough to dig with a shovel.

The level of volcanic activity in the northern Rocky Mountains while the Renova formation was accumulating does not seem nearly great enough to explain all the ash in the Renova formation. Besides, the volcanic ash in the Renova formation is not the sort of deposit that might have been laid down by local eruptions. Most of it must have come from elsewhere. One likely source of ash was in the Western Cascades of Oregon and Washington. Large eruptions there produced enormous volumes of rhyolite precisely while the broad valleys of western Montana were filling with Renova formation sediments. Imagine the overwhelming clouds of pale rhyolitic ash blotting out the sun for days at a time as they drifted east after large eruptions. Great blankets of ash settled heavily, like gray snow, on the mountains of western Montana, then washed into the valleys.

The Renova formation also contains beds of limestone that must have formed in shallow lakes and coal seams that began as deposits of peat laid down in marshes. Both are exactly what one would expect in a sedimentary deposit laid down during a dry climatic period. Dry regions tend to contain lakes with marshes around their borders simply because the weak streams don't drain the landscape effectively. The Great Salt Lake is an oversized modern example: limestone is forming in the lake, and the marshes around its borders contain deposits of peat that will someday become coal. Similar situations exist on a smaller scale throughout the dry regions of western North America.

Beautiful petrified wood abounds in the Renova formation, and the beds of gray volcanic ash commonly contain well preserved fossil leaves. Most of the wood and leaves are from the Dawn Redwood, a close relative of the modern Sequoia tree that was thought to be extinct until survivors were found growing in China. Evidently the climate of Montana during Renova time was dry enough that streams could not carry eroded sediment and volcanic ash away to the ocean, but neither too dry nor too cold to keep redwood trees from growing on the mountains.

Although fine grained sediments like those that form most of the Renova formation may contain large volumes of water, they hold it so tightly that little can move to a well. Drilling for water in the Renova formation is a risky business that succeeds only if the well happens to intersect one of the occasional beds of sand or gravel in which the pore spaces are large enough to permit water to flow freely. Many residents of western Montana lose a lot of money drilling dry holes into the Renova formation.

Most of the volcanic ash in the Renova formation is degraded into expandable clay minerals that swell up when they get wet, by as much as 40 percent of their original volume. Furthermore, expandable clay becomes extremely weak and slippery when it is wet. Landslides are common wherever the Renova formation absorbs large quantities of water. Several western Montana towns, including Missoula, Bozeman, and Deer Lodge, are now spreading onto benches underlain by the Renova formation. As more people build on those high benches, and in so doing encourage more water to soak into the Renova formation, landslides may become a serious problem.

THE TROPICAL TIME

People stick "Native of Tropical Montana" bumper strips on their cars, never dreaming that such creatures did once exist. Starting sometime around the middle of Miocene time, perhaps about 20 million years ago, and continuing for approximately 10 million years, Montana really did have a climate suitable for a lush island in the Caribbean. No one knows why that happened, but evidence that it did happen is abundantly clear.

Tropical Soils

In many places in western Montana, a buried layer of red soil lies on top of the Renova formation, on older rocks in a few places. Watch for splashes of red in roadcuts and hills within the broad valleys filled with Renova formation. That buried soil has the chemical and mineral composition of the modern red soils in the wet tropics, laterites. The tropical climate prevailed over the entire Pacific Northwest almost exactly when enormous volcanic eruptions built the Columbia Plateau of Washington, Oregon, and nearby parts of Idaho and California. Red layers of buried laterite soils like those in western Montana make bright stripes sandwiched between the black basalt lava flows.

Some of the big lava flows on the Columbia Plateau dammed rivers to impound large lakes. Sediments deposited in those lakes are also sandwiched between lava flows. They contain fossil leaves preserved as perfectly between layers of white volcanic ash as though they were pressed between the pages of a dictionary. Many are leaves of tropical hardwood trees like those that live today in Florida and in the Caribbean. Similar trees must have flourished in Montana during the tropical interlude.

Late Miocene Streams

When the dry period of Renova time ended and heavy tropical rains began to fall, streams began to flow through Montana. Some were large rivers. Imagine them flowing through heavily forested, hilly landscapes in eastern Montana and green mountains cloaked in tropical and subtropical hardwoods in the western part of the state. The trees are gone now, except for occasional deposits of petrified wood in the Columbia Plateau, but parts of those old valleys survive in the modern landscape of western Montana. Miocene valleys filled with younger gravel deposits almost certainly exist in eastern Montana.

THE SECOND LONG DRY SPELL

Sometime late in Miocene time, or early in Pliocene time, perhaps about 10 million years ago, the humidly tropical period ended and a new dry period began. All the evidence suggests that the climate was much more arid during this second dry period than in the first. Montana was as truly a desert then as Death Valley is now, and was to remain so through Pliocene time, until the first ice ages began about 2.5 million years ago. Again, no one knows what caused the climate to change.

When the rains stopped, the streams that had drained all of Montana during the wet years of Miocene time dried up. With no streams draining them, the broad mountain valleys of western Montana once again became undrained desert basins like those in the modern deserts of the southwestern United States. Meanwhile, an enormous desert plain developed east of the main mountain front as the shifting desert streams spread a thick blanket of coarse gravel across the region.

Geologists call the sediments deposited during the second dry period the Six Mile Creek formation in western Montana, the Flaxville formation east of the main mountain front. Neither contains volcanic ash. Both consist largely of coarse gravel, as one would expect of an extremely arid desert where very few plants shelter the ground and keep the soil open so it can absorb water. Most of the little rain that falls runs off the hills almost as though they were tin roofs. The surface runoff pours down the normally dry stream beds as torrential flash floods that typically carry heavy burdens of coarse gravel and then spread them across the landscape as the stream dries up.

In western Montana, the new generation of valley filling sediment began once again to fill the mountain valleys, this time with layers of brown gravel with beds of sand and mud sandwiched between them. In most places, the gravelly Six

Mile Creek formation lies on the reddish remains of the laterite soil that developed during the tropical interlude. The accumulating gravel first filled the valleys the late Miocene streams had eroded into the Renova formation, then buried the hills between the valleys.

The Six Mile Creek gravel commonly lies on tilted layers of the Renova formation, but is itself generally undisturbed. Evidently, crustal movements continued after the Renova formation was deposited, but had almost stopped by the time deposition of the Six Mile Creek formation began.

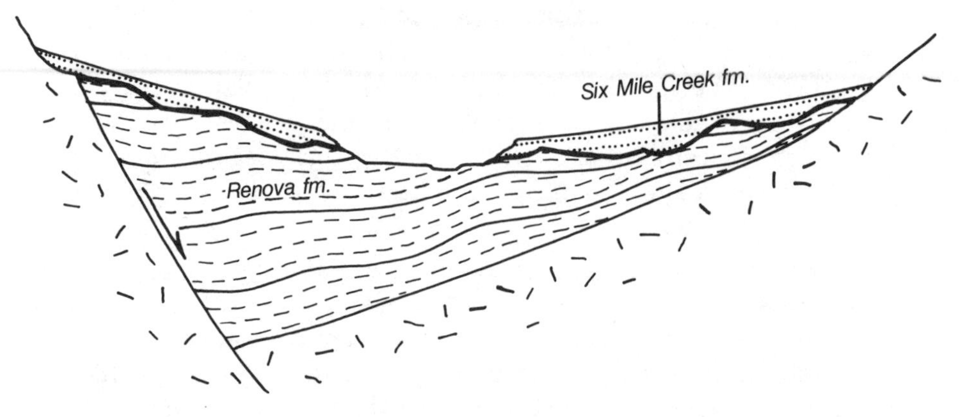

Gravel of the Six Mile Creek formation buried the landscape that had been eroded into the folded Renova formation.

Unlike the Renova formation, the Six Mile Creek gravels make excellent aquifers. Wherever the formation lies below the water table, wells drilled into it predictably produce large quantities of good water. Drilling a deeper well through the gravel and into the Renova formation beneath it rarely improves the water supply.

The Flaxville formation of eastern Montana also consists mostly of gravel interlayered with lesser amounts of fine sediment. South of Montana, the same gravel is called the Oglalla formation. The Flaxville and Oglalla gravels extend from the Rocky Mountain front east at least as far as a north-to-south line drawn through central North Dakota and central Texas. Deposition of the Flaxville gravels doubtless began in the old stream valleys that had formed during the wet period of Miocene time. As the valleys filled, the gravel deposits buried the hills between them, and finally spread to make a nearly continuous blanket across the vastness of the High Plains. Now, we find the thickness of the Flaxville gravels ranging from a few to several hundred feet, no doubt because they bury an older landscape.

Like the Six Mile Creek formation in the Rocky Mountains, the Flaxville and Oglalla gravels provide excellent aquifers filled with large quantities of good water. Occasional wells drilled in eastern Montana penetrate an uncommonly thick section of gravel, and produce an extraordinary volume of water. Those wells were probably located, by sheer good luck, on one of the buried valleys. If geologists someday piece together a map of the large streams that flowed through eastern Montana during the wet part of Miocene time, it would make it possible to locate such wells intentionally.

Gravels of the Six Mile Creek and Flaxville formations contain very little petrified wood, no fossil leaves. Here and there, the Pliocene gravels contain bones and teeth of mammals of all sizes from mice to elephants. The fossils include the remains of early models of animals such as horses and camels that thrive in dry regions today. The sparse vegetation of dry regions is more nutritious than the lusher growth of wetter climates because desert soils naturally contain large amounts of fertilizer nutrients. That is why herds of large grazing animals are common in dry country today, and why it is no surprise to find the bones of such animals in gravels deposited during Pliocene time.

ICE AGES AND THE MODERN STREAMS

The youngest fossils known from the Six Mile Creek and Flaxville gravels are the remains of animals that lived just as Pliocene time ended and Pleistocene time began. Evidently, that was when deposition of the Six Mile Creek and Flaxville gravels ceased, presumably because the climate was no longer as dry as it had been. Age dates of about 2.5 million years on volcanic ash associated with the oldest known glacial deposits suggest that Pleistocene time, the period of the great ice ages, began about then. That date marks the beginning of the last major phase in development of the landscape of Montana.

The Streams Begin to Flow

As the climate became wetter, the umbrella of green plants spreading over Montana greatly reduced the rate of soil erosion. We can easily imagine clear streams beginning to flow east down the gentle slope of the vast deposits of Flaxville gravel that stretched from the front of the Rocky Mountains into the Dakotas. With more rain to fill them and less sediment to carry, the invigorated streams must have flowed continuously, instead of only during wet weather. They began to carve deep valleys into the formerly smooth plains of eastern Montana.

Large and small remnants of the old Pliocene desert surface survive almost untouched by erosion throughout the eastern two thirds of Montana. Recognize them as flat uplands between streams that wear a cap of Flaxville gravel and generally support a generous sweep of wheat fields. Geologists call those remnants the High Plains surface. They probably owe their survival to the ability of the permeable Flaxville gravel to absorb surface water, thus preventing erosion of the surface by surface runoff.

Streams could not begin to flow quite so easily in western Montana where the broad valleys had become basins of internal drainage with no outlets. But it is easy to imagine how the problem solved itself. First, those isolated desert valleys filled with water like so many bird baths. The lakes then overflowed across the lowest points on the divides between them. Flow through the spillways eroded them into the canyons that now connect the valleys of western Montana. Watch as you drive through western Montana for the tendency of the streams to flow for a few miles through one of the broad valleys floored with basin fill deposits and then cut through a mountain range in a narrow canyon.

When the modern streams began to flow in western Montana about 2.5 million years ago, many followed old valleys that earlier streams had eroded during the wet interval of late Miocene time. If the modern stream is too small to fit the old valley in which it flows, remnants of the old valley floor remain as bedrock benches along the sides of the valley. In many other places, the modern streams did not flow through the old Miocene valleys, probably because crustal movements along

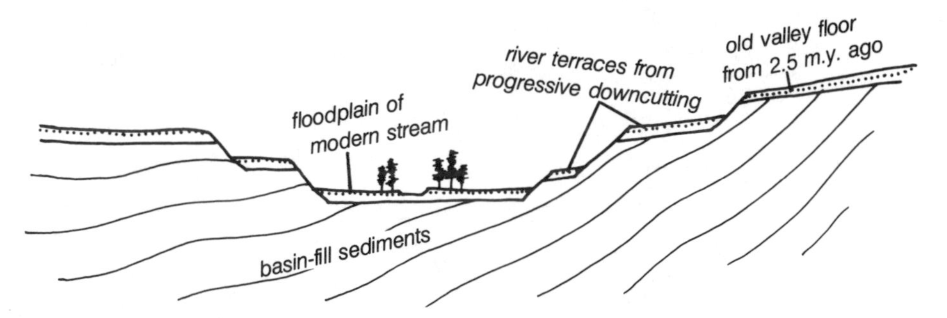

Modern streams are slowly excavating the basin fill sediments that accumulated in the big valleys of western Montana during the long periods when the climate was dry. The modern valley floors are hundreds of feet lower than the valley floor of 2.5 million years ago.

faults during the Pliocene dry period had raised the Miocene valley too high. Now we see sections of those old Miocene valleys preserved in the modern landscape as obvious stream valleys, complete with tributary valleys, in which no streams flow.

As the desert basin fill sediments accumulated in the big valleys of western Montana, they buried all the older landscape on the valley floor, including entire hills. Here and there, the modern streams uncover hills that had been buried as they excavate the valley fill sediments. Now we can walk on hillslopes that must look almost exactly as they did before they were buried tens of millions of years ago, the same slopes walked by animals long since extinct.

Ice Ages and Glaciated Landscapes

Of all the geologic events we associate with Pleistocene time, ice ages are the most dramatic. Time after time, enormous glaciers formed in high mountains everywhere, and on flat land in northern Europe and central Canada. As each ice age continued, the glaciers grew until the climate changed, and the ice melted.

No one really knows what ice age climates were like. We can be sure that the weather was considerably wetter than that we know, and it seems safe to assume that it was somewhat colder. Exactly how wet or how cold remain matters of debate. Most estimates place the drop in mean annual temperature during the last ice age in the range of a few degrees—noticeable, but

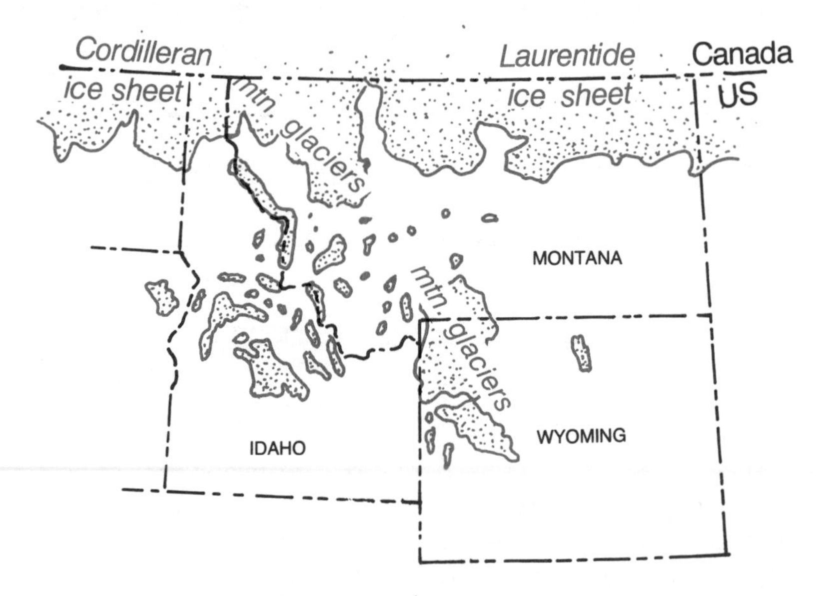

The approximate maximum extent of glacial ice in the northern Rockies.

nothing that our wardrobes could not handle. All of us are, after all, direct descendants of people like ourselves who camped out all their lives, generation after generation, during the last ice age.

Most geologists assume that the future will bring more ice ages simply because they have been happening in the geologically recent past. If a new ice age were to start, the immediate effect on Montana would probably be an increase in snowfall and rainfall accompanied by stormier and cooler weather. Almost every crop now grown in Montana could flourish in wetter and cooler weather, so the change in climate might well increase the agricultural productivity of the state. A new ice age would have to continue for thousands of years before the glaciers could grow large enough to cover any of the places where many people live.

Much of our modern landscape shows in one way or another the mark of the ice. Our high mountain peaks owe their jagged form to glacial sculpture, and the lower valleys in those mountains, as well as large expanses of nearby low areas, contain deep deposits of glacial sediments. Continental ice creeping out of central Canada covered most of eastern Montana north of the Missouri River, and left deposits of sediment, moraines, that mark its exact outline.

Like any carving tool, glaciers make a distinctive cut. Glaciers straighten mountain valleys and deeply gouge them into something approximating a semi-circular profile. The heads of those valleys end in deep hollows, called cirques, that look from a distance as though some mythical giant wielding an ice cream scoop had gouged a helping of rock out of the top of the peak.

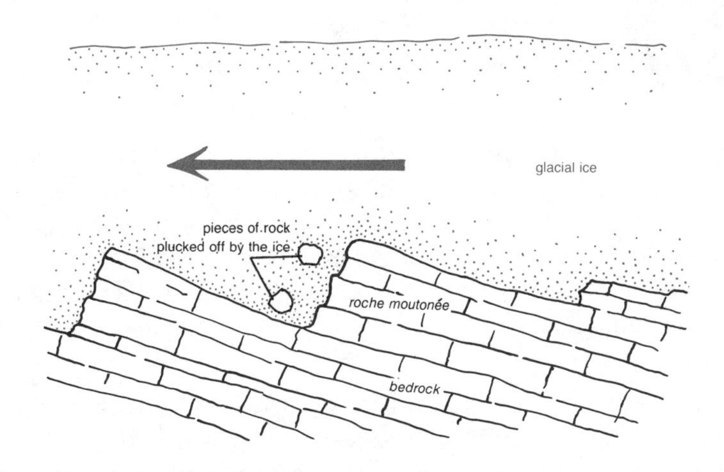

Roches moutonée, the large tilted steps eroded by the plucking action of ice in a glacier.

Where several glaciers gouge their cirques into the top of a mountain, they reduce the peak to a gnarled pinnacle of rock called a horn. Heavy ice age glaciation converted the higher ranges of western and central Montana into serrated rows of craggy horn peaks, each one dropping off into cirques that lead downward into deeply gouged valleys. Many of those glaciated valleys are so straight that you can stand in their lower reaches and look directly at the distant peaks that spawned the glacier. River valleys never provide such long views.

As mountain glaciers descend in their valleys, and as continental glaciers move farther south, they advance into warmer climates. Eventually, they reach an area where the climate is warm enough to melt the ice as fast as it moves

Glacially scoured bedrock, once covered by till on the west side of Flathead Lake.

forward. That balance establishes the end of the glacier. As glaciers reach warmer climates, the ice thins and begins to deposit its load of sediment. Debris dumped directly from glacial ice makes a distinctive sediment called till, a chaotic mass of all sizes of material mixed together. Till looks like something that a bulldozer might have scraped up and dumped.

Any deposit of glacial till is a moraine. Mountain glaciers deposit a bench of lateral moraine along the valley wall near their lower edges and a hummocky terminal moraine across their front. Together, the lateral and terminal moraines make a continuous hairpin ridge that encloses the lower end of the glacier and records its precise outline. Continental glaciers like the one that covered much of the northern part of central and eastern Montana are not confined within valley walls, so they leave only terminal moraines that trace the old ice front for hundreds of miles.

The typical arrangement of a mountain glacier, its moraines, and their associated outwash plain

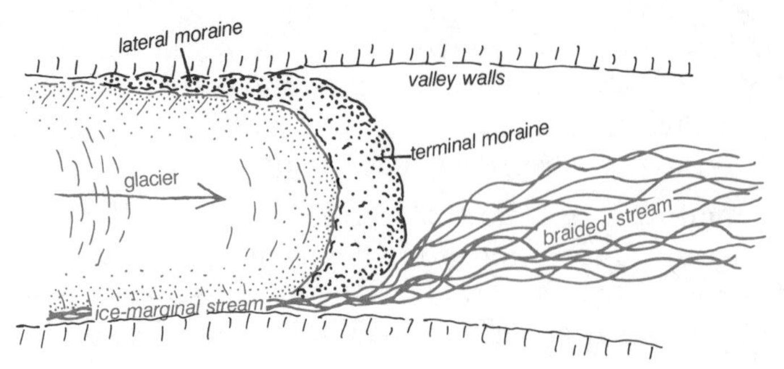

As long days full of sunshine mellowed the ice age summers, vast torrents of water poured off the melting glaciers, carrying enormous loads of sediment melted out of the ice. Meltwater rivers crossed the moraines and dumped their loads of sediment downstream as deposits of sand and gravel called outwash. Now we find outwash flooring the lower parts of glaciated mountain valleys and spreading away from terminal moraines as great alluvial fans.

For many years, most geologists agreed that Pleistocene time brought four great ice ages separated by interglacial episodes in which most of the land ice melted. Evidence recently obtained from deep sea cores shows that there were many ice ages, perhaps as many as 20. In Montana, we see clear evidence of just two major episodes of glaciation, both late in Pleistocene time. Geologists call the older and larger the Bull Lake glaciation, the younger the Pinedale glaciation. Widely scattered evidence of earlier ice ages exists, but their record is so fragmentary and hard to read that almost nothing is known.

Some geologists believe that the Bull Lake glaciers reached their maximum sometime about 70,000 years ago; others place that event closer to 130,000 years ago. Both dates are shaky because Bull Lake glaciation happened in a time span for which we have no reliably effective radioactive age dating techniques. Glaciers of the Pinedale ice age reached their maximum about 15,000 years ago and melted about 10,000 years ago. Those are firm figures that rest on numerous reliable radiocarbon dates.

THE CURRENT SCENE

Although the pace may be slower than it was at times in the past, Montana is still an active place. Igneous activity continues in and near Yellowstone National Park. Faults continue to move rapidly enough in several parts of western Montana to chop old mountain ranges into pieces and raise new ones.

Sometime after the events of 50 million years ago, a new generation of faults began to dissect the tortured rocks of the northern Rockies into blocks that moved up to form mountain ranges, down to form broad basins. Most of those faults trend

The distribution of earthquakes defines the intermountain seismic zone.

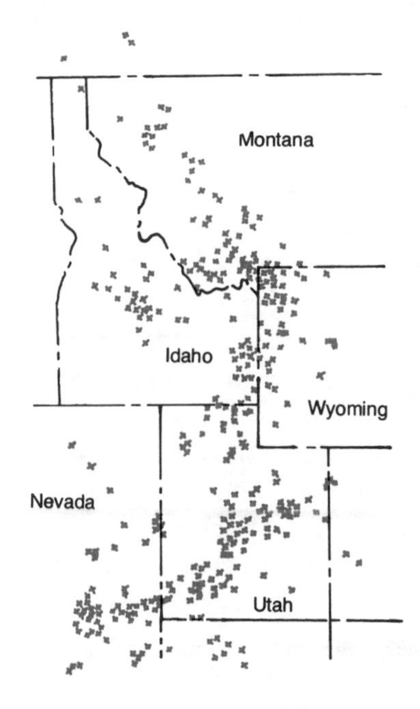

northwest, a few northeast. We can be sure that many of them are still active because they generate earthquakes as they suddenly move a mountain block a bit higher or drop a valley a few feet lower.

Plots of the locations of earthquakes in the northern Rockies show them widely scattered over the western part of Montana. But the greatest concentration defines a broad zone south from Helena through Yellowstone National Park to the vicnity of Salt Lake City. At Helena, the greater part of that zone branches northwest to cross Flathead Lake, where it finally fades. The remainder more quickly fades off to the northeast. Geologists call that branching zone of most intense earthquake activity the Intermountain seismic zone, but can not explain its existence.

The Yellowstone Volcano

Yellowstone National Park is one of the largest and most violent volcanoes in the world. There is no reason to assume it is extinct, every reason to expect further eruptions. Past eruptions of the Yellowstone volcano have been hundreds of times greater than the eruption of Mount St. Helens in 1980, making that minor outburst look like a shot from a toy popgun by comparison. Any future eruptions in Yellowstone National

Park may well be equally impressive.

Geologists call volcanoes like that in Yellowstone National Park resurgent calderas because they form an enormous collapse crater called a caldera when they erupt, fill it with new volcanic rocks, and then erupt again. The Yellowstone volcano has done that three times at intervals of about 600,000 years, most recently about 600,000 years ago. If there is any periodicity to such things, another eruption must be about due. Only a few resurgent calderas are now active worldwide, and none has erupted within the period of recorded human history, so we have no eyewitness accounts. To judge from the rocks, there may be some difficulty finding surviving eyewitnesses if the volcano does erupt again.

Large volumes of molten magma exist at shallow depth beneath Yellowstone National Park. The simplest evidence of its presence is the extremely high temperature of many of the hot springs, which strongly suggests magma at depth. Furthermore, earthquake waves that pass beneath the park lose the kind of shear wave motion that does not pass through a liquid, and that also happens at shallow depth. Virtually all the volcanic rock in Yellowstone National Park is rhyolite, which has the same composition as granite, so it is safe to assume that the molten magma beneath the park will become a mass of granite if it does not erupt.

If the magma that now lies a few thousand feet below the surface in Yellowstone National Park were to crystallize into granite without further eruption, the result would be a batholith lying beneath a cover of volcanic rocks of the same composition. That is exactly what we see in the Boulder batholith, which appears to have been a resurgent caldera during its emplacement.

What Next?

It isn't over yet. The mountain ranges of southwestern Montana that are now rising along active faults will continue to rise for some time to come. Those faults will continue to generate earthquakes, including occasional large earthquakes that we will all feel. The Yellowstone volcano may erupt almost anytime, and if that happens, the whole country will share the experience.

Roads covered in this section.

II
Northwestern Montana—Where Nothing is Nailed Down

Northwestern Montana west of the Rocky Mountain front and north of the Lewis and Clark fault zone is, more than anything else, a region in which the bedrock moved east in great slabs. In fact, it is no exaggeration to say that every rock in the province, without exception, moved east along faults for distances of some tens of miles from where it formed. First, let us look at the rocks where they formed, and then move them to where we now see them.

PRECAMBRIAN SEDIMENTARY ROCKS—THE BELT FORMATIONS

People who travel through northwestern Montana see mostly Belt rock, enormously thick sections of Precambrian mudstone, sandstone, and limestone. Age dates show in various ways that the Belt rocks accumulated as deposits of mud, sand and lime mud between about 1500 and 800 million years ago. That is impressively old, even for rocks. Especially for sedimentary rocks.

The Belt rocks provide us a marvelous window opened wide to a view of an early period in the history of our planet. If we find it hard to understand what we see through that window, that is understandable, because the Belt rocks are as foreign to us as though they were survivors of a distant planet now long since lost in the corridors of time. The first step toward an interpretation of the peculiar Precambrian sedimentary rocks is to attempt to imagine the earth as it was then.

The Precambrian World

Although no one really knows what the earth was like a billion and more years ago, it is possible to offer reasonable conjecture. We can be sure that the atmosphere then was nothing like the air we breathe now. We can be reasonably confident that the Precambrian atmosphere contained very little oxygen because some sedimentary rocks that formed then contain minerals that can not exist in the presence of oxygen. And we can also be reasonably sure that the early atmosphere contained an enormous amount of carbon dioxide. Evidence to support that idea exists in every outcrop of limestone or dolomite.

The earth's crust contains enormous volumes of the common carbonate sedimentary rocks limestone and dolomite, which consist of calcium carbonate and calcium-magnesium carbonate respectively. In the modern world those rocks form as deposits of sticky gray lime mud laid down in shallow water in such places as Florida Bay. No matter where or when they accumulate, the carbonate in limestone and dolomite forms from carbon dioxide taken from the atmosphere. So formation of the carbonate sedimentary rocks involves transfer of carbon dioxide.

Such things are hard to estimate, but it seems likely that if all the carbon dioxide now tied up in limestone and dolomite were magically released, the atmosphere might then contain something like 20 molecules of carbon dioxide for every molecule of gas of any other kind that it now contains, perhaps more. Anything even remotely resembling such a carbon dioxide concentration would make an extremely dense atmosphere, one that would instantly suffocate any form of animal life. And it would make the earth into a planetary sauna bath.

Carbon dioxide traps heat, just as glass does, to create a "greenhouse effect." So the Precambrian atmosphere with all its carbon dioxide was the equivalent of a giant greenhouse that covered the entire planet. That must have made the earth uncomfortably warm, but we can be sure that the temperature was below the boiling point of water because many Precambrian sedimentary rocks contain fossils of blue-green algae that could not have survived in boiling water. High temperatures increase the rate of evaporation from the oceans, so the

Precambrian atmosphere must have contained large amounts of water vapor, which probably maintained a dense cloud cover.

Imagine then the Precambrian world. The earth's steaming surface was shrouded in dense clouds that dropped showers of hot rain. Pools of scalding water evaporated, and then filled again with the next rain. The scarcity of oxygen in the Precambrian atmosphere must have prevented formation of an ozone layer in the upper atmosphere that might protect the surface from ultraviolet radiation. It was a hostile world; hot, without breathable air, bombarded by intense ultraviolet radiation. No place for animals of any kind. It is easy to understand why geologists have never found a single animal fossil in any of the Belt sedimentary rocks.

But primitive plants flourished.

Precambrian Plants

Belt rocks are full of the fossil remains of blue-green algae, extremely primitive plants that still abound. Modern blue-green algae, which look exactly like their Precambrian ancestors, thrive in any moist place where they are reasonably secure from grazing vegetarian animals such as snails. They flourish in the steaming hot springs of Yellowstone Park. During Precambrian time, the blue-green algae probably thrived in every place that provided moisture and light. Their fossil remains are so abundant and diverse that they convey an impression of a world scummy with algae. Look at those humble plant fossils with respect; they converted the suffocating early atmosphere into something more like the air we breathe.

Two of the many forms of fossil blue-green algae.

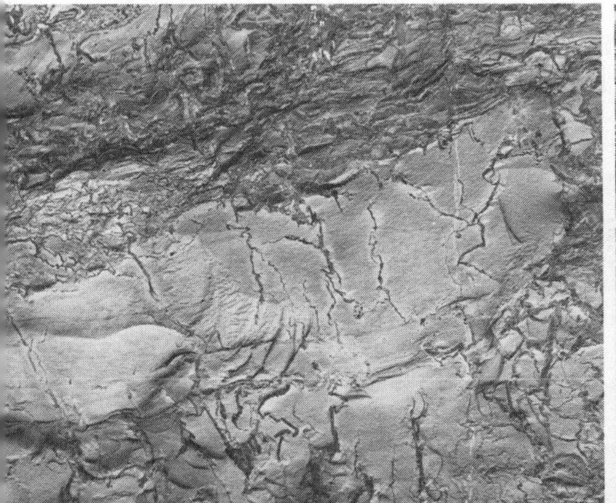

Breathable Air

Like all green plants, growing blue-green algae absorb carbon dioxide from the atmosphere, use the carbon to build their tissues, and release free oxygen. In the normal course of events, the carbon and oxygen recombine into carbon dioxide as the plant tissues finally decay. But if accumulating sediments bury dead plants before they can decay, the oxygen remains in the atmosphere, and the carbon becomes part of the sedimentary deposit. Many of the Belt rocks contain carbon, some in such abundance that it stains the rock black. That carbon is certain evidence that oxygen was accumulating in the atmosphere as the Belt rocks formed. By the end of Precambrian time, the atmosphere finally lost enough carbon dioxide and acquired enough oxygen to make it possible for primitive animals to breathe.

Abundant and varied animal fossils first appear in Cambrian sedimentary rocks deposited about 600 million years ago. Quite quickly, all sorts of small animals flourished. No one knows for sure why animals appeared when they did. Maybe their appearance marks the time when the earth finally had an atmosphere that contained enough oxygen for small animals to breathe, enough to maintain an ozone layer in the upper atmosphere to block the killing ultraviolet radiation. The very first known animals appeared at least a few million years before Cambrian time began, but left little trace of themselves because they lacked hard shells.

Belt Rock

Some of the Belt formations are full of bedding surfaces covered with mudcracks, ripple marks, even the dainty little craters left by a spattering of rain drops on soft mud. Those rocks formed from mud that dried in the air. On land. Other Belt formations contain no evidence of air drying, and evidently consist of sediments that accumulated under water. In the absence of fossils, little in those rocks can tell us whether that was sea water.

Look at almost any outcrop or pebble of Belt rock, and you will see delicate features such as sedimentary layers hardly thicker than a sheet of paper, ripple marks, mud cracks, in some places tiny cubic indentations left by salt crystals. Muds and sands deposited in the modern world rarely contain such exquisitely preserved details because burrowing and scurrying animals destory them while the rock is still soft mud or sand. Only in the Precambrian world was such preservation so generally possible.

Some of the Belt formations, especially those in the eastern part of the region, consist of great thicknesses of bright red and green mudstones full of ripple marks, suncracks and other features that speak of wetting and drying. At first glance, those look like something that might have formed on tidal flats, where the mud is alternately wet and dry. But those layers extend for miles and miles, for distances vastly greater than the width of any modern tidal flat. Futhermore, it is hard to imagine how deposition so closely linked to sea level could pile up thousands of feet of sediment. Deposition on land seems far more likely.

In the absence of leafy plants to protect the ground, the rate of soil erosion must have been high during Precambrian time. So it seems reasonable to suspect that the surface runoff after rains was muddy. If the climate really was very hot, surface water probably evaporated quickly, perhaps so quickly that it could not gather into a network of streams that could carry the eroded sediment to the ocean. If so, the drying streams might well have flowed into shallow temporary lakes like those in modern deserts that fill with water after a rain, then dry up a few days later. As the runoff ponded, it would lay down its burden of sediment, then the suncracks would develop as the water evaporated. That kind of deposition could build a deep pile of sedimentary layers regardless of sea level.

Belt formations deposited on land also include thick sandstones full of the distinctive internal layers typical of sands deposited from running water. Those sandstones appear to have accumulated on nearly level plains that stretched for miles, on water washed surfaces similar to those that form in modern deserts. The similarity between land surfaces of Precambrian time and modern deserts lies in the lack of plant cover, not necessarily in the climate. Although blue-green algae lived in Precambrian time, they do not form the kind of plant cover that might spread an umbrella of leaves across the ground. Rain water probably ran off the shadeless landscape of Precambrian time, just as it does in the barren deserts of today.

Despite those enormous blankets of sand that must have been exosed to the air, no Belt formation contains the distinctive internal layering of sand dunes. There was plenty of sand. All the suncracked mud surfaces in other formations clearly show that the ground did dry out in Precambrian time. So where was the wind? To judge from what we can see in the rocks, whatever wind blew in the Precambrian world did not blow hard enough to move sand.

In most of the more western part of northwestern Montana, the Belt formations consist of dark rock in somber shades of gray and brown, rock that shows every evidence of having been deposited under water. Some geologists contend that those formations accumulated in sea water, others that they were deposited in some sort of deep lake. At this point, it is impossible to be sure. Remember that much, if not all, of the sediment that became those formations had probably been laid down before the continental rift established a new sea coast in eastern Washington. So the area was well within the continent, and that helps an argument in favor of a lake. But there is nothing unreasonable about suggesting that an interior part of a continent was below sea level. Nor is it unreasonable to suggest that an early rift like the modern Red Sea may have developed within the continent.

The Precambrian formations of northwestern Montana include several enormously thick sections of limestone and dolomite that must have accumulated under water. Again, it is not clear whether the water was fresh or salty. But most of those rocks are full of fossil blue-green algae, so it is at least certain that the water was shallow enough to pass sunlight to the bottom to support green plants.

Many of the Precambrian sedimentary formations of western Montana consist of alternating thick or thin layers of red and green mudstone. The red mudstones contain iron oxides whereas the green layers get their color from silicate minerals that contain iron in its more reduced, ferrous state. The difference in color seems to reflect differences in the original mud, not something that has happened to it since it was deposited. Evidently the red layers accumulated under conditions in which iron was more oxidized than in the green layers. Rain freshly fallen through an atmosphere rich in carbon dioxide would probably have been acidic enough to keep the iron in its reduced state so the green silicate minerals could form. As a pool of water evaporated, it would become more alkaline, and then the iron oxide minerals that color the red layers could form.

PALEOZOIC AND MESOZOIC FORMATIONS

Virtually all of the Paleozoic and Mesozoic sedimentary rock in northwestern Montana is now in the eastern part of the Sawtooth Range. When those rocks formed, they probably spread their blanket over much, if not all, of northwestern Montana.

Paleozoic Formations

Cambrian sedimentary rocks exist in the Sawtooth Range, also here and there near a line from Libby south through Thompson Falls to Alberton. Although they include some thick limestones, including one that forms the Chinese Wall in the Bob Marshall Wilderness, the Cambrian formations do not contribute much to the landscapes visible from the highways of northwestern Montana. The same can be said of the Devonian sedimentary rocks that also appear in the Sawtooth Range.

Everywhere it comes to the surface, the Madison limestone, which accumulated during Mississippian time, is a conspicuous part of the landscape. In northwestern Montana, it forms the forbidding white wall that is the eastern front of the Sawtooth Range all the way from just north of Helena to just south of Glacier Park.

Mesozoic Formations

Mesozoic formations are visible in northwestern Montana only in the eastern part of the Sawtooth Range, and in the hills just east of that mountain front. They consist mostly of sandstones and shales deposited near the shoreline during Cretaceous time, while the region was shallowly flooded.

The Overthrust Belt

If geologists generally agree that the rocks of northwestern Montana all migrated east, that doesn't mean that they also agree about why, how, and when those rocks moved. We will present our own analysis, starting with that extremely peculiar feature, the Rocky Mountain trench. It seems to split northwestern Montana into two geologic provinces.

Eureka, Whitefish, Columbia Falls, Kalispell, Polson, Ronan, and St. Ignatius all lie in the same continuous valley. Montana people call their different parts of it the Tobacco, Flathead, and Mission valleys, and it continues north through British Columbia, where people have their own names. Geologists call the entire 900 mile length of valley, from St. Ignatius to the Yukon, the Rocky Mountain trench. The word trench fits.

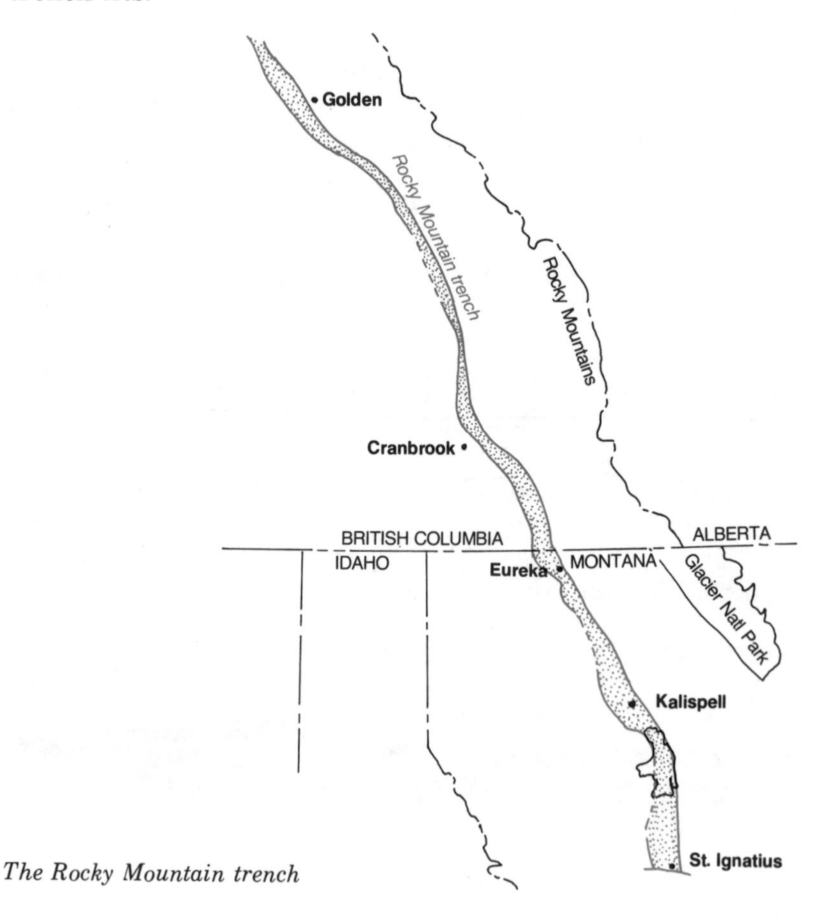

The Rocky Mountain trench

Along much of its length, the Rocky Mountain trench looks like a long trough confined between high mountain walls. That trough traces a course almost precisely parallel to the eastern front of the overthrust belt of the Canadian Rockies and its southern continuation in the Montana Sawtooth Range. And the rocks along the west side of the Rocky Mountain trench have been deeper in the earth's crust than those on the east side. Both the overthrust belt and the Rocky Mountain trench end on the line of the east-trending St. Mary fault, near St. Ignatius.

Its continuity as a valley doesn't necessarily mean that the Rocky Mountain trench is geologically the same along its entire length. In fact, there is good reason to suppose that different parts may have formed in different ways. Fortunately, the title of this book doesn't obligate us to consider the Canadian part of the Rocky Mountain trench.

The Montana part of the Rocky Mountain trench marks the western margin of the part of the overthrust belt that consists of a series of great slabs, relatively thin compared to their size, that moved east and came to rest stacked on each other like shingles on a roof. Those slabs may well have been at the earth's surface when they moved, and certainly were not at any great depth. Geologists call that kind of fault movement thin-skinned tectonics. Although the bedrock west of the Rocky Mountain trench also consists of large slabs that moved east along faults, the structure there appears to involve enormous and relatively thick slabs that moved at depth along faults that tilt very gently down to the west. That is not thin-skinned tectonics.

We think it reasonable to suppose that development of the bedrock structure of northwestern Montana proceeded in two stages. The first stage began with the collision between the Pacific Ocean and North America along the former trench in eastern Washington. We think that encounter shoved the deep slabs west of the Rocky Mountain trench east, thus telescoping the sedimentary veneer, perhaps also the continental crust. Shoving those slabs of rock east must have thickened the crust and raised the earth's surface. In the second stage of deformation, the slabs that now lie east of the Rocky Mountain trench spread east from the raised western edge of the continent under the pull of gravity. In other words, movement along the deep faults west of the Rocky Mountain trench created the conditions for later movement along the shallow faults to the east.

Nearly all the rocks in the area west of the Rocky Mountain trench are either Precambrian sedimentary rocks, Belt formations, or diabase sills that intruded them during the big continental split of Precambrian time. The only exceptions are a discontinuous belt of Cambrian formations that trends from near Libby south through the Thompson Falls area to Alberton and the west end of the Missoula Valley. The Cambrian formations between the Libby and Alberton areas are sandwiched between slabs of Belt rocks that were shoved east along thrust faults.

Bedrock exposed east of the Rocky Mountain trench includes a wide variety of sedimentary rock formations deposited through long periods of geologic time. There are several Precambrian Belt formations, all probably younger than those exposed west of the Rocky Mountain trench, all exposed mostly in the western part of the region. Paleozoic formations deposited in sea water between 600 and 200 million years ago form most of the bedrock in the eastern part of the region. Mesozoic sedimentary rocks dominate the easternmost part of the region.

Most of the northern Montana part of the overthrust belt is in the Sawtooth Range, geologically the southern end of the Canadian Rockies. It extends from the line of the St. Mary fault in the area just north of Helena to Glacier Park, and consists essentially of a series of long ridges that trend from north to south between parallel valleys. Each ridge is the upturned edge of a tilted layer of resistant rock; each valley follows the trend of a layer of weak rock. The extraordinary arrangement of those layers of resistant and weak rocks in a series of overlapping displaced slabs leads geologists to call the Sawtooth Range the overthrust belt.

The disturbed belt is a parallel band of similarly deformed rocks that extends in most areas about 10 miles east of the Sawtooth Range. The main difference between the overthrust and disturbed belts is in the strength of the rocks. The older sedimentary formations in the overthrust belt consist largely of hard rocks that remained fairly stiff during deformation.

Some of those rocks resist weathering and erosion well enough to stand up in the bold ridges that make the Sawtooth Range. The much younger formations in the disturbed belt consist mostly of weak rocks, sandstones and shales, that deformed easily into very tight folds. Those rocks now erode into low hills hardly more striking than those eroded in the same formations farther east, where the layers still lie flat. The change in resistance of the rocks, more than the way they are deformed, defines the front of the Sawtooth Range.

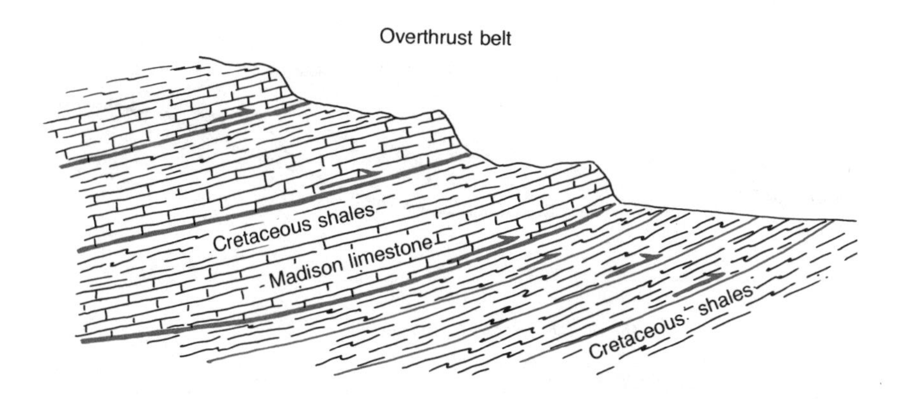

Typical cross section through the overthrust and disturbed belts showing overthrust slabs piled on each other, and becoming younger from west to east. The easternmost slabs of resistant Madison limestone define the mountain front.

The Sawtooth Range consists of many slabs of sedimentary rock that tilt gently down to the west, and overlap each other like the shingles on a roof. Each slab is a long way from home, having moved to where we now see it by sliding east on a fault that lies almost flat, an overthrust fault. In general, the rock formations in the successive overthrust slabs become progressively older from east to west, so older rocks lie on top of younger ones.

The original stack of sedimentary formations accumulated from bottom to top as younger formations were laid down on their older predecessors. What process shuffled them into their present reversed arrangement with the older rocks lying on the younger ones? Various geologists imagine several ways that could have happened.

The simplest explanation that seems to fit the rocks requires us to imagine a slab of rocks moving east toward some obstacle, perhaps a ridge. Now imagine that the moving layer encounters the obstacle, climbs the ramp of its side, and shears off. That will leave the sheared off slab on the side of the ramp waiting to do the same thing to the next part of the moving slab that attempts to climb over. As that process repeats itself, the sheared off slabs will stack overlapped on each other in a pattern like that in the overthrust belt.

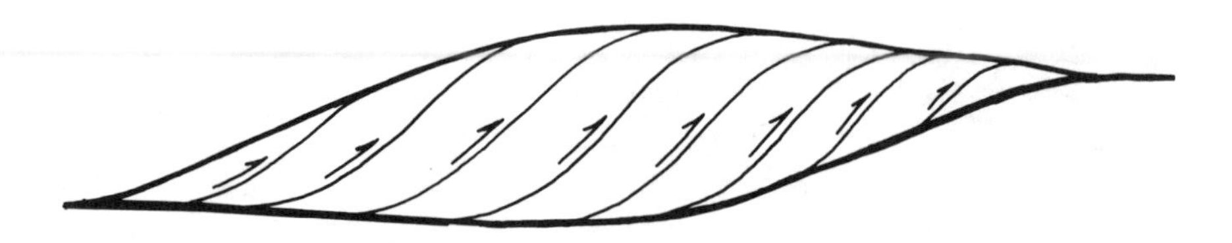

One way the overthrust belt may have formed

Rocks in the disturbed belt east of the Sawtooth Range include Cretaceous formations that were deposited in shallow water as recently as 70 million years ago. Obviously, the deformation that crumpled those rocks must have happened since 70 million years ago. The North Fork Valley along the west side of Glacier Park, near the west side of the overthrust belt, contains basin fill sediments that may be as much as 45 million years old. So the North Fork Valley must have existed by then. Age dates on minerals heated by friction along the moving faults in the disturbed belt give figures in the range from 65 to 70 million years—that is probably when it all happened.

Does Northwest Montana Contain Oil and Gas?

We write this at a bad time just as exploration for oil and gas in the northern Montana overthrust belt begins and no one knows what will be found. It would be easier and safer to wait until the end of the century, when everyone will know how things turned out.

It is proverbial that someone looking for elephants should begin by going to elephant country. Even if not proverbial, it is almost equally true that someone looking for oil should begin by going to oil country. The northern Montana overthrust belt has never been oil country, or even very close to oil country. Neverthless, many people think that may be its destiny. If so, several geologic realities may conspire to limit that destiny.

First, consider the rock structures. The Sawtooth Range contains extremely tight folds and faults almost beyond counting. Together, those faults and folds certainly conspire to create many closed structures of the kind that would trap any oil and gas that may exist in the rocks. Likely places to drill certainly exist. Unfortunately, the complexity of the deformation also makes it likely that the closed structures will be small, as will any reserves of oil and gas they may contain.

The same rock formations that exist tightly folded and faulted in the overthrust belt lie almost flat beneath the high plains to the east. Those formations produce abundant oil and gas in the high plains of Alberta, so it is no surprise to find that they also contain gas in the Alberta part of the overthrust belt. Efforts to extend the oil country of Alberta into the high plains immediately east of the mountain front in Montana have yielded relatively little south of the big oil fields near Cut Bank. That is about where the oil country seems to stop. Perhaps the rocks farther south contain less organic matter, or possibly they have never been quite hot enough to cook the organic matter they do contain into oil. Whatever the reason, the scarcity of oil and gas in the high plains of Montana near the Sawtooth Range suggests that the potential of the Sawtooth Range may also decrease southward.

Different problems exist west of the Rocky Mountain trench. Exposed bedrock there consists almost entirely of Precambrian formations, which never contain petroleum. One look at the geologic map is enough to discourage almost anyone. But ordinary geologic maps show only the kinds of rocks exposed at the surface, not what lies at depth.

There is every reason to believe that the Precambrian formations at the surface rest on faults; considerable reason to suspect the existence of younger sedimentary formations beneath at least some of those faults. If the younger formations exist, they are probably Paleozoic rocks similar to those exposed in the Sawtooth Range, and could contain oil and gas. Nevertheless, the chances of large production from the area west of the Rocky Mountain trench seem remote.

The most obvious problem is the great depth to the potentially productive rock formations. High drilling costs will discourage exploration, and also mean that wells must produce fairly large quantities of oil and gas to return a reasonable profit. Small discoveries will not turn northwestern Montana into oil country. Great depth poses another problem. Temperature increases downward, so deep burial may cook oil and gas out of the rock. The rocks exposed at the surface west of the Rocky Mountain trench are slightly metamorphosed, so there is no doubt that they were very hot. If the rocks at depth were equally hot, they certainly lost any petroleum they may have contained.

THE ICE AGES

Northwest Montana is the most heavily and most picturesquely glaciated part of the state. All the higher mountains had large glaciers in them during the ice ages; quite a few have small glaciers in them now. Those glaciated mountains are easy to recognize by their jagged skylines, parades of gnarled peaks that drop off into deeply gouged valleys. In many northern parts of the region, the ice age glaciers covered the mountains so broadly that they coalesced to make a nearly continuous sheet of ice, instead of isolated valley glaciers. Northwestern Montana really was ice bound.

Throughout the region, two conspicuous sets of glacial features clearly record two ice ages. The largest moraine, the one farthest down the valley, is also the oldest. It formed during

Scratches left in the bedrock under a glacier.

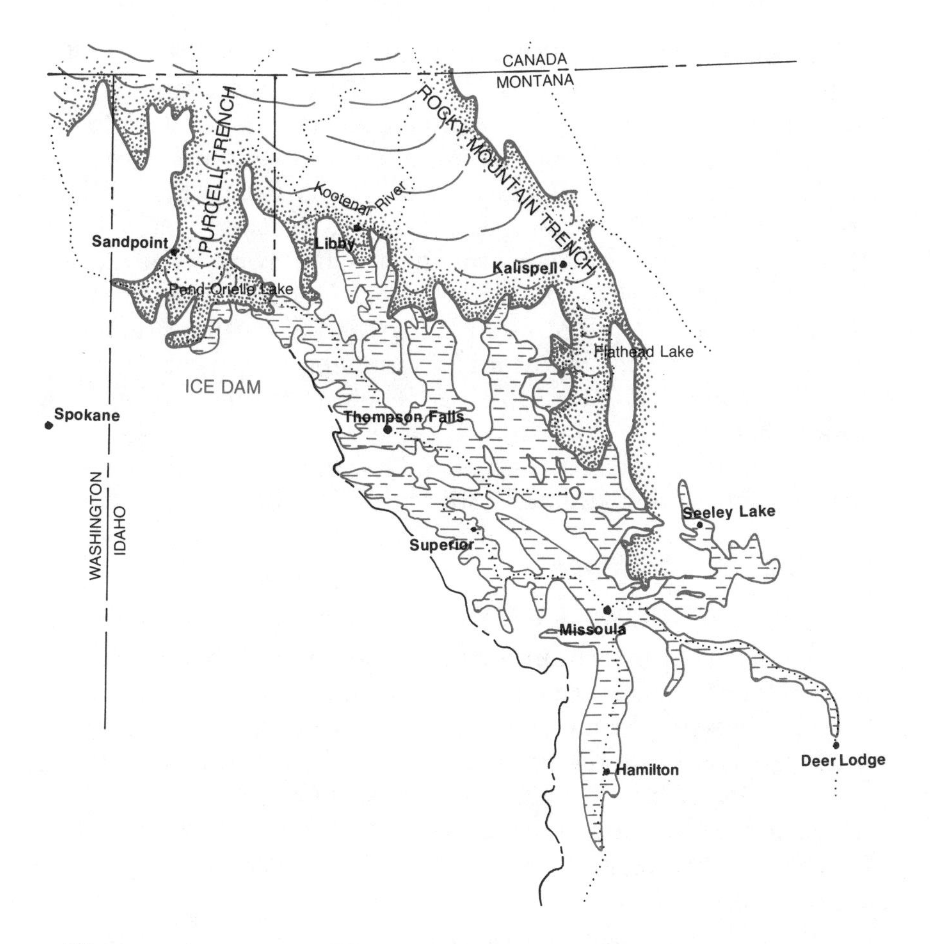

The approximate extent of ice age glaciers in northwestern Montana. Leading edges of ice in dotted pattern; Glacial Lake Missoula in horizontal dashes. Highways dotted.

the Bull Lake ice age, which reached its maximum sometime between 70,000 and 130,000 years ago. The smaller moraine not quite so far down the valley records the maximum advance of the Pinedale glaciers.

During the ice ages, glaciers flowing out of the mountains of British Columbia completely filled the Rocky Mountain trench to form an enormous valley glacier that was about 6000 feet deep where it crossed the Canadian border to enter Montana. As it moved into the Flathead Valley, the Rocky Mountain trench glacier split on the wedge of the low northern end of the Mission Range, sending one branch down the Swan Valley, the other down the Mission Valley. During the Bull Lake ice age, the glacier buried the Mission Range as far south as Ronan, and left its terminal moraines at Ninepipes Reservoir in the Mission Valley, slightly south of Clearwater Junction in the Swan Valley. The Pinedale ice age version of the Rocky Mountain trench glacier reached only as far south as Polson in the Mission Valley, Salmon Lake in the Swan Valley.

Glacial Lake Missoula

The first geologist to visit the Missoula Valley, a member of one of the early western surveys, reported in 1878 that a glacial lake had once flooded the area. The evidence was unmistakeably obvious in the numerous horizontal shoreline benches faintly grooved into the grassy slopes of the mountains at the eastern edge of the valley. Only a series of old shorelines could form a set of absolutely horizontal, perfectly parallel lines on a hillside. Nothing else in nature appears as a perfectly horizontal line on a hillside.

As the ice age glaciers approached their maximum extent, mountain glaciers emptying into the broad trench of the Purcell Valley of British Columbia filled it with ice to form an enormous glacier that finally pushed south into northern Idaho. When that large glacier, which was at least 20 miles wide, crossed the valley of the Clark Fork River at the present site of Pend Oreille Lake, it became an ice dam that impounded the river to form Glacial Lake Missoula. It also impounded the Kootenai River to form another ice dammed glacial lake that may have connected with Glacial Lake Missoula. That probably happened about 15,000 years ago, give or take a

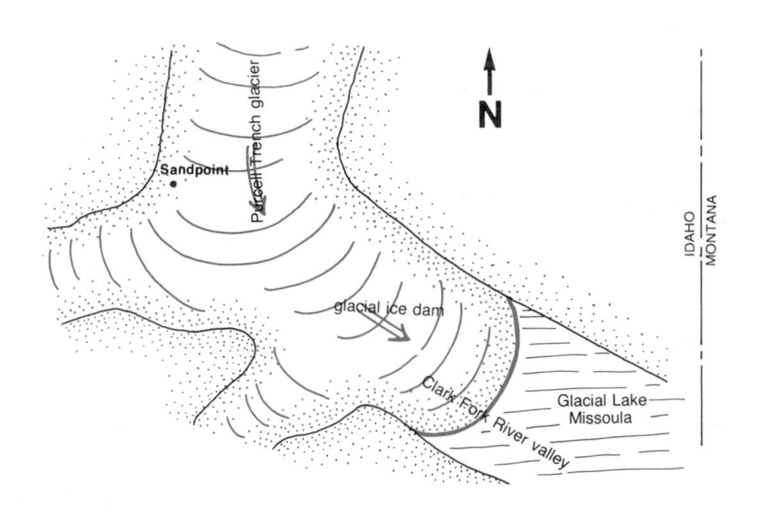

The Clark Fork River was dammed by a large glacier filling the Purcell trench about 15,000 years ago.

thousand years. The same thing must have happened during the earlier Bull Lake ice age, probably on a much larger scale.

The entire flow of the Clark Fork River, which was probably considerably greater in the wet climates of the ice ages than it is today, backed up behind the glacier dam. Year after year,

Snow melting on an afternoon in March emphasizes the old glacial lake shorelines on Mount Sentinel, at the east end of the Missoula Valley.

Glacial Lake Missoula flooded deeper into the valleys of the Clark Fork drainage, spread farther into their headwaters. At its maximum during the last ice age, the lake level reached an elevation of about 4350 feet. The water was then at least 2000 feet deep at the ice dam, and the volume of the lake was about 500 cubic miles—comparable to that of modern Lake Ontario.

Ice is disastrously poor material for a large dam because it floats before the water behind it gets deep enough to spill over the top. When that happens, the glacier breaks up, and the dam fails catastrophically. When the ice dam in northern Idaho floated, Glacial Lake Missoula emptied in a matter of days, releasing the greatest flood of known geologic record.

After the ice dam washed out and the lake drained, the glacier continued to flow south through the Purcell Valley until it established a new ice dam, perhaps within a period of a few years. Then the impounded Clark Fork River again flooded the mountain valleys of western Montana to form a new Glacial Lake Missoula that deepened year after year until it, too, floated its ice dam, and drained.

Evidence of those catastrophic drainages of Glacial Lake Missoula exists in many parts of northwestern Montana, most conspicuously as ponds and giant ripples in places where the water flowed through narrow places. Ponds formed where the rush of water scoured holes in the bedrock. Giant ripples look from the air about like the sand ripples that form in the beds of streams. But their scale is so overwhelming that it is difficult to perceive them from the ground.

Every time the ice dam in northern Idaho floated and broke, it emptied Glacial Lake Missoula in a sudden flood that poured west and southwest from the Spokane area across eastern Washington to the Columbia River. Those floods, geologists call them the Spokane floods, started as walls of water as much as 2000 feet high at the ice dam. They completely filled the normal stream valleys of eastern Washington and poured across the divides between them to erode a vast system of high stream channels, now dry, called the channeled scabland. After having throughly scoured a large part of eastern Washington, the floodwaters continued down the Columbia River to the Pacific. Boulders of Belt rock brought from western Montana in floating icebergs that stranded along the sides of the Columbia

Gorge lie as much as 1000 feet above the river level, showing that the floodwaters there were at least that deep. Similar boulders litter the floor of the Willamette Valley as far south as Salem, Oregon, showing that the floodwaters ponded there, too.

The Spokane floods created many temporary lakes in eastern Washington as they backed up behind bottlenecks such as the Columbia Gorge, and reversed the flow of streams tributary to the Columbia River. Those temporary lakes probably lasted no longer than a few days, but they were extremely muddy and deposits of sediment remain to record their brief existence. Geologists working in eastern Washington have found as many as 41 layers of sediment laid down one upon the other in places that held temporary lakes during the Spokane floods. They record at least 41 Spokane floods. In western Montana, deposits of lake sediment record at least 36 fillings of Glacial Lake Missoula. The difference between 36 and 41 is not important because some fillings that may not have been deep enough to leave a record of themselves in the higher valleys of western Montana would still have flooded the valleys of eastern Washington when they drained. Glacial Lake Missoula probably filled and drained at least 41 times.

Glacial lakes keep their archives in thin layers of light and dark sediment called varves. Glacial meltwater is typically milky with finely ground rock flour pulverized as the rocks embedded in the moving glacier grind each other. Rock flour accumulates on the floors of glacial lakes during the summer when large quantities of ice melt. Meanwhile, algae and microscopic animals flourish in the sunlit surface waters of the lake. The coming of winter ends the melting, thus cutting off the supply of rock flour, and the algae and animals that thrived during the long summer days die with the freeze. Their remains settle to the lake floor during the winter to become a layer of dark sediment. Each pair of light summer and dark winter layers records the seasons of one year of the glacial lake's existence.

Counting the varves in Glacial Lake Missoula sediments left during the last ice age in western Montana shows that the earliest filling of the lake lasted for 58 years. Each successive filling was for a shorter period of time, with the latest lasting only nine years. In all of its 36 fillings during the last ice age in western Montana, the lake existed for a total of just under 1000 years. Varved sediments deposited from the Bull Lake ice age version of Glacial Lake Missoula also exist, but no one has counted their varves.

The lake must have grown deeper and larger with every year the ice dam lasted. Because each successive filling lasted fewer years, each shoreline must be younger than the one above it, older than the one below. Therefore, the highest shoreline must correspond to the earliest filling, which lasted for the longest time. The flood deposits in the valleys of eastern Washington reflect the same pattern. The oldest are thickest and they become thinner upward, thus suggesting that the earlier versions of the Spokane flood were the greatest.

Evidently the glacier of the last ice age did not push far enough south in the Purcell Valley to dam the Clark Fork River until nearly the end of the ice age. Then the climate began to change, and the glacier began to get thinner. As the glacier thinned, it floated in shallower water, so each successive filling of the lake drained at a lower level, after fewer years.

GNP

Ice-age glaciers sculptured these craggy peaks along the crest of Glacier National Park.
—National Park Service photo

GLACIER NATIONAL PARK

Glacier National Park is a geologic as well as a scenic spectacular. This is where geologists first recognized the existence of large overthrust faults back in the 1890s. The park also contains some of the finest exposures of Precambrian Belt formations in the country. And the landscape is a magnificent example of what large valley glaciers can do to a mountain range.

Lewis Overthrust

Most of Glacier Park is an enormous slab, some two miles thick, of Precambrian Belt rocks that moved east at least 35 miles, probably considerably more. Now those rocks, a billion or more years old, lie on top of Cretaceous sedimentary rocks only 65 to 100 million years old. Although they are not exposed, there is little reason to doubt that Paleozoic formations also lie beneath the slab. The Precambrian rocks moved on a surface called the Lewis overthrust fault.

Watch along the eastern front of the mountains for the abrupt change from gently rolling hills eroded in the soft Cretaceous formations to steep cliffs above in the more resistant Precambrian rocks. That is the line of the Lewis overthrust. Because the fault lies almost flat, it makes a raggedly embayed trace along the mountain front, dodging far back into the big valleys, and then swinging east around the bases of the mountain ridges.

North Fork Valley

The North Fork Valley, a long trough that extends well into British Columbia, lies along the western margin of Glacier Park. According

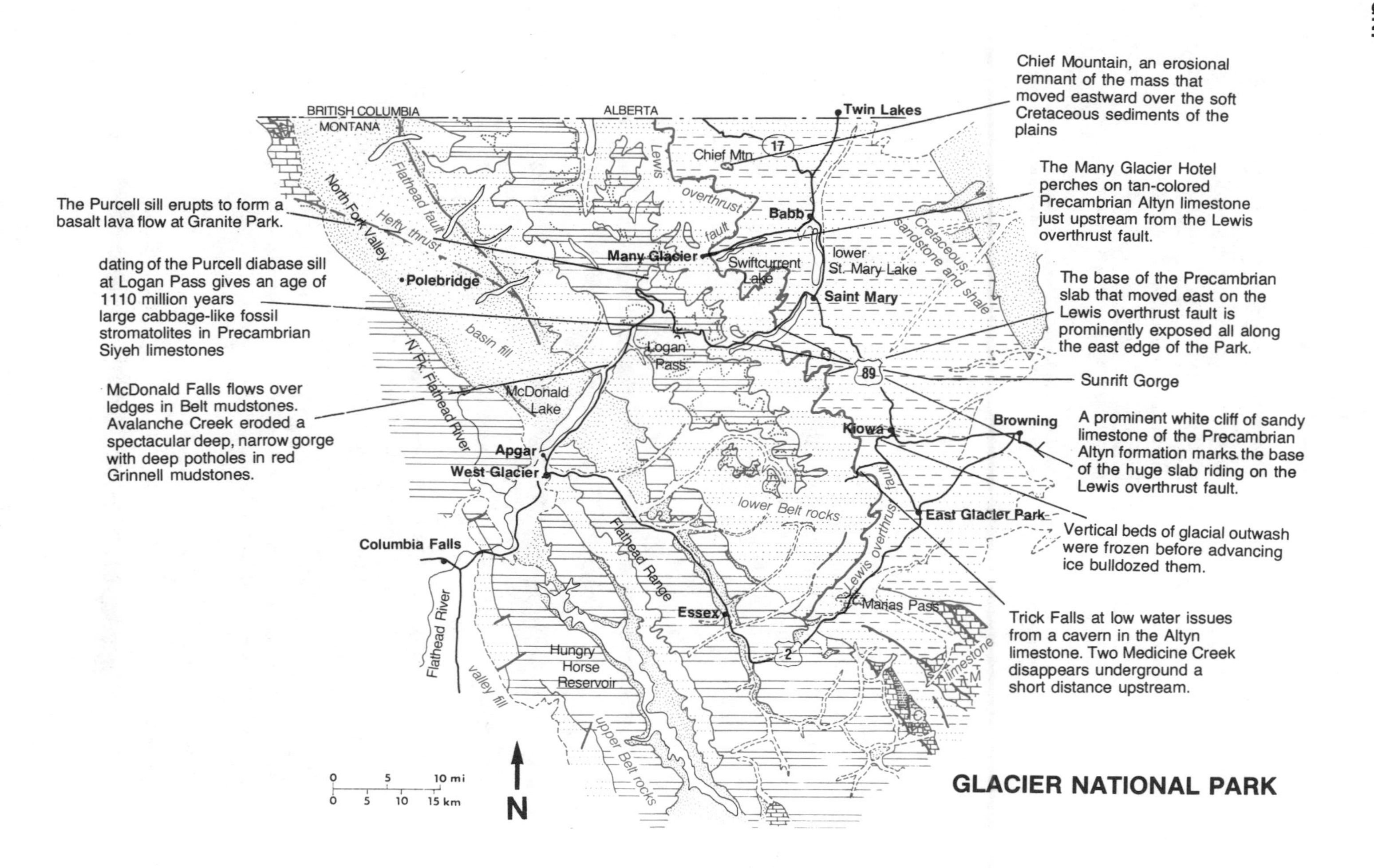

GLACIER NATIONAL PARK
Chief Mountain, an erosional remnant of the mass that moved eastward over the soft Cretaceous sediments of the plains
The Many Glacier Hotel perches on tan-colored Precambrian Altyn limestone just upstream from the Lewis overthrust fault.
The base of the Precambrian slab that moved east on the Lewis overthrust fault is prominently exposed all along the east edge of the Park.
Sunrift Gorge
A prominent white cliff of sandy limestone of the Precambrian Altyn formation marks the base of the huge slab riding on the Lewis overthrust fault.
Vertical beds of glacial outwash were frozen before advancing ice bulldozed them.
Trick Falls at low water issues from a cavern in the Altyn limestone. Two Medicine Creek disappears underground a short distance upstream.
The Purcell sill erupts to form a basalt lava flow at Granite Park.
dating of the Purcell diabase sill at Logan Pass gives an age of 1110 million years
large cabbage-like fossil stromatolites in Precambrian Siyeh limestones
McDonald Falls flows over ledges in Belt mudstones. Avalanche Creek eroded a spectacular deep, narrow gorge with deep potholes in red Grinnell mudstones.
BRITISH COLUMBIA
MONTANA
ALBERTA
Twin Lakes
Chief Mtn
17
Lewis
overthrust
fault
Babb
Many Glacier
Swiftcurrent Lake
lower St. Mary Lake
Cretaceous sandstone and shale
Saint Mary
89
Logan Pass
North Fork Valley
Flathead fault
Hefty thrust
Polebridge
basin fill
N. Fk. Flathead River
McDonald Lake
Apgar
West Glacier
Browning
Kiowa
East Glacier Park
Lewis overthrust fault
lower Belt rocks
Flathead Range
Columbia Falls
Flathead River
Essex
Marias Pass
limestone
Hungry Horse Reservoir
valley fill
upper Belt rocks
2
0 5 10 mi
0 5 10 15 km
N

to some geologists, it may have opened as the earth's crust stretched after the overthrust belt formed. It contains a deep fill of basin sediments often called the Kishenehn formation even though they appear to be the same as the Renova formation of other valleys. Deep deposits of glacial debris blanket most of the valley floor, making exposures of the older fill hard to find.

Early settlers noticed that they could recognize bear skins from the North Fork Valley because they smelled like kerosene. They solved that mystery during the 1890s by finding several oil seeps, in which the bears wallowed. A well drilled at the head of Kintla Lake in 1901 penetrated only Belt rock, and found no trace of oil. Several wells drilled in the valley since then did find shows of oil and gas, but no actual production. And geologists have found at least several hundred feet of oil shale in the Renova formation. It is not clear whether the oil is coming from the oil shale or, as seems more likely, from older rocks deep beneath the valley floor.

Regardless of its source, an oil seep certainly tells us something, and many geologists expect to see the North Fork Valley produce oil. If the oil shales are the source of the seeps, commercial production seems unlikely. The Renova formation has not been hot enough to convert much of its organic matter into oil.

Belt Formations

Glacier Park is one place where anyone can quickly learn to recognize rock formations. You can tell most of them apart by their colors.

The Altyn limestone is the oldest Precambrian Belt formation in the park. It is tan on a weathered outcrop, white in the fresh roadcuts, and everywhere full of sand grains. Look for the Altyn limestone just above the Lewis overthrust fault. The easiest places to see it are in roadcuts beside St. Mary Lake just west of the Rising Sun Campground, and outcrops around the parking lot of the Many Glacier Hotel.

The Appekunny formation, some 3500 feet of green mudstone, lies next above the Altyn limestone. It appears most conspicuously in the east side of the park, especially in big roadcuts near the western end of St. Mary Lake. The equivalent mudstones on the west side of the park are almost black, and look more like another rock unit that geologists call the Prichard formation.

Some 2500 feet of barn red mudstones, the Grinnell formation, lie above the green Appekunny mudstones. The two formations grade into each other through an interval several hundred feet thick in which layers of red and green mudstone alternate. Many of the layers

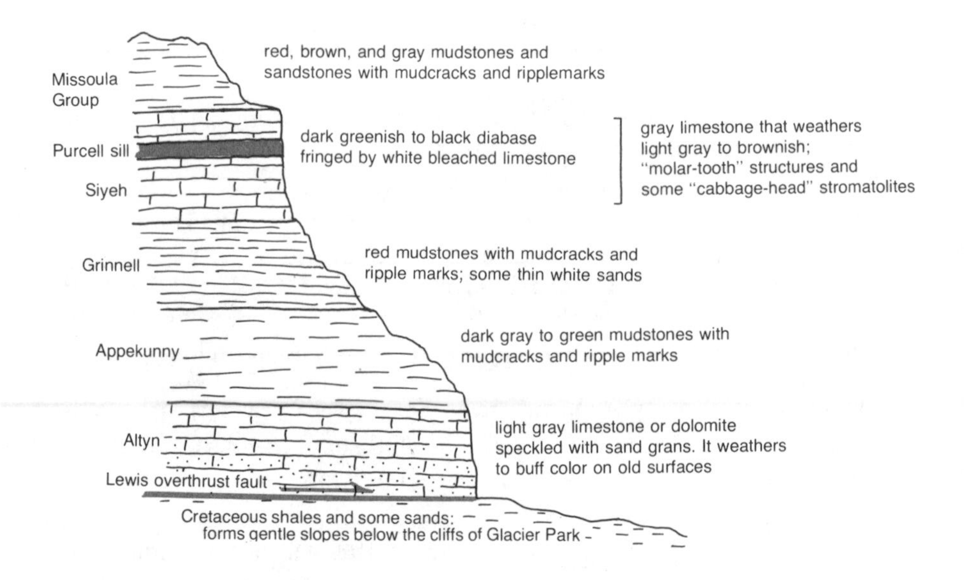

Precambrian rock formations exposed along the Going to the Sun Road

are full of beautifully preserved mudcracks, ripple marks, rain drop imprints, and other such structures. The Grinnell formation also contains layers of snow-white sandstone that contrast beautifully with the red mudstone.

About 3500 feet of gray limestone, the Siyeh formation, lies on the red Grinnell mudstones. Look for the Siyeh formation in the pale gray cliffs high in the park, and in much darker gray roadcuts on both sides of Logan Pass.

Mudcracks on a bedding surface of the Grinnell mudstone.

GNP

A reef of blue-green algal structures called stromatolites exposed in a roadcut in the Siyeh formation west of Logan Pass.

Many exposures of the Siyeh formation contain curious structures called stromatolites, that look almost like cabbages, but are actually the remains of extremely primitive blue-green algae, the most abundant and advanced form of life a billion years ago. To judge from the abundance of fossil algae in the Siyeh formation, it must have accumulated in shallow water, scummy with algae. Those primitive green plants absorbed carbon dioxide from the atmosphere, used the carbon to make their tissues, and released free oxygen. Some of the algae were buried in the accumulating lime muds before they could decay, and their incompletely decomposed tissues give the fresh exposures of the Siyeh limestone their dark color. For every atom of carbon in the rock, there is one molecule of oxygen in the atmosphere—that is how the earth acquired its breatheable air.

The youngest Precambrian rocks in the park belong to the Shepard and Kintla formations, which consist mostly of mudstones full of mudcracks and ripple marks. The Shepard formation appears from a distance as an inconspicuous tan section above the gray Siyeh limestone. The Kintla formation is bright red, and impossible to overlook. Watch for high peaks along the continental divide that wear a bright red cap of Kintla mudstone. The easiest place to see both formations at close range is along the trail from Logan Pass to Hidden Lake.

The Purcell Sill

The Purcell sill is a thick slab of black diabase, a coarse grained variety of basalt, that squirted as molten magma between rock layers in the upper part of the Siyeh formation. Heat from the magma baked

The Purcell sill is faintly visible in the distant cliffs overlooking Many Glacier.

the dark organic matter out of the nearby Siyeh limestone, bleaching it almost snow white. Watch in the high cliffs for the layer of dark diabase fringed above and below with thin ribbons of sparkling white marble.

At Granite Park there is no granite. The magma that formed the Purcell sill erupted there to become a basalt lava flow. Therefore, we can be sure that the sill was injected into the Siyeh formation while it was still layers of lime mud accumulating in shallow water. Age dates on the Purcell sill yield a figure of some 1200 million years, one of the best indications we have of the age of the Precambrian Belt rocks. We can be sure that all of the rock beneath the Purcell sill is even older.

Glaciation

It should have been called "Glaciated Park." The mighty glaciers of the ice age are gone, and those that now exist in Glacier Park are puny and shrinking, hardly more than oversized snow patches. This is a poor place to see glaciers in action. But every part of the park bears the marks of enormous ice age glaciers that filled the valleys with groaning and creaking rivers of ice thousands of feet thick.

During peak ice age glaciation, the cover of ice on Glacier Park almost became a continuous ice cap. To visualize the park as it then was, imagine a vast sea of ice with isolated peaks jutting above it as islands of exposed bedrock.

On the west side of the park, ice poured down the major valleys to pond in the North Fork Valley, which was completely full of ice. The glacier that moved down McDonald Creek joined another in the valley of the Flathead River, which poured through Bad Rock Canyon to join the vast sea of ice that filled the Flathead Valley. Ice pouring down the major valleys on the east side of the park ponded on the flat country east of the mountains to make a big glacier that spread east across the plains beyond Browning, where it met the continental glacier moving southwest from central Canada. Moraines left by that piedmont glacier cover most of the country between Glacier Park and Browning.

View west up St. Mary Lake to peaks on the continental divide.

The most conspicuous marks of glaciation are the straight valleys that permit long views to distant peaks like those up McDonald and St. Mary lakes. Unglaciated river valleys curve too much to permit such sweeping vistas. And the sides of glaciated valleys tend to rise more steeply than those of river valleys. The upper parts of glaciated valleys have deeply gouged bedrock floors; the lower parts have flat floors underlain by deep deposits of sediment.

As a glacier moves, it pulls away from the head of its valley, opening a gap there between ice and bedrock. Then snow and water fill the gap to form new ice that freezes fast to the bedrock. As the glacier continues to move, the ice again pulls away from the bedrock, taking blocks of rock as it goes. So the upper end of a glacier actively erodes the head of its valley. Now that the ice has melted, we see a deep basin called a cirque broadly scooped into the bedrock at the head of each glaciated valley. Notice that the headwall cliffs of cirques have a roughly quarried look. Most cirques cradle lakes in their floors.

Where several glaciers headed in the same mountain, they left little of the original peak except a gnarled pinnacle of rock, called a horn, that drops off into deep cirques on all sides. Horn peaks abound in Glacier Park. They give glaciated mountain ranges their typical ragged skyline.

In the lower parts of the glaciated valleys, melting ice left deposits of till heaped into moraines that neatly mark the outline of the glacier. Blanket deposits of till floor the lower parts of most glaciated valleys, helping to make the valley floor flat. Glacial meltwater left more sediment in the form of outwash deposits that trail long distances downstream from the moraines. Those deposits are what make the lower valley floors so flat.

Going to the Sun Road: West Glacier—St. Mary

53 mi./85 km.

West Glacier is in the southern part of the North Fork Valley, and St. Mary is on Cretaceous rocks east of the Lewis overthrust fault. So the road across Logan Pass crosses the entire slab of Precambrian rocks that moved east. It provides good views of all the Precambrian formations, as well as of the glaciated landscapes.

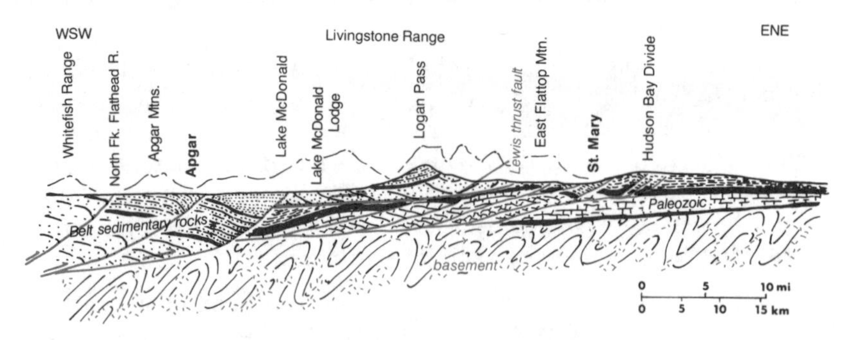

Section approximately along the line of the Going to the Sun Road

Lake McDonald

Lake McDonald, on the west side of the park, fills a deep basin glacially gouged into the soft Renova formation in the floor of the North Fork Valley. Those rocks appear as beds of sticky clay in a few places along the shore near the upper end of the lake when the water is low. Imagine the glacier coming off the hard Precambrian bedrock at the lower end of McDonald Creek, and plunging deeply into the soft bottom of the North Fork Valley. Evidently, a mass of ice lingered where the lake now is as the ice age ended, and so prevented accumulation of outwash that would otherwise have filled the basin. Deep deposits of till and outwash beyond the lower end of Lake McDonald help impound the water.

The long ridges that follow either side of Lake McDonald are lateral moraines deposited along the margins of the big glacier that filled this valley when the last ice age ended. Large glacially transported boulders carried down from the high peaks litter their surfaces.

A few roadcuts and the outcrops in the bed of McDonald Creek immediately east of Lake McDonald expose dark green mudstones. Farther east, and higher in the section, the green gives way to bright red Grinnell mudstones. Red and green pebbles eroded from those mudstones make colorful gravel in the bed of McDonald Creek. Above McDonald Creek, the road traverses the steep valley wall in a long grade, passing a long succession of large roadcuts in gray Siyeh limestone.

Logan Pass

Logan Pass exists where two glaciers eroded their cirques back-to-back to cut a low saddle in the continental divide. During the ice ages, a continuous mass of glacial ice went right over the pass, part moving west, the rest moving east. Look on the cliffs high above the road to see the line where glacially scoured and relatively smooth bedrock surfaces pass upward into ragged outcrops that bear the mark of ice wedging blocks of rock loose. That line marks the former upper surface of the ice.

Roadcuts on both sides of Logan Pass expose the Purcell sill. But the best place to see it is where the south end of the trail to Granite Park passes along the face of the steep cliff above the road. Rock cuts there provide fresh exposures of the black sill, and of the bleached Siyeh limestone above and below it. Fossil blue green algae show up especially well in the bleached limestone.

Good exposures of the Shepard mudstones full of mudcracks and

GNP

The walls and floor of Sunrift Gorge show no sign of stream erosion.

ripple marks abound along the path from Logan Pass to Hidden Lake. Bright red Kintla mudstones cap some of the high peaks around Logan Pass. Red blocks fallen from those cliffs litter the ground in places along the trail to Hidden Lake, providing a convenient view of the Kintla formation for those who prefer not to climb.

Sunrift Gorge

Sunrift Gorge is in the transition zone between the Appekunny and Grinnell formations where roadcuts and cliffs reveal red and green mudstones interfingering in alternating layers. Roadcuts west of Sunrift Gorge expose red Grinnell mudstones; those farther east are in green Appekunny mudstone and tan Altyn limestone.

Sunrift Gorge is a geologic curiosity, a small valley that did not form through erosion. Notice that the gorge is perfectly straight, and that its sides and floor have none of the fluted carving typical of rock shaped by running water. Compare them to the intricately carved rock along the trail to Avalanche Lake. Evidently, the block of rock that lies beneath the hillside south of Sunrift Gorge moved a few feet downhill after the ice melted, and the stream took advantage of the narrow gap behind it.

GNP

Streams carved bedrock in McDonald Creek.

St. Mary Lake

As the last ice age ended, muddy meltwater pouring off a shrinking glacier in a tributary valley deposited the big alluvial fan that impounds upper St. Mary Lake. The St. Mary entrance gate and Visitor Center stand on that fan.

Along most of St. Mary Lake, the glacier cut through the Precambrian formations into soft Cretaceous bedrock beneath the Lewis overthrust. Cliffs of pale Altyn limestone beside the road west of the Rising Sun Campground mark where the road crosses the fault. In most places the Cretaceous rocks east of the fault lie beneath a deep blanket of glacial debris, so it is hard to find exposures.

Many Glacier Road: US 89—Swiftcurrent Lake

12 mi./19 km.

The road between US 89 and Swiftcurrent Lake follows the valley of Swiftcurrent Creek, crossing the Lewis overthrust fault about a mile east of Many Glacier. Except in the Many Glacier area, Cretaceous sedimentary rocks form the bedrock beneath the valley floor; Precambrian sedimentary rocks form the valley walls.

All the Precambrian rocks exposed along the road to Many Glacier are tan limestones that belong to the Altyn formation. Green and red mudstones of the Appekunny and Grinnell formations make broad color bands on the higher slopes. Distant views of the high cliffs in the

Siyeh limestone show the black ribbon of the Purcell sill with its white borders of bleached limestone.

Glaciated cliffs of Siyeh limestone loom above the Many Glacier area. The horizontal black band is the Purcell sill.

The Big Copper Boom

In 1892, prospectors found copper minerals in and near diabase dikes in the Altyn limestone around Swiftcurrent Lake. The area was then part of the Indian reservation, so developers had to wait and agitate until the boundaries were changed in 1898 before they could start mining. Then the little town of Altyn sprang up on the flats about three quarters of a mile below the Many Glacier Hotel.

Many of the diabase dikes and sills in the Precambrian sedimentary formations throughout western Montana deposited small amounts of copper, nowhere enough to support a mine. As with similar occurrences elsewhere, mining around Swiftcurrent Lake wasn't as good as the promoters expected, so Altyn never grew beyond a population of 100. The town was deserted by 1910, and nothing remains but a few crumbling foundations near the head of Sherburne Reservoir.

Oil in Glacier National Park

In 1902, a miner prospecting for copper near Swiftcurrent Lake found oil seeping into his workings. Drilling began in 1903, and a well sunk to a depth of 550 feet in 1905 actually produced small amounts of oil, the first ever found in Montana. Another well finished in 1909 produced natural gas, which was used until 1914 to heat one house, the only customer the field ever had. Other efforts to establish production were even less successful. The boom had already fizzled when Glacier National Park was established in 1910. Sherburne Reservoir flooded the site in 1919.

Two Medicine Road: US 89—Two Medicine Lake
8 mi./13 km.

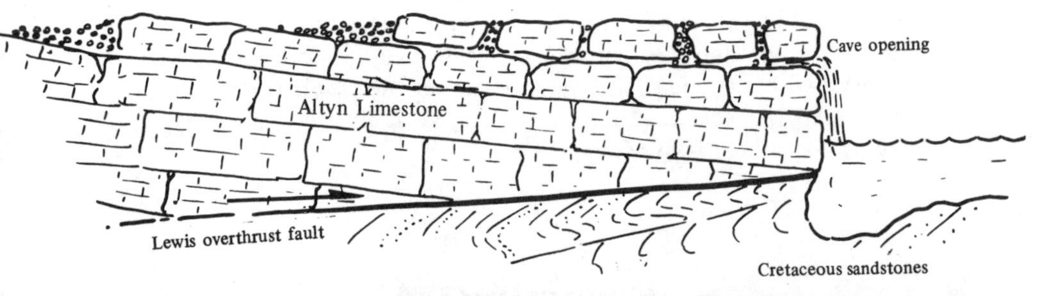

Trick Falls tumbles down a ledge of Altyn limestone just above the Lewis overthrust fault.

The road to Two Medicine Lake follows a valley that ice age glaciers gouged into a spectacular deep trough. The road crosses the Lewis overthrust fault and climbs into the hard Precambrian rocks at

Two Medicine Creek. Cretaceous formations form the bedrock east of that area, Precambrian rocks to the west.

Two Medicine Lake lies where three tributary glaciers joined to form a single, much larger glacier that gouged a deep basin. Many glacial lakes exist in such places, immediately below the confluence of two or more glaciers. Swiftcurrent Lake is another example. After the ice age ended, the thick mass of ice that filled this more deeply gouged area of valley floor lingered after thinner parts of the glacier had melted, and so saved the basin from filling with outwash sediments.

Cliffs above Two Medicine Lake are an especially good place to see the former depth of the ice. Look high for the line that separates glacially smoothed rock below from much rougher outcrops above the former level of the ice. The rough outcrops owe their ragged form to the quarrying action of ice expanding as it freezes in cracks, and prying loose blocks of rock.

US 89: Browning—Alberta Border
51 mi./82 km.

Between Browning and the Canadian line, the highway follows the east front of Glacier National Park. Bedrock near the road is Cretaceous sedimentary formations, mostly buried under glacial deposits. The mountains consist entirely of Precambrian sedimentary formations.

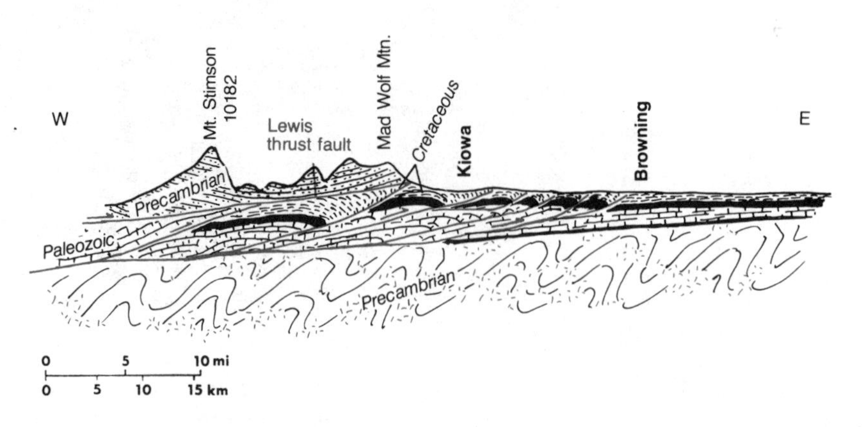

Section north of US 89 between Kiowa and Browning

Glacial till exposed in a roadcut near St. Mary.

From St. Mary to the area several miles north of Babb, the road follows the valley of St. Mary Creek, past lower St. Mary Lake and along several miles of valley deeply floored in glacial outwash. Lower St. Mary Lake appears to flood an area where a large mass of stagnant glacial ice lay unmelted for some time after other parts of the valley were clear. Meltwater could not deposit outwash where the ice lingered, so that part of the valley floor is now a deep basin that holds water.

Chief Mountain

No one who lives within sight of Chief Mountain ever forgets that great gnarled stump west of highway 17 between Babb and the Canadian border. Chief Mountain is an outpost of the great slab of

Section showing the structure of Chief Mountain

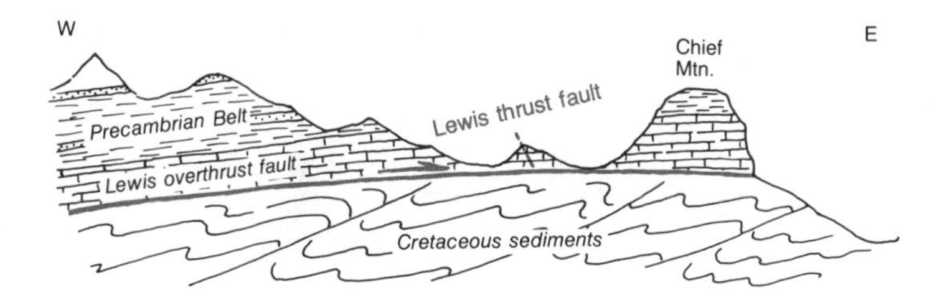

Chief Mountain, a pillar of Precambrian Belt rocks standing on the much younger Cretaceous formations that lie beneath this part of the high plains. Cretaceous rocks lie beneath the forested lower slopes—under the Lewis overthrust.

Precambrian Belt rock that moved east on the Lewis overthrust fault, and now forms the mountains of Glacier National Park. All geologists learn about Chief Mountain as undergraduates because it is such an outstanding example of an outlying bit of an overthrust slab isolated by erosion.

The steep cliffs that surround most of Chief Mountain rise above the Lewis overthrust fault. Rocks exposed in those cliffs are Precambrian sedimentary formations, Altyn limestone and Appekunny mudstone, the formations that form the steep mountain front along most of the eastern edge of Glacier Park. Softer Cretaceous sedimentary formations about 80 or so million years old lie beneath the gently rolling slopes below the cliffs.

Interstate 90: Lookout Pass—Missoula

104 mi./168 km.

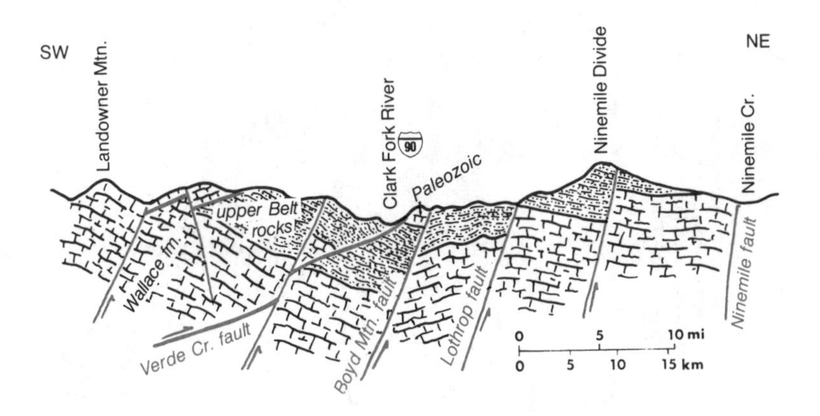

Section across the highway between Tarkio and Superior. Except for the small area of Paleozoic formations on the crest of the ridge north of the highway, all the rocks are Precambrian Belt formations. The Ninemile fault is part of the Lewis and Clark fault zone.

The highway follows the St. Regis River between Lookout Pass and St. Regis, the Clark Fork River between St. Regis and Missoula, through remarkably straight canyons eroded along the Lewis and Clark fault zone. Between Frenchtown and Missoula, the road follows the north side of the Missoula Valley. Except for a few exposures of Glacial Lake Missoula sediments and some Cambrian rocks scattered between the west end of the Missoula Valley and Tarkio, all the rocks are Precambrian sedimentary formations, Belt rock.

Precambrian Rocks

Many of the Precambrian formations exposed between Alberton and St. Regis are colorful red and green mudstones that accumulated on land. The mudstones contain beautifully preserved ripple marks, mudcracks, raindrop imprints, even the little square and triangular dents that the faces and corners of cubical salt crystals make as they imprint soft mud. A careful search of bedding surfaces reveals nearly everything imaginable that can leave an impression in soft mud

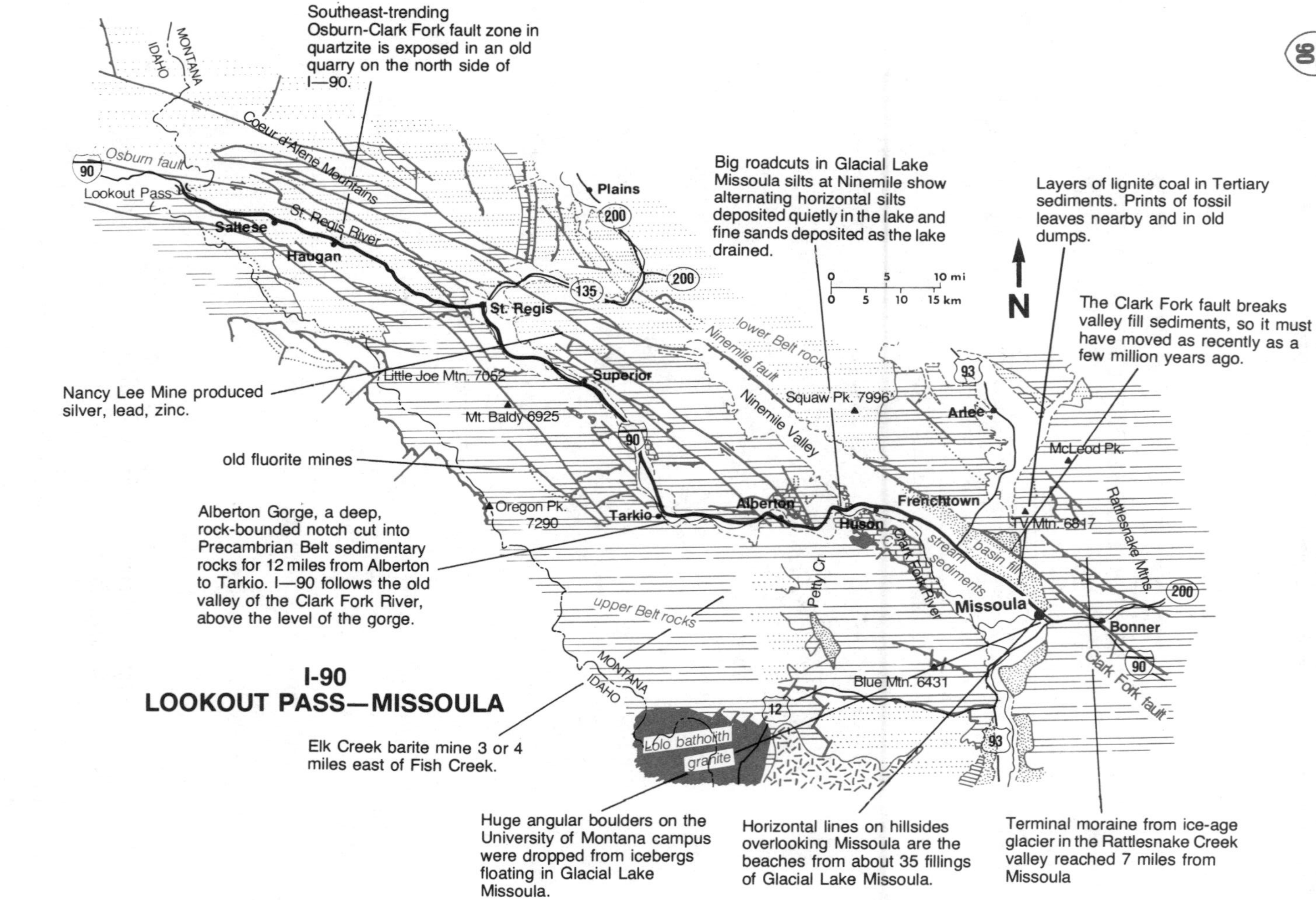

90
Southeast-trending Osburn-Clark Fork fault zone in quartzite is exposed in an old quarry on the north side of I—90.
MONTANA
IDAHO
Coeur d'Alene Mountains
Osburn fault
90
Lookout Pass
Saltese
St. Regis River
Haugan
Plains
200
200
135
St. Regis
Superior
Big roadcuts in Glacial Lake Missoula silts at Ninemile show alternating horizontal silts deposited quietly in the lake and fine sands deposited as the lake drained.
0 5 10 mi
0 5 10 15 km
N
Layers of lignite coal in Tertiary sediments. Prints of fossil leaves nearby and in old dumps.
The Clark Fork fault breaks valley fill sediments, so it must have moved as recently as a few million years ago.
lower Belt rocks
Ninemile fault
Ninemile Valley
Squaw Pk. 7996
93
Arlee
McLeod Pk.
Nancy Lee Mine produced silver, lead, zinc.
Little Joe Mtn. 7052
Mt. Baldy 6925
90
old fluorite mines
Alberton
Frenchtown
Huson
TV Mtn. 6817
Rattlesnake Mtns.
Oregon Pk. 7290
Tarkio
Alberton Gorge, a deep, rock-bounded notch cut into Precambrian Belt sedimentary rocks for 12 miles from Alberton to Tarkio. I—90 follows the old valley of the Clark Fork River, above the level of the gorge.
Petty Cr.
Clark Fork River
stream sediments
basin fill
Missoula
200
Bonner
upper Belt rocks
90
Clark Fork fault
MONTANA
IDAHO
Blue Mtn. 6431
12
93
Lolo batholith
granite
Elk Creek barite mine 3 or 4 miles east of Fish Creek.
Huge angular boulders on the University of Montana campus were dropped from icebergs floating in Glacial Lake Missoula.
Horizontal lines on hillsides overlooking Missoula are the beaches from about 35 fillings of Glacial Lake Missoula.
Terminal moraine from ice-age glacier in the Rattlesnake Creek valley reached 7 miles from Missoula

I-90
LOOKOUT PASS—MISSOULA

Ripple marks on a slab of Precambrian mudstone near Alberton

except for any sign of animal life. In fact, the exquisite preservation of all that detail is one reason to believe that no animals were around to stir the mud.

Cambrian Rocks

Some roadcuts and outcrops along the river between Missoula and Alberton expose steeply tilted Cambrian sedimentary formations. Cambrian formations on the ridge overlooking Alberton must be the source of the tumbled blocks of Hasmark dolomite that stand north of the road at the Alberton exit. These Cambrian rocks lie at the southeast end of a long trend that extends north through Thompson Falls to near Libby. The Cambrian formations were deposited on top of the Precambrian rocks about 500 million years ago, then sandwiched between them along faults that probably moved about 100 million years ago.

The Lewis and Clark Fault Zone

Under different names in different areas, including the Osburn and Clark Fork fault zones, the Lewis and Clark fault zone extends from the Okanogan Valley of Washington at least as far as the area of the Boulder batholith west of Helena. The Ninemile fault, one of many in the zone, defines the straight north side of the Missoula Valley. It sliced some of the Renova formation into Mount Jumbo on the northeast side of Missoula, so it must have been moving as recently as 25 million years ago, but there is no evidence that any of the faults have moved within the last few million years.

The Lewis and Clark fault zone exposed just north of the highway near St. Regis. The chunks of solid rock are in a matrix of crushed rock bleached to pale colors by hot water circulating through the fractured rock.

Ore Deposits

The Superior area appears to be the easternmost extremity of the trend of the heavily mineralized Coeur d'Alene district that follows the Lewis and Clark fault zone through northern Idaho. Numerous mines have produced lead and silver from the hills around Superior, along with small quantities of gold.

The Iron Mountain Mine, on Flat Creek about six miles north of Superior, was one of the largest in the district. Prospectors found outcrops of a vein about 40 feet wide there in 1888, and the mine, along with the little town of Pardee, went into full operation when the railroad came in 1891. Ore went down the mountain to a mill in Superior on an aerial tramway of huge buckets running on steel cables. The mine closed in 1897, then opened again a few years later to limp along off and on for years. Pardee struggled along until about 1930.

Cedar Creek, south of Superior, was once the host of a regular gold rush. Prospectors found placer gold in the creek in late 1869, inspiring a midwinter stampede in which the entire gulch was staked in one day, and almost as quickly acquired a population of some 3000 men, who apparently had little else to do. Some estimates place the total production during those years at around two million dollars, when an ounce of gold was worth about 20 dollars. Considering the number of man years invested in digging gravel in Cedar Creek, that return could not have made many people rich.

Missoula Valley

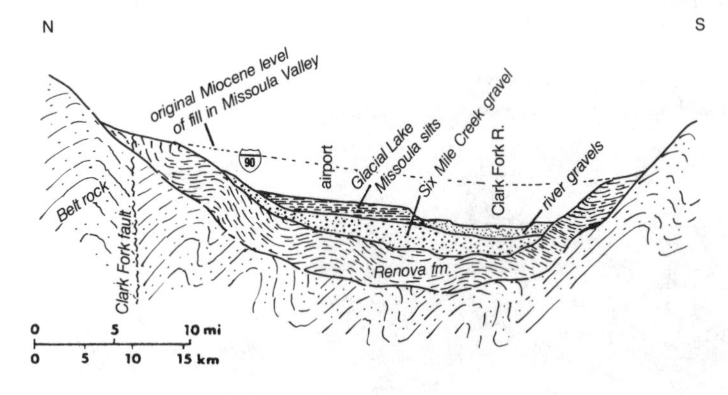

Section across the Missoula Valley

Between Missoula and Frenchtown, the highway follows the full length of the Missoula Valley, a broad structural basin formed during the crustal movements that created the Rocky Mountains. West of Frenchtown, the Missoula Valley continues as the Ninemile Valley, the same basin. Although it probably owes its existence at least in part to the Ninemile fault, the exact origin of the valley is not clear.

The high and glacially carved Rattlesnake Mountains north of the Missoula Valley contain complexly folded Precambrian sedimentary formations, Belt rocks, broken along several large faults. The much lower and unglaciated Sapphire Mountains that rise along the southeast end of the Missoula Valley are the trailing edge of the Sapphire detachment block. They contain less deformed Precambrian formations.

Roadcuts along the highway between Missoula and the area near Frenchtown expose valley fill sediments, along with a few outcrops of the Precambrian bedrock that lies beneath the valley. Watch for the pale colors of the silty Renova formation, the first generation of valley filling sediment. Some roadcuts also contain colorful splashes of reddish brown laterite, the tropical soil that developed during late Miocene time when the region enjoyed a warm climate with abundant rainfall. Brownish gravel exposed near the tops of some roadcuts is the Six Mile Creek formation, which covered the red soil during Pliocene time, the second period of valley filling.

Glacial Lake Missoula

The great rushes of water that drained Glacial Lake Missoula poured down the Clark Fork Valley. Watch in the narrower stretches

of valley between Frenchtown and St. Regis for rough outcrops of bedrock in the valley floor, and for lower valley walls with nearly all the soil scrubbed off.

Dark winter layers alternating with pale summer layers in a section of glacial lake sediment

The large roadcut in soft silts just east of the bridge across the river at Ninemile contains a remarkably complete record of Glacial Lake Missoula. The section there consists of sequences of lake bed deposits alternating with river sediments. There are 36 sequences of lake bed deposits, each recording a separate filling of Glacial Lake Missoula. The river deposits that separate them accumulated while the lake was drained.

Yaak Falls; the river slides down a bedding surface in green Precambrian mudstone.

US 2: Kalispell—Idaho Border
121 mi./195 km.

All the bedrock between Kalispell and Libby is Belt rock, Precambrian sedimentary formations deposited more than a billion years ago. Along much of the route, that bedrock lies beneath deposits of glacial debris left when the glaciers melted at the end of the last ice age.

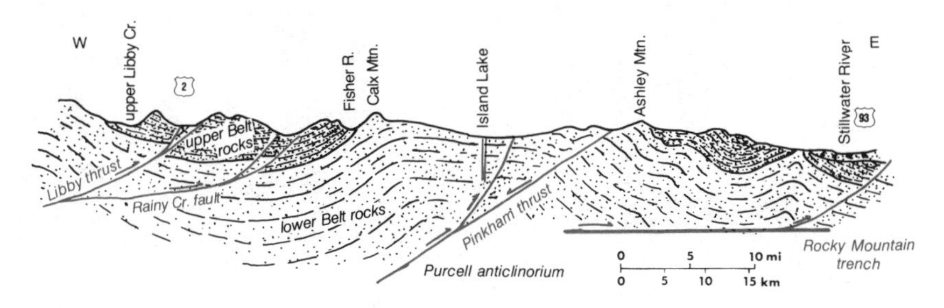

Section along a line north of US 2 from about 15 miles south of Libby to Kalispell. Except for the thin wedge of Cambrian formations near Libby, all the rocks are Precambrian Belt formations. They were warped into a broad fold, as they were shoved east along deep thrust faults.

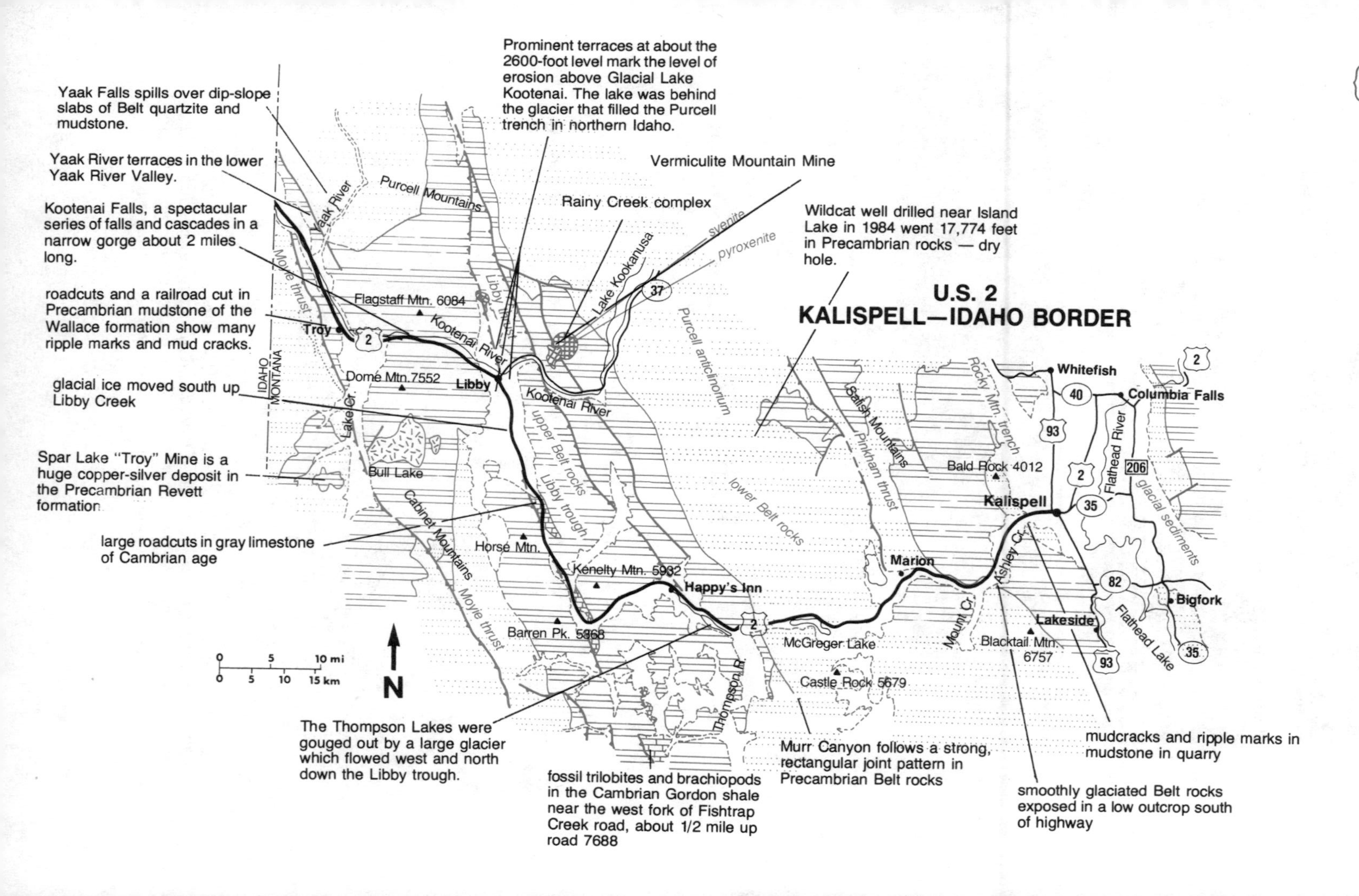
U.S. 2
KALISPELL—IDAHO BORDER
Yaak Falls spills over dip-slope slabs of Belt quartzite and mudstone.
Yaak River terraces in the lower Yaak River Valley.
Kootenai Falls, a spectacular series of falls and cascades in a narrow gorge about 2 miles long.
roadcuts and a railroad cut in Precambrian mudstone of the Wallace formation show many ripple marks and mud cracks.
glacial ice moved south up Libby Creek
Spar Lake "Troy" Mine is a huge copper-silver deposit in the Precambrian Revett formation.
large roadcuts in gray limestone of Cambrian age
Prominent terraces at about the 2600-foot level mark the level of erosion above Glacial Lake Kootenai. The lake was behind the glacier that filled the Purcell trench in northern Idaho.
Vermiculite Mountain Mine
Rainy Creek complex
Wildcat well drilled near Island Lake in 1984 went 17,774 feet in Precambrian rocks — dry hole.
The Thompson Lakes were gouged out by a large glacier which flowed west and north down the Libby trough.
fossil trilobites and brachiopods in the Cambrian Gordon shale near the west fork of Fishtrap Creek road, about 1/2 mile up road 7688
Murr Canyon follows a strong, rectangular joint pattern in Precambrian Belt rocks
mudcracks and ripple marks in mudstone in quarry
smoothly glaciated Belt rocks exposed in a low outcrop south of highway
IDAHO
MONTANA
Yaak River
Purcell Mountains
Moyie thrust
Troy
Flagstaff Mtn. 6084
Kootenai River
Libby thrust
Dome Mtn.7552
Libby
Lake Cr.
Bull Lake
Cabinet Mountains
Horse Mtn.
upper Belt rocks
Libby trough
Kenelty Mtn. 5932
Happy's Inn
Barren Pk. 5368
Lake Kookanusa
syenite
pyroxenite
37
Purcell anticlinorium
lower Belt rocks
Thompson R.
McGregor Lake
Castle Rock 5679
Salish Mountains
Pinkham thrust
Rocky Mtn. trench
Bald Rock 4012
Marion
Mount Cr.
Ashley Cr.
Kalispell
Blacktail Mtn. 6757
Lakeside
Whitefish
Columbia Falls
Flathead River
glacial sediments
Bigfork
Flathead Lake
2
40
93
206
35
82
0 5 10 mi
0 5 10 15 km
N

The road crosses a series of large slices of Precambrian formations that moved east on big thrust faults. At least some of that movement, perhaps all of it, probably happened when the western edge of the continent collided first with the Pacific Ocean floor, then with the Okanogan micro-continent. These thrust faults greatly thickened the western margin of the continent more than 100 million years ago. Some millions of years later, the younger formations that once covered this area moved east into the overthrust belt.

Thrust faults tantalize and tempt geologists because they tend to place older formations over younger ones, leaving us to wonder what they may conceal. Many geologists speculate that the Precambrian formations we see at the surface in northwest Montana may lie on top of much younger Paleozoic rocks. That is an interesting idea because Paleozoic rocks might contain oil and gas. The Precambrian rocks are too old and have been too hot to offer much hope. It is true that Cambrian formations, the oldest Paleozoic rocks, do peek out from under one of the thrust plates along a trend from near Libby southeast to Alberton.

The broad arch of the Purcell anticline looked especially attractive because it is the kind of structure that should trap oil and gas if they exist in the rocks. Unfortunately, the only deep hole drilled there so far, the Island Lake well, found no Paleozoic rocks. Instead it penetrated a thick section of Precambrian formations with numerous large diabase sills sandwiched between the sedimentary layers—no hope in that sort of rock. That result does not finally eliminate all possibility that oil or gas may exist in northwestern Montana, but it does enormously worsen the odds.

Kootenai Falls

About midway between Troy and Libby, the Kootenai River plunges over two beautiful waterfalls and down a spectacular series

Kootenai Falls

of steep rapids. It is difficult to compare waterfalls because they vary in both height and volume, and of course in their general scenic effect. The Kootenai River is a large stream, and it drops about 200 feet over the two cascades in an almost unspoiled setting. By any standard of comparison, these are the largest undammed falls in the Northern Rockies, one of the largest in the country. Many people also think them among the most beautiful in the country.

The Rainy Creek Complex

The largest vermiculite mine in the country began producing in 1925 from a big open pit in the Rainy Creek stock, high on a mountain about seven miles northeast of Libby. Vermiculite is an altered variety of mica that expands like popcorn when it is heated, to occupy about 15 times the volume of the original mineral. The yellowish brown puffs find many uses, mostly for fireproof insulation, acoustical plaster, and soil conditioner.

The Rainy Creek stock is an extraordinary igneous intrusion, most of it a complex assortment of exotic black rocks composed mainly of black pyroxene. The core of that assemblage is a mass of rock that originally consisted almost entirely of black biotite mica, now altered to yellow vermiculite. A cap of white syenite, a rock composed mostly of feldspar, covers part of the complex. The whole assemblage is strikingly similar to the Skalkaho stock east of Hamilton, and to a much more famous intrusion at Magnet Cove, in Arkansas.

Magmas that crystallize into rocks composed mostly of pyroxene can come only from somewhere in the mantle, from deep beneath the continental crust. Several age dates suggest that the Rainy Creek stock formed about 90 million years ago, a time that does not fit well into the general geologic picture. But that may not be a problem because such rocks are notoriously difficult to date. In this case, we suspect that the true age of the Rainy Creek stock may be closer to 50 million years old, that it is part of the widespread alkalic igneous activity of that time.

Gold on Libby Creek

Early prospectors found placer gold in Libby Creek about 20 miles south of Libby in 1886, and a town, Old Libby, sprang up in the next year. The miners built a high flume to deliver water under pressure to hydraulic nozzles that could wash down the higher gravel banks, and flush their contents through sluices to recover the gold. The gravel was rich for a distance of several miles. Placer mining in Libby Creek continued off and on until fairly recently, but production since the turn of the century has been pitifully small considering the effort expended, probably less than 100,000 dollars.

The bedrock source of the gold was discovered in 1887 near the contact of a granite intrusion into the Belt rocks in the Cabinet Range west of Libby Creek. Some of the mines worked for several decades, and produced silver and lead, as well as gold. The district has been essentially dead for many years, after having produced something between four and five million dollars worth of gold. Even though those were turn of the century dollars, the production was a small return for the investment of labor and capital it required.

Mining around Troy

Troy has a long history and probably a long future in mining. In 1888, prospectors found deposits of copper, silver, lead, and gold at Sylvanite, on the Yaak River nearly 20 miles north of Troy. A narrow gauge railroad hauled ore from the mines at Sylvanite to a concentrating mill in Troy. The mines produced a total of more than 4 million dollars worth of metal before they closed down after the mill burned in 1927. The Sylvanite area has seen very little activity since then.

A large mine at Spar Lake, south of Troy, produces silver from a rich ore body in Precambrian sedimentary rocks, the Revett formation. The ore minerals are scattered through beds of sandstone, and they appear to have arrived at about the same time as the sand grains—most silver deposits are considerably younger than the rocks that contain them. Discovery of the Spar Lake deposit during the 1960s surprised everyone because the ore is quite unlike any known elsewhere. Geologists have since paid careful attention to every outcrop of Revett sandstone.

Glaciation

Between Libby and the Idaho line, the highway follows the Kootenai River, which separates the Purcell Mountains north of the road from the Cabinet Mountains to the south. Although the Precambrian sedimentary formations north and south of the river are identical, the jagged Cabinet Mountains look quite different from their more rounded counterparts north of the Kootenai River. That is a matter of glaciation.

When the ice ages were at their maximum, glaciers pushing south out of British Columbia so nearly buried the Purcell Mountains that only the highest peaks rose above the ice cap. Ice covered and eroded nearly the entire range, leaving the mountains looking generally smooth and rounded. Meanwhile, the Cabinet Mountains escaped

A deposit of bouldery glacial till, the raw material of glacial moraines.

such general coverage. Their glaciers gouged out the valleys, leaving a dramatically carved alpine landscape of ragged peaks and ridges.

When the ice age was at its maximum, one long tongue of ice, probably a prong of the Flathead Valley glacier, pushed south past Rexford down the valley of the Kootenai River that Lake Koocanusa, the reservoir behind Libby Dam, now floods. Other long tongues of ice came down the valley of Libby Creek well south of Libby, and past Troy as far south as Bull Lake. Smaller glaciers flowed down the mountain valleys on either side of those lobes of ice, and joined them as tributaries. Dense forests almost certainly cloaked the mountainsides above the ice-filled valleys, just as they do today in Alaska.

The Purcell Valley of northern Idaho filled with an enormous glacier almost comparable to the one that filled the Rocky Mountain trench in Montana. That ice blocked the drainage of the Kootenai River long after much of the ice in the smaller valleys of northwestern

Wanless Lake huddles in the floor of a glacial cirque in the Cabinet Range.
—U.S. Forest Service photo by L. Bernston

Montana had melted back far enough to free its valley. Then a lake flooded the Kootenai River drainage and left deep deposits of glacial lake sediments. Watch for roadcuts and stream banks in pale gray silt.

So little research has been done on the glacial geology of northwestern Montana that it is impossible to say whether the glacial lake in the Kootenai drainage connected with Glacial Lake Missoula, as seems likely. There are valleys, such as that of the Bull River, low enough to connect them. The problem is that no one knows whether those valleys were free of ice at the right time.

Almost the entire route of US 2 passes through valleys full of lakes that tell of vanished glaciers. Watch along the way for hummocky moraines littered with boulders, deposits that tell as clearly as anything can that glaciers once lay there. Between Libby and the divide, the road follows the valleys of Libby Creek and the Fisher River to the east end of the Thompson Lakes. The glacier that filled that valley flowed northwest. East of the divide, the road follows the equally glaciated valley of Ashley Creek past McGregor Lake, the former route of a glacier that flowed into the great expanse of ice that once filled the Flathead Valley.

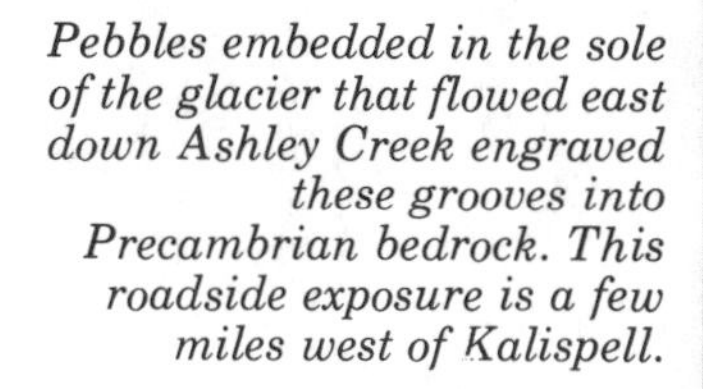

Pebbles embedded in the sole of the glacier that flowed east down Ashley Creek engraved these grooves into Precambrian bedrock. This roadside exposure is a few miles west of Kalispell.

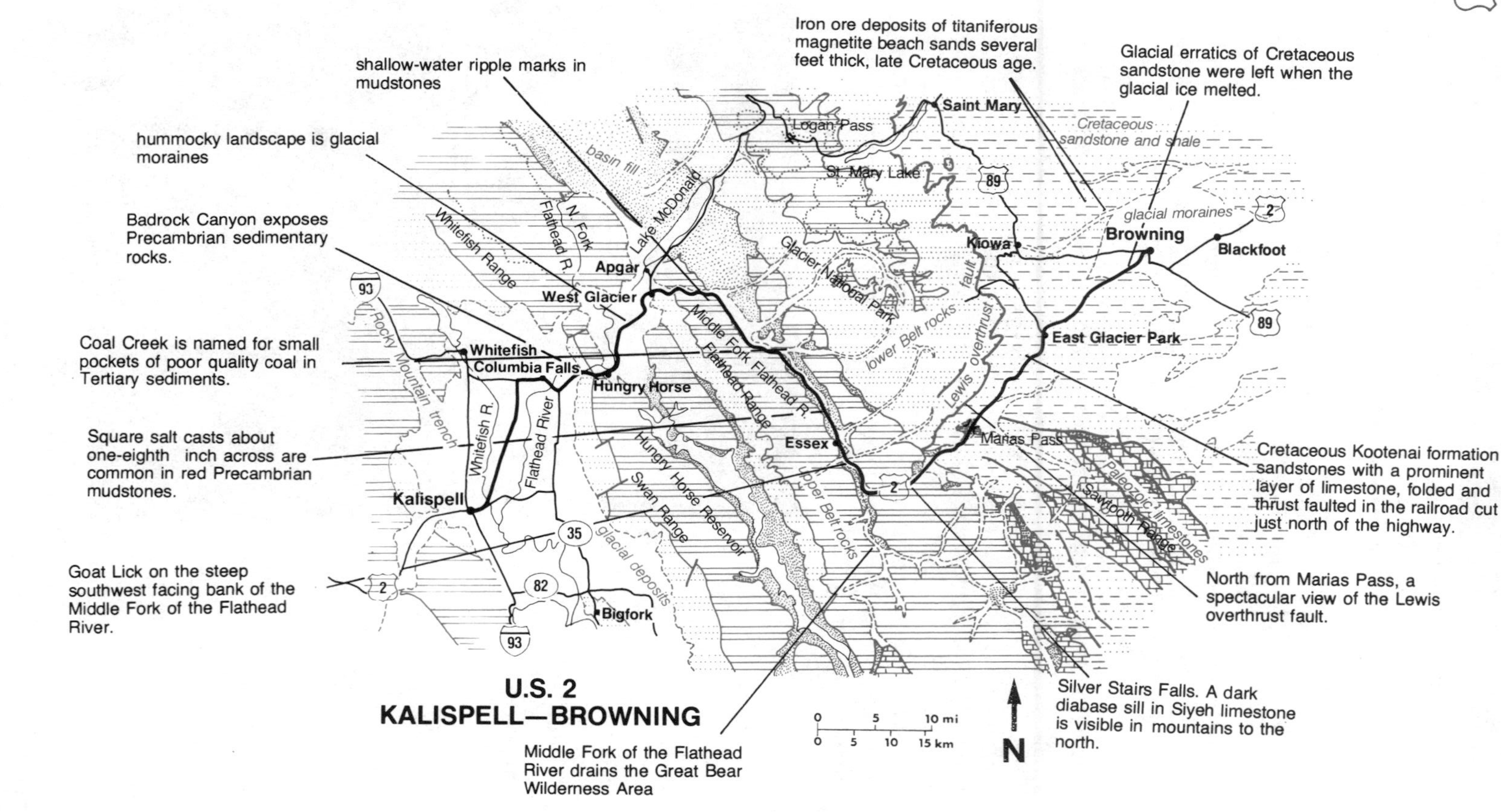
shallow-water ripple marks in mudstones
Iron ore deposits of titaniferous magnetite beach sands several feet thick, late Cretaceous age.
Glacial erratics of Cretaceous sandstone were left when the glacial ice melted.
hummocky landscape is glacial moraines
Badrock Canyon exposes Precambrian sedimentary rocks.
Coal Creek is named for small pockets of poor quality coal in Tertiary sediments.
Square salt casts about one-eighth inch across are common in red Precambrian mudstones.
Goat Lick on the steep southwest facing bank of the Middle Fork of the Flathead River.
Cretaceous Kootenai formation sandstones with a prominent layer of limestone, folded and thrust faulted in the railroad cut just north of the highway.
North from Marias Pass, a spectacular view of the Lewis overthrust fault.
Silver Stairs Falls. A dark diabase sill in Siyeh limestone is visible in mountains to the north.
Middle Fork of the Flathead River drains the Great Bear Wilderness Area
Saint Mary
Logan Pass
St. Mary Lake
Cretaceous sandstone and shale
glacial moraines
Browning
Blackfoot
Kiowa
East Glacier Park
Glacier National Park
lower Belt rocks
Lewis overthrust fault
Marias Pass
Paleozoic limestones
Sawtooth Range
basin fill
Lake McDonald
N. Fork Flathead R.
Whitefish Range
Apgar
West Glacier
Rocky Mountain trench
Whitefish
Columbia Falls
Hungry Horse
Middle Fork Flathead R.
Flathead Range
Essex
upper Belt rocks
Whitefish R.
Flathead River
Hungry Horse Reservoir
Swan Range
glacial deposits
Kalispell
Bigfork
2
35
82
89
93
0 5 10 mi
0 5 10 15 km
N

U.S. 2
KALISPELL—BROWNING

Looking across the North Fork Valley to the mountains of Glacier Park

US 2: Kalispell—Browning
98 mi./157 km.

Between Kalispell and Columbia Falls the highway crosses the Flathead Valley, the southern end of the Rocky Mountain trench. Low mountains west of the valley are part of the Salish Ranges, composed of Belt formations. High mountains on the east side of the

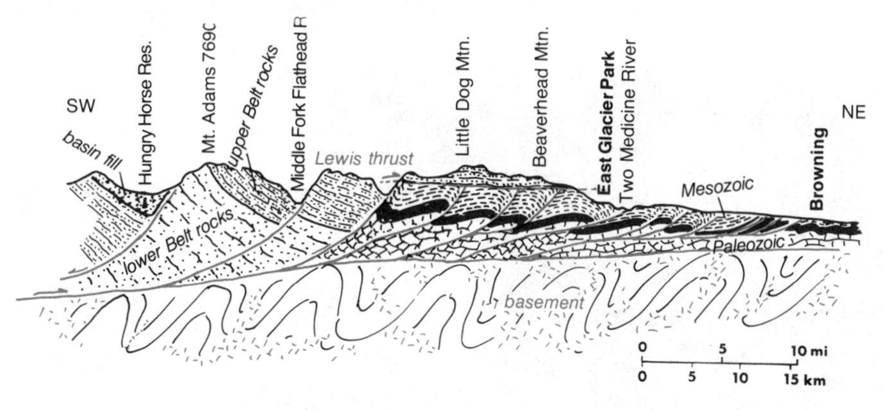

Section through the southern edge of Glacier National Park, just north of US 2

valley are the Whitefish and Swan ranges, also composed of Belt formations. Those ranges are in the western edge of the overthrust belt.

The Overthrust Belt

The route between Columbia Falls and East Glacier crosses the northern Montana overthrust belt, skirting the southern edge of Glacier National Park along much of the way. Rocks exposed along the highway between the area just east of Columbia Falls and the top of Marias Pass are the same Precambrian sedimentary formations, Belt rocks, that exist in Glacier Park.

A pullout at the top of Marias Pass is the best place to get a good view of the Lewis overthrust fault that moved the ancient Precambrian formations onto the much younger Cretaceous rocks. Look north at the mountains in the southern end of Glacier Park to see the straight trace of the fault angling very gently up to the east, like a long railroad grade. The talus slopes below the fault nearly cover the bedrock beneath, but in a few places you can glimpse exposures of black Cretaceous shale.

Rocks above the Lewis overthrust are a great slab of Precambrian sedimentary formations deposited a billion or more years ago. That slab slid at least 35 miles east over younger rocks, probably about 65 million years ago. Rocks below the fault east of Marias Pass are Cretaceous sedimentary formations that accumulated in and near a shallow inland sea about 80 or so million years ago. This was the area where geologists first recognized the existence of overthrust faulting, back in the 1890s.

All the bedrock along the highway between Marias Pass and East

The view north toward Glacier Park from Marias Pass. The Lewis overthrust fault trends gently upward to the right along the base of the pale Altyn formation, about half way up the mountain slope.

Glacier is Cretaceous sedimentary rock. Watch for occasional inconspicuous exposures of black shale and brown sandstone. Some of the black shales contain small fossil ammonites that look at first glance almost like big snails, but were actually more like an octopus that lived in a shell. They were close relatives of the modern pearly nautilus. Ammonites flourished in great numbers and many varieties in the shallow seas that flooded much of Montana during Cretaceous time, then died in the same calamity that annihilated the dinosaurs.

Glaciation

During the ice ages, the glacier that filled the Rocky Mountain trench pushed south through the Flathead Valley and into the Mission Valley. When it melted, that glacier left a deep fill of sediment in the floor of the Flathead Valley. Meanwhile, the valley of the Middle Fork of the Flathead River filled with ice coming down from the mountains on either side. That glacier ground its way west through Bad Rock Canyon to join the huge glacier that filled the Rocky Mountain trench.

Glacial deposits on the mountain sides show that the ice that filled the Flathead Valley was about 6000 feet thick at the Canadian border. The ice thinned so rapidly southward that the glacier ended not far south of Flathead Lake. Evidently the ice age summers were warm enough to melt enormous volumes of ice, and must have released torrents of muddy meltwater.

Hummocky moraines littered with boulders between East Glacier and Browning record an enormous piedmont glacier. It formed as glaciers flowing east from the big canyons in the Rocky Mountains spread across the high plains, and ponded there to form a large sheet of ice that reached several miles east of Browning. Because they spread across such a large area, piedmont glaciers move so slowly that the ice is nearly stagnant. Their moraines are vast fields of irregular humps and hollows that typically show little of the shaping influence of moving ice.

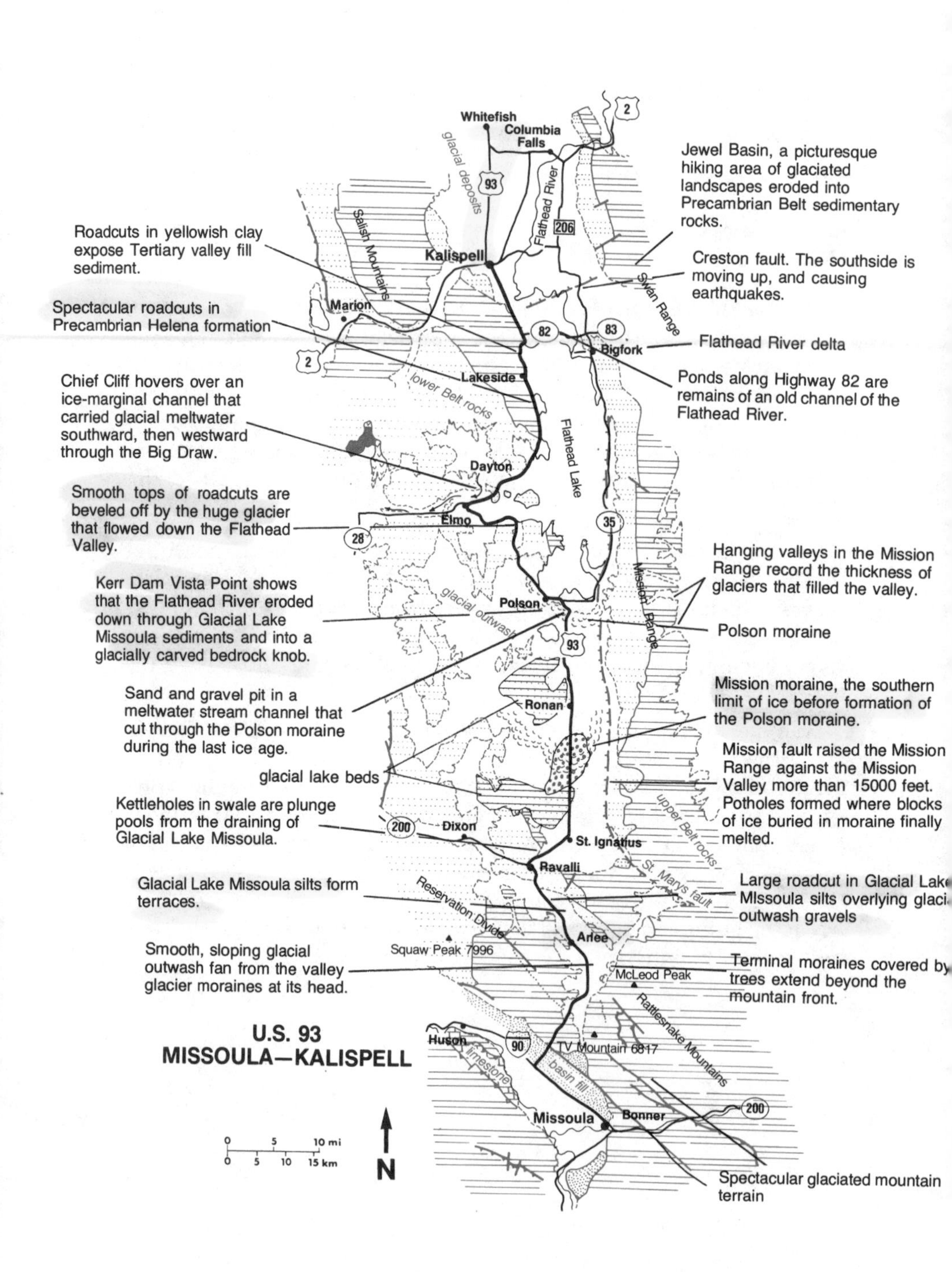
Whitefish
Columbia Falls
glacial deposits
Flathead River
Jewel Basin, a picturesque hiking area of glaciated landscapes eroded into Precambrian Belt sedimentary rocks.
Roadcuts in yellowish clay expose Tertiary valley fill sediment.
Salish Mountains
Kalispell
Creston fault. The southside is moving up, and causing earthquakes.
Swan Range
Marion
Spectacular roadcuts in Precambrian Helena formation
Bigfork
Flathead River delta
Lakeside
lower Belt rocks
Ponds along Highway 82 are remains of an old channel of the Flathead River.
Chief Cliff hovers over an ice-marginal channel that carried glacial meltwater southward, then westward through the Big Draw.
Flathead Lake
Dayton
Smooth tops of roadcuts are beveled off by the huge glacier that flowed down the Flathead Valley.
Elmo
Hanging valleys in the Mission Range record the thickness of glaciers that filled the valley.
Kerr Dam Vista Point shows that the Flathead River eroded down through Glacial Lake Missoula sediments and into a glacially carved bedrock knob.
glacial outwash
Polson
Mission Range
Polson moraine
Sand and gravel pit in a meltwater stream channel that cut through the Polson moraine during the last ice age.
Ronan
Mission moraine, the southern limit of ice before formation of the Polson moraine.
glacial lake beds
Mission fault raised the Mission Range against the Mission Valley more than 15000 feet.
Kettleholes in swale are plunge pools from the draining of Glacial Lake Missoula.
Dixon
upper Belt rocks
Potholes formed where blocks of ice buried in moraine finally melted.
St. Ignatius
Ravalli
Glacial Lake Missoula silts form terraces.
Reservation Divide
St. Marys fault
Large roadcut in Glacial Lake Missoula silts overlying glacial outwash gravels
Arlee
Squaw Peak 7996
Smooth, sloping glacial outwash fan from the valley glacier moraines at its head.
McLeod Peak
Terminal moraines covered by trees extend beyond the mountain front.
Rattlesnake Mountains
U.S. 93
MISSOULA—KALISPELL
Huson
limestone
TV Mountain 6817
basin fill
Missoula
Bonner
0 5 10 mi
0 5 10 15 km
N
Spectacular glaciated mountain terrain
2
93
206
82
83
28
35
200
90

US 93: Missoula—Kalispell
128 mi./205 km.

All the hard bedrock exposed along the road and in the mountains within sight of the road between Missoula and Kalispell is Belt rock, Precambrian sedimentary formations. Unconsolidated Tertiary valley fill sediments and an assortment of glacial debris—till, outwash, and lake sediments—cover the bedrock in long sections of the route.

The Rattlesnake Mountains

In its joint segment with Interstate 90 between Missoula and the junction 10 miles west of town, the highway follows a line just south of the Ninemile fault, part of the Lewis and Clark fault zone. It is difficult to know what to call the high mountain range north of the Missoula Valley. Although they are labeled the Jocko Range on some maps, few people call them that. Most people know them as the "Rattlesnake Mountains," the "Missoula Hills," or, most commonly, "those mountains over there." We will call them the Rattlesnake Mountains. By whatever name, they consist of very tightly folded Precambrian formations broken by many faults. Ice age glaciers carved the higher peaks into a magnificent alpine landscape.

A few miles north of its junction with Interstate 90, the highway crosses the Ninemile fault onto hard bedrock. Watch the distribution of outcrops. North of the fault, the highway follows Evaro Canyon through the Rattlesnake Mountains to the Jocko Valley, one of the smaller structural basins of the northern Rocky Mountains.

The Jocko Valley

Craggy peaks of the glaciated Rattlesnake Mountains make a bold wall along the southern margin of the Jocko Valley. Gently rounded hills that never contained glaciers: the Jocko Hills on the east, the

The low wooded hill in the middleground is a terminal moraine of a valley glacier on the northwest side of the Rattlesnake Mountains, as seen from US 93 at the south end of the Jocko Valley. It probably dates to the Bull Lake ice age.

Reservation Divide on the west, contain the other sides. All consist entirely of Precambrian Belt rock.

Large exposures of brownish gravel north of Evaro Canyon are on the divide between the Missoula and Jocko valleys. They appear to be Pliocene deposits laid down during the last long period of dry climate that ended about two or three million years ago. Evidently, the Missoula and Jocko valleys then joined in this area, and the big sheets of gravel in their floors merged. Since then, streams have hauled much of the gravel out of both valleys, leaving this high remnant.

The Mission Range

Between St. Ignatius and Polson, the deeply glaciated Mission Range looms to the east, its jagged skyline as dramatic as any in the region. The St. Mary fault abruptly chops off the southern end of the range just as it reaches its greatest elevation.

View across St. Ignatius to the high southern end of the Mission Range

The landscape within the southern part of the Mission Range is virtually identical to that in Glacier Park. Bedrock in both areas is quite similar, and both were deeply carved by large valley glaciers. Tracts of low and hummocky hills across the mouths of the big valleys in the southern part of the Mission Range are glacial moraines.

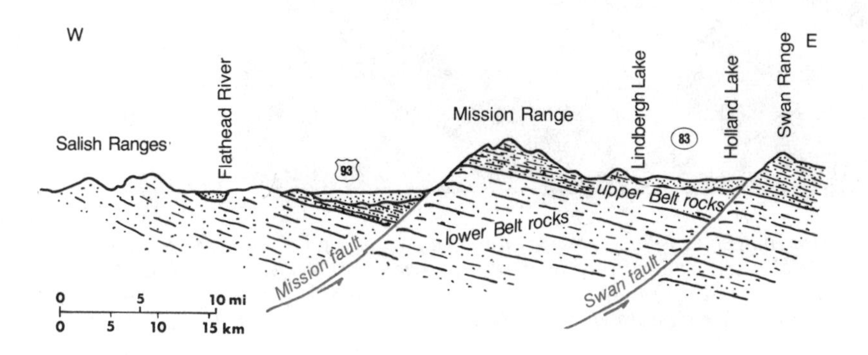

Section across the Mission and Swan valleys north of St. Ignatius. The Mission and Swan ranges consist basically of great slabs of Precambrian sedimentary formations, Belt rocks, tilted gently down to the east. Rocks in the Salish Ranges are older Belt formations.

The Mission Range is the westernmost of the displaced slabs that make up the overthrust belt. Rocks in the much lower Salish Ranges west of the Mission Valley are Belt formations much older and formerly more deeply buried than those in the Mission Range. It seems likely that the slabs now in the overthrust belt slid off the rocks we see in the Salish Ranges, thus exposing the older rocks at the surface.

Glaciation of the Mission Valley

During the ice ages, enormous glaciers filled the Rocky Mountain trench of British Columbia and slowly moved southward through the Flathead Valley. As the ice ages climaxed, ice reached into the Mission Valley, the southern end of the Rocky Mountain trench. The landscape between St. Ignatius and Polson contains clear evidence of major glaciation during the Bull Lake ice age that probably reached its maximum sometime between 70,000 and 130,000 years ago. Glaciers advanced again during the Pinedale ice age that climaxed about 15,000 years ago.

Hummocky hills in the foreground are part of the Polson moraine. The part of the Mission Range in the background was completely buried under glacial ice during the Bull Lake ice age. During the Pinedale glaciation, valley glaciers eroded the range down to the level of the ice that then filled the valley.

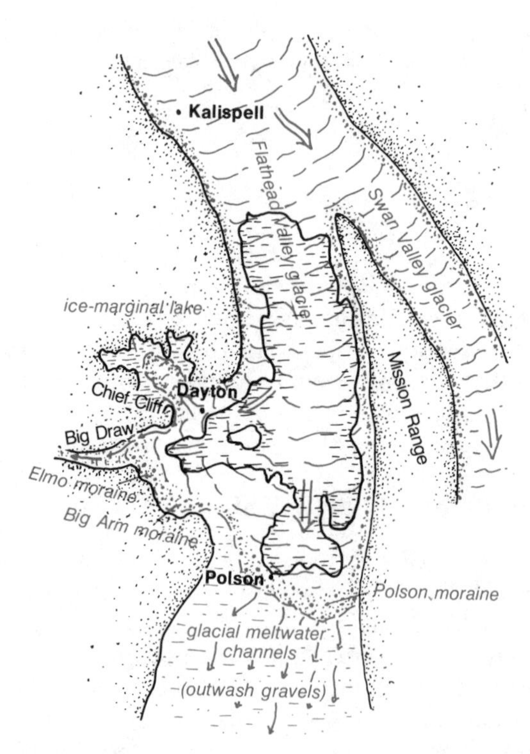

Ice flowing south through the Rocky Mountain trench split on the north end of the Mission Range to send separate glaciers down the Mission and Swan valleys.

The older and larger glacier reached the Ninepipes Reservoir area, left an enormous moraine there, and an apron of glacial outwash on the floor of the Mission Valley south of Ninepipes Reservoir. That moraine and outwash plain show their age in their generally melted down appearance. Watch in the Ninepipes area for numerous kettle ponds, each one marking a spot where a large chunk of ice buried in the moraine left a hole when it melted.

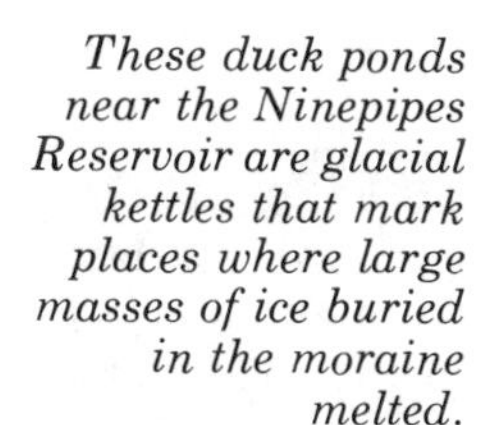

These duck ponds near the Ninepipes Reservoir are glacial kettles that mark places where large masses of ice buried in the moraine melted.

That same older glacier buried the Mission Range about as far south as Ronan, where the profile of the range changes abruptly. North of Ronan, the overriding ice left softly rounded peaks along the crest of the range. Bedrock outcrops on the tops of those rounded mountains are covered with deep scratches gouged by the passing ice, and littered with boulders that the ice left as it melted. Mountain glaciers gouged the valleys south of Ronan while the mountains farther north were buried. The glaciers left a legacy of ice-carved craggy peaks and sharp ridges high in the mountains, hummocky moraines at the mouths of the valleys.

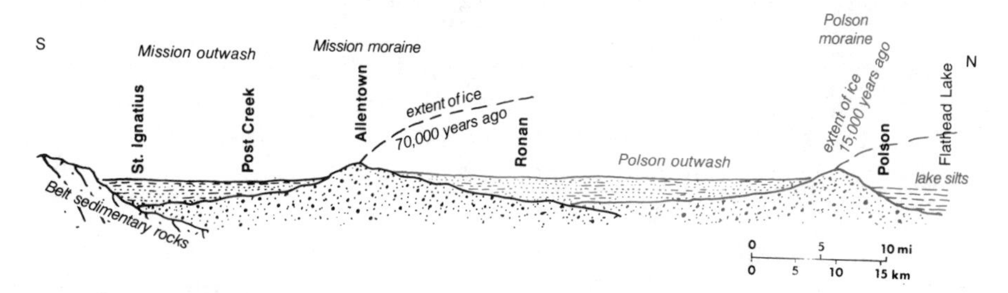

Profile of the Mission Valley between St. Ignatius and Polson showing moraines and outwash of the most recent Pinedale ice age, shown in color, partially burying older deposits left by the Bull Lake glacier

The younger glacier of the last ice age reached only as far south as Polson, and left its terminal moraine as the low ridge that crosses the valley just south of Polson. People driving north get their first glimpse of Flathead Lake just as they come over the crest of that Polson moraine. Big roadcuts in the Polson moraine expose stream deposited outwash, clean sand and gravel, instead of glacial till, the disorderly mixture of material most common in moraines. Evidently, the road follows the path of a glacial meltwater stream.

Most of the smooth surfaces between Ninepipes and Pablo are outwash plains formed as meltwater streams draining off the younger glacier spread deposits of sand and gravel across the valley floor. Little hills covered with pine trees in the Pablo area are old sand dunes that marched across the outwash plain during the last ice age, driven before the cold wind that constantly drained off the glacier.

During the last ice age, mountain glaciers gouged the valleys in the northern part of the Mission Range that the earlier glacier had left smoothly rounded. Look for the narrower valleys shallowly gouged into the very broad valley floors that the older glacier left. The younger valley glaciers eroded down to the upper level of the ice that then filled the floor of the Flathead Valley.

When Glacial Lake Missoula flooded to its higher stands, water filled the Mission Valley, and lapped onto the ice north of Polson. The lower end of the glacier floated during those times, and icebergs calved off to drift south across the lake. Many of the large rocks that litter the valley floor south of Polson probably dropped from those icebergs as they began to melt in the summer sun.

Kerr Dam

Kerr Dam stands near the north end of a narrow gorge that the Flathead River cut through a buried knob of hard Precambrian sedimentary rock, Belt rock. The north side of the bedrock knob is smoothly rounded; its south side is jagged. That is the typical shape of a hill sculptured beneath a glacier. Pieces of rock embedded in the glacier sandpapered the north side smooth while the ice froze to the rock and plucked blocks loose to create the ragged profile of the south side. Obviously, a glacier rode over that buried hill, probably the glacier that left the Mission moraine in the Ninepipes area south of Ronan.

Deep deposits of sediment that buried the bedrock knob form the gray bluffs on the far side of the river. It is easy to imagine the Flathead River spilling as glacial meltwater across the Polson moraine, beginning to flow across the sediments. As the river cut its

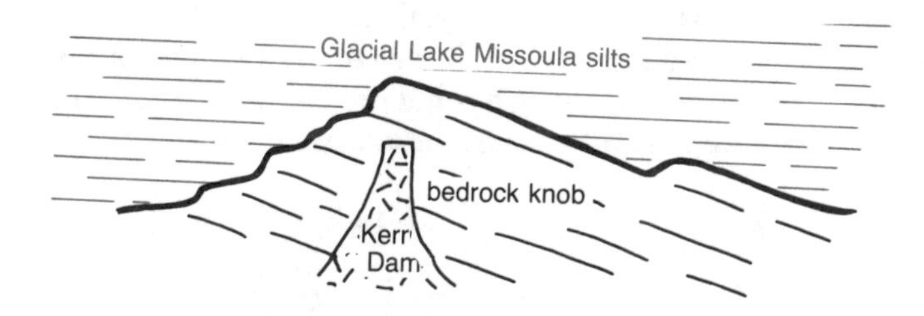

Section showing the geologic situation visible from the visitor's overlook at Kerr Dam just west of Polson

channel deeper, it let itself down onto the buried bedrock knob, and eventually carved the gorge that holds Kerr Dam. It is inconceivable that the river could have cut such a deep gorge in such hard rock during the 10,000 years since the last ice age ended, so we must assume that all this happened during and after some earlier ice age.

Flathead Lake

Most of the route between Polson and Kalispell follows the west side of Flathead Lake. The lake fills a basin that appears to have formed where a large mass of stagnant ice lingered as the last ice age ended. Glacial deposits filled the valley floor elsewhere, but could not accumulate where the ice still lay. When the ice finally melted, it left a depression in the deposits of sediment that filled the valley floor.

Near the north end of Flathead Lake, the highway passes a number of road cuts that expose deposits of yellowish clay and silt, probably Tertiary valley fill sediment, the Renova formation. These are among the few such exposures in the Flathead Valley where glacial deposits generally bury all the older valley sediments.

Lake Mary Ronan and Chief Cliff

Lake Mary Ronan is west of the highway, several miles up the road from Dayton. It lies behind a moraine that the Flathead Valley glacier left on the valley wall. The moraine blocks the drainage to impound the lake in a part of the valley that was never glaciated.

The ice age ancestor of Lake Mary Ronan was a much larger lake trapped between the glacier and the hills to the west. Overflow from that lake, a substantial river, poured south along the edge of the ice, eroding a deep channel in the bedrock along the edge of the glacier.

The ice is gone now, but long stretches of that abandoned channel still gouge the slopes west of the highway. Chief Cliff is the most striking souvenir of that vanished stream. It was eroded as a bluff above the meltwater stream, and the dry channel eroded in bedrock still winds along its base, like a giant irrigation ditch, now dry.

Chief Cliff, a bluff rising above a dry abandoned stream channel

Earthquakes

Minor earthquakes and swarms of earthquakes frequently rattle the valley that holds Dayton and Proctor. This area on the west side of Flathead Lake is the extremity of the northwestern branch of the Intermountain Seismic Zone that continues south from Helena through Yellowstone Park to the Wasatch Front east of Salt Lake City. The frequent small tremors probably mean that the faults are not storing enough energy between movements to unleash a large and dangerous earthquake.

US 93: Kalispell—Canadian Border

174 mi./248 km.

Only glacial deposits, almost no bedrock, are exposed along this route through the Rocky Mountain trench. The mountains on either side of the great valley consist almost entirely of Precambrian sedimentary formations, Belt rock. Except for a small patch of Cambrian limestone north of Eureka, the only exceptions are occasional diabase sills, probably emplaced into the Belt formations during the great continental rift of 800 million or so years ago.

The Rocky Mountain Trench

The big problem with our cross section is that it doesn't seem to explain the Rocky Mountain trench. In that respect, it does not differ from most other interpretations of the area. Such a long valley must reflect bedrock structure. Unfortunately, it is extremely difficult to

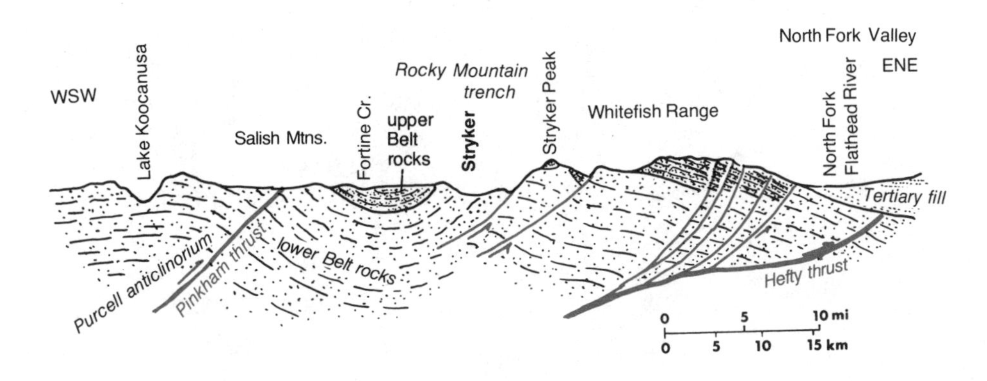

Section across US 93 at Stryker. Formations in the upper part of the Belt section are exposed along Fortine Creek, and in the Whitefish Range.

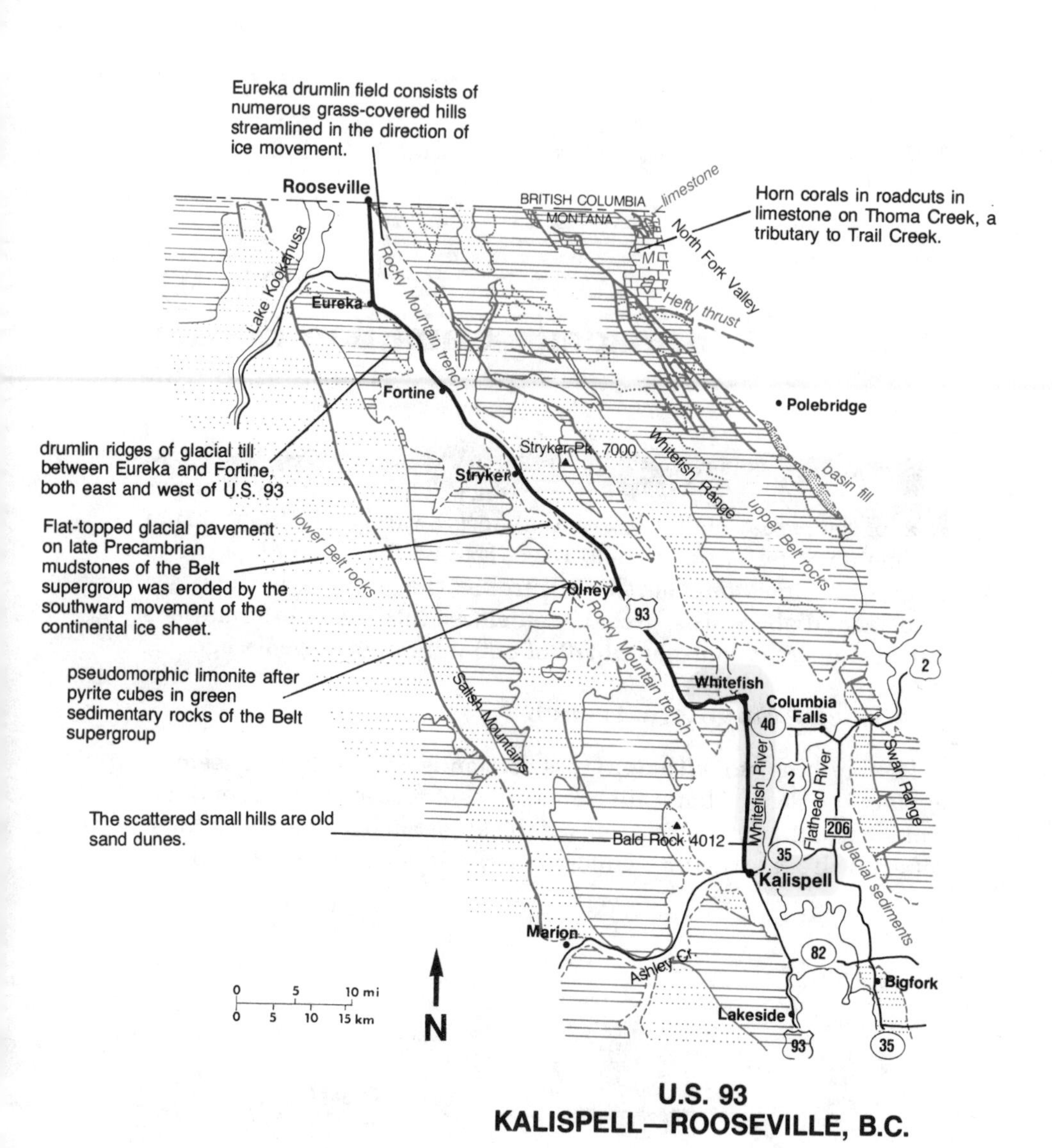
Eureka drumlin field consists of numerous grass-covered hills streamlined in the direction of ice movement.
Horn corals in roadcuts in limestone on Thoma Creek, a tributary to Trail Creek.
drumlin ridges of glacial till between Eureka and Fortine, both east and west of U.S. 93
Flat-topped glacial pavement on late Precambrian mudstones of the Belt supergroup was eroded by the southward movement of the continental ice sheet.
pseudomorphic limonite after pyrite cubes in green sedimentary rocks of the Belt supergroup
The scattered small hills are old sand dunes.
Rooseville
BRITISH COLUMBIA
MONTANA
limestone
North Fork Valley
Hefty thrust
Lake Koocanusa
Eureka
Rocky Mountain trench
Fortine
Polebridge
Stryker Pk. 7000
Stryker
Whitefish Range
basin fill
upper Belt rocks
lower Belt rocks
Olney
93
Rocky Mountain trench
Salish Mountains
Whitefish
Columbia Falls
40
2
Whitefish River
Flathead River
Swan Range
206
Bald Rock 4012
35
Kalispell
glacial sediments
Marion
Ashley Cr.
82
Bigfork
Lakeside
93
35
0 5 10 mi
0 5 10 15 km
N

U.S. 93
KALISPELL—ROOSEVILLE, B.C.

determine the structure of bedrock that comes in extremely thick formations that tend to look alike and contain no fossils. All that makes it difficult to trace rocks of the same age from one place to another, and there is no other way to start working out their structure.

Glaciation

The road follows the route of the great Rocky Mountain trench glacier that pushed south as far as the Mission and southern Swan valleys. Near the Canadian border, the ice was deep enough to bury most of the mountains, and spread across the landscape as an almost continuous ice cap. But it thinned southward, drawing away from the mountains and becoming a great river of ice with tributary rivers of ice entering from the mountain valleys on either side.

Rocks embedded in a glacier gouged these scratches in an outcrop of Precambrian mudstone.

Vast fields of peculiar hills called drumlins cover long stretches of the valley floor, most conspicuously in the area near Eureka. These are the only large drumlin fields in Montana. Drumlins are deposits of glacial till molded into streamlined forms beneath the flowing ice. Their high end is blunt, and faces into the direction of ice flow, and they trail off into thin tails in the downflow direction. From the air, drumlins suggest schools of giant tadpoles all lined up with their big

A drumlin north of Eureka—the high end to the left faces north upstream into the direction of ice flow.

heads facing upstream, their slender tails pointing downstream.

Common as they are in many glaciated regions, the origin of drumlins is poorly understood, the subject of considerable controversy. The basic problem is that drumlins form beneath thousands of feet of moving ice, where direct field observation is impossible. Some geologists contend that drumlins form as the ice plasters glacial till around cores of bedrock. That seems unlikely in this area where many of the drumlins must rest on a base of deep valley fill sediments hardly solid enough to feel like bedrock to a glacier.

Whatever the exact mechanics of their origin, drumlins generally form somewhere near the lower end of the glacier where the ice is thinning, but well back from the terminal moraine. That is fairly easy to understand. Thin ice near the lower end of the glacier slides harmlessly over deposits of glacial debris, whereas the much thicker ice closer to the source of the glacier probably bulldozes such soft material. Somewhere in between, the ice is heavy enough to mold soft sediments in the valley floor without bulldozing them. That is where drumlins form.

Glacial sculpture on an outcrop of Precambrian mudstones south of Eureka.

Montana 28: Plains—Elmo
147 mi./235 km.

All the bedrock exposed along the road is Precambrian sedimentary formations, Belt rock. Most is somber gray and dark brown rock, Ravalli and Prichard formations of the lower part of the Precambrian section. But the main attractions are the spectacular relics of the ice ages, and of Glacial Lake Missoula.

Rainbow Lake

When Glacial Lake Missoula suddenly drained, part of the water that filled the Little Bitterroot Valley poured over the divide at the head of Boyer Creek, and rushed south along the line of the highway towards Plains. That surge of water eroded a long channel in the bedrock floor of the upper part of the valley. Now, Rainbow Lake fills the lower part of that floodway. The entire lake basin was eroded in hard Precambrian bedrock during several catastrophic drainings of Glacial Lake Missoula, probably within a total period of a few incredible hours.

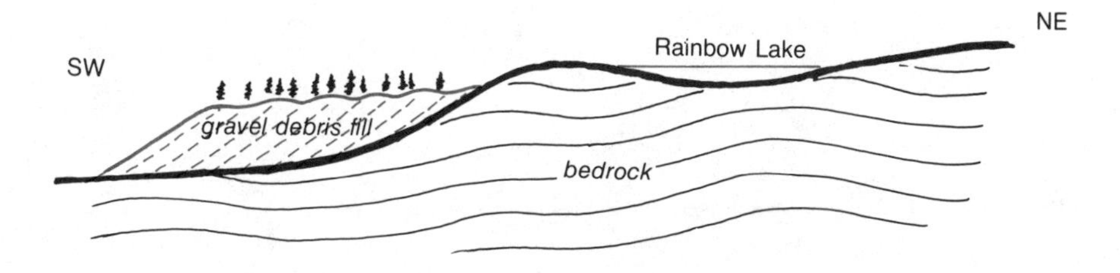

Schematic section showing the eroded basin near the drainage divide, and the deep debris fill in the lower part of Boyer Creek

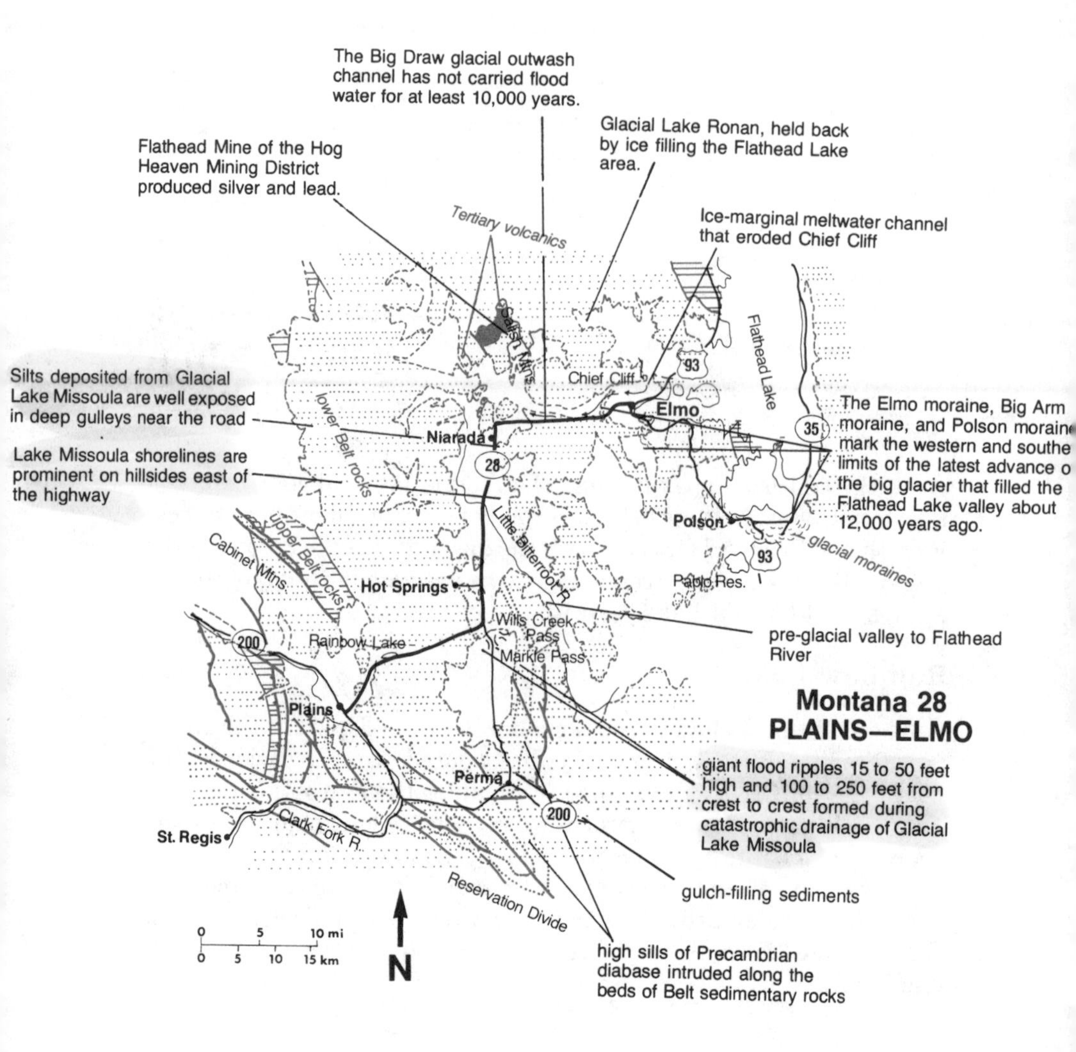

The Big Draw glacial outwash channel has not carried flood water for at least 10,000 years.
Glacial Lake Ronan, held back by ice filling the Flathead Lake area.
Flathead Mine of the Hog Heaven Mining District produced silver and lead.
Tertiary volcanics
Ice-marginal meltwater channel that eroded Chief Cliff
Flathead Lake
Salish Mtns
Chief Cliff
93
Elmo
Silts deposited from Glacial Lake Missoula are well exposed in deep gulleys near the road
lower Belt rocks
Niarada
28
35
The Elmo moraine, Big Arm moraine, and Polson moraine mark the western and southern limits of the latest advance of the big glacier that filled the Flathead Lake valley about 12,000 years ago.
Lake Missoula shorelines are prominent on hillsides east of the highway
Little Bitterroot R.
Polson
glacial moraines
Cabinet Mtns
upper Belt rocks
Hot Springs
Pablo Res.
Wills Creek Pass
Markle Pass
pre-glacial valley to Flathead River
200
Rainbow Lake
Montana 28
PLAINS—ELMO
Plains
giant flood ripples 15 to 50 feet high and 100 to 250 feet from crest to crest formed during catastrophic drainage of Glacial Lake Missoula
Perma
St. Regis
Clark Fork R.
gulch-filling sediments
Reservation Divide
0 5 10 mi
0 5 10 15 km
N
high sills of Precambrian diabase intruded along the beds of Belt sedimentary rocks

Rock debris swept out of the basin that holds Rainbow Lake and the marshes above it accumulated as a deep fill in the lower part of the valley of Boyer Creek. People driving north see the lower end of that debris fill stretching before them like a wall across the valley a few miles north of Plains. The road crosses the face of that fill in a steep grade, and passes for several miles across its top in the area below Rainbow Lake.

The Little Bitterroot Valley

Between Hot Springs and Niarada, the road crosses the Little Bitterroot Valley, obviously a large structural basin even though the bedrock structure remains unexplained. The Little Bitterroot Valley probably contains the usual deep deposits of sediments that date from the two dry periods of Tertiary time, but they are well hidden. A top layer of sediments deposited in Glacial Lake Missoula now covers most of the valley floor.

In the northern part of the valley, the lake sediments are thick and well exposed in a large area of old badlands topography, now degraded and largely overgrown with sagebrush. As nearly everywhere, the lake sediments are fine-grained, and light gray—slightly pinkish in wet weather. A bit of digging into fresh material reveals the thin light and dark layers deposited during the summers and winters of the lake's existence.

The Big Draw

Northeast of Niarada, Montana 28 follows the Big Draw, a pre-glacial valley of the Flathead River. The valley floor contains deep deposits of glacial outwash gravels left there by meltwater

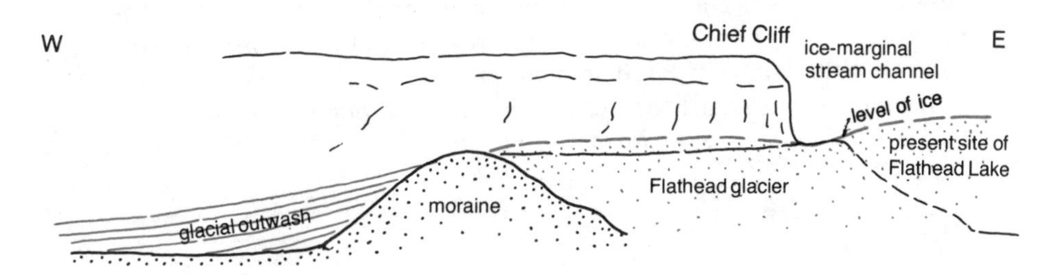

Schematic cross section showing the moraine and glacial outwash that partially fill the Big Draw

View from the air of the moraine at the head of the Big Draw and the meltwater stream channel downstream from Chief Cliff.
—Rick Hull photo, Kalispell Daily News

pouring off the glacier that filled the Flathead Valley. Even though the outwash deposits accumulated at least 10,000 years ago, the meltwater stream channels are still so clearly visible that they look as though they might have formed last year. The Montana Highway Department was so impressed that it built culverts where the road crosses several of those channels, even though they have not carried water since the last ice age.

At the head of the Big Draw, the road leaves the outwash, and tops the crest of a moraine left by a lobe of the glacier that filled the Flathead Valley. The moraine blocked the old valley of the Flathead River, forcing it to find its present course west and then south from Flathead Lake.

Trees grow above a nearly horizontal line on the hillside north of the road between the crest of the moraine and Elmo. Although it is not easy to see from the road, that line is an old channel carved by a meltwater stream that flowed where the ice lapped onto the hillside. Now that the ice is gone, the old bedrock channel remains high on the hillside. This is the same channel that winds around the base of Chief Cliff, visible from US 93. The slope below the old stream channel contains several small terraces with flat tops, deposits of sediment laid down in ponds that formed along the edge of the glacier as it began to melt.

The Hog Heaven Hills

The forested Hog Heaven Hills in the distance north of Niarada are the eroded remains of an old volcano, the only one in the region.

Geologists know very little about this volcano except that it erupted mostly light-colored rocks during Tertiary time. Many of the rocks erupted from the Hog Heaven volcano resemble some of the alkalic rocks in parts of central Montana, so it seems possible that the Hog Heaven volcano is the westernmost outpost of that activity.

A prospector who discovered outcrops full of barite in 1913 located the Flathead Mine, a silver producer, near the center of the old volcano. The property produced large quantities of silver, and has much more in reserve. If the price of silver permits, the mine could again produce for many years. It is barely visible from the road as a large cut on the side of one of the hills.

The Giant Ripples of Camas Prairie

The 16 mile drive along Montana 382 between Perma and Hot Springs passes through the famous giant ripples of Camas Prairie, one of Montana's geologic spectaculars. Camas Prairie is also a bit of a relic.

Like all the large valleys in the Northern Rockies, Camas Prairie acquired a deep accumulation of basin fill sediments during the two

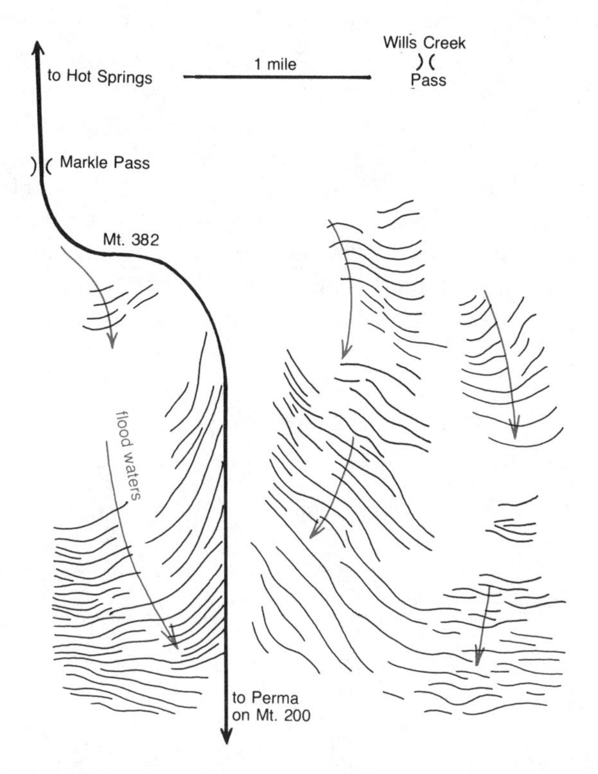

The pattern of giant ripples in Camas Prairie

long periods of dry climate. But Camas Prairie is unique among those valleys in not having contained a stream during the last several million years. It is a perfectly preserved fossil, a desert valley that still looks almost exactly as it must have at the end of Pliocene time, some two to three million years ago.

When Glacial Lake Missoula drained, some of the water that filled the Little Bitterroot Valley spilled south through Markle and Wills Creek passes into Camas Prairie, then on into the Flathead River. The torrent scoured holes into the bedrock in both passes. A deep funnel-shaped hole east of the road in Markle Pass contains a little pond that dries up most summers. Several much larger, but shallower, holes in Wills Creek Pass hold the Schmitz Lakes, which always contain water. The angular rubble eroded from the passes is now in the giant ripples south of them.

Water pouring south across Camas Prairie dumped its load of sediment in two incredible trains of enormous current ripples one below each pass. Except for their incredible scale, they look exactly like the little sand ripples that form in the bed of a stream, or in a gutter after a heavy rain. The giant ripples are easy to perceive from the air, because they look from a distance exactly like what they are. They are obvious from the ground too, if you know what to expect.

The best view of the giant ripples is from just south of Markle Pass, where you can see them spread out below you over miles of valley floor. From that vantage, it is clear that the two sets of ripples meet approximately along the line of the road. Ripples immediately below the passes are as much as 35 feet high, and about 300 feet from crest to crest. Although they become lower and harder to see southward, the ripples remain clearly visible almost to the lower end of Camas Prairie. They appear most vividly when the sun is at a low angle, or in the spring when the grass is just beginning to grow.

Bedding surfaces in Precambrian sedimentary rocks slope down towards Island Lake in the Mission Range.

Montana 83: Clearwater Junction—Bigfork
91 mi./146 km.

Montana 83 follows the length of the heavily forested Swan Valley, crossing glacial debris all the way. Bedrock everywhere is Precambrian sedimentary formations, but outcrops appear only near the ends of the route, around Salmon Lake and near Bigfork.

The Swan Valley

The Swan Valley lies between the Mission Range on the west and the Swan Range on the east. Both ranges are great slabs of nearly identical Precambrian sedimentary formations similarly tilted down to the east. Both contain magnificent alpine landscapes, expressed from a distance in their crosscut saw profiles, the mark of heavy ice-age glaciation. Many vantage points reveal marvelous views of the Swan Range, but the trees permit only a few glimpses of the more distant Mission Range.

The Swan Valley Glacier

As the great ice ages wore on, an enormous mass of ice slowly filled the Rocky Mountain trench of British Columbia, and moved

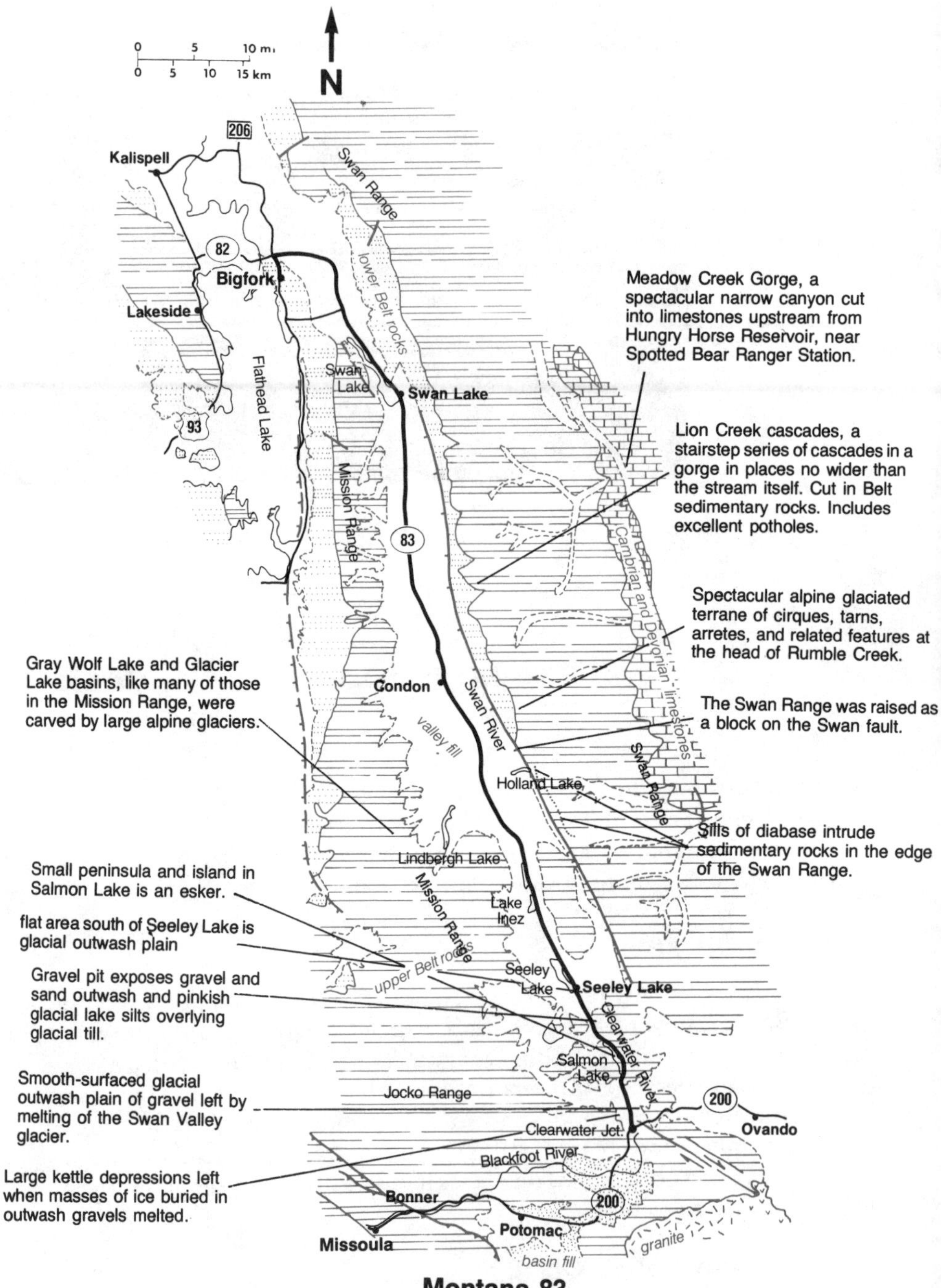
0 5 10 mi
0 5 10 15 km
N
206
Kalispell
82
Bigfork
Lakeside
93
Flathead Lake
Swan Range
lower Belt rocks
Swan Lake
Swan Lake
Mission Range
83
Meadow Creek Gorge, a spectacular narrow canyon cut into limestones upstream from Hungry Horse Reservoir, near Spotted Bear Ranger Station.
Lion Creek cascades, a stairstep series of cascades in a gorge in places no wider than the stream itself. Cut in Belt sedimentary rocks. Includes excellent potholes.
Cambrian and Devonian limestones
Spectacular alpine glaciated terrane of cirques, tarns, arretes, and related features at the head of Rumble Creek.
Gray Wolf Lake and Glacier Lake basins, like many of those in the Mission Range, were carved by large alpine glaciers.
Condon
Swan River
valley fill
The Swan Range was raised as a block on the Swan fault.
Swan Range
Holland Lake
Sills of diabase intrude sedimentary rocks in the edge of the Swan Range.
Lindbergh Lake
Mission Range
Small peninsula and island in Salmon Lake is an esker.
Lake Inez
flat area south of Seeley Lake is glacial outwash plain
upper Belt rocks
Seeley Lake
Seeley Lake
Gravel pit exposes gravel and sand outwash and pinkish glacial lake silts overlying glacial till.
Clearwater River
Salmon Lake
Smooth-surfaced glacial outwash plain of gravel left by melting of the Swan Valley glacier.
Jocko Range
200
Ovando
Clearwater Jct.
Blackfoot River
Large kettle depressions left when masses of ice buried in outwash gravels melted.
Bonner
200
Potomac
Missoula
basin fill
granite

Montana 83
CLEARWATER JUNCTION—BIGFORK

McDonald Peak rises above a glacially sculptured landscape in the Mission Range.

ponderously south. As the ice ages approached their maximum, the ice finally crept into the Flathead Valley. There, the low northern end of the Mission Range split the glacier, sending one branch down the wide open northern end of the Swan Valley. During the Bull Lake ice age, that branch of the glacier left moraines that form many of the hills just south of Clearwater Junction.

The road between Clearwater Junction and Salmon Lake crosses the smooth surface of the outwash plain below a younger moraine deposited around the lower end of Salmon Lake during the Pinedale ice age. About 15,000 years ago, at the height of that last ice age, that smooth surface must have been a broad expanse of watery sand and gravel that flooded with torrents of glacial meltwater on warm

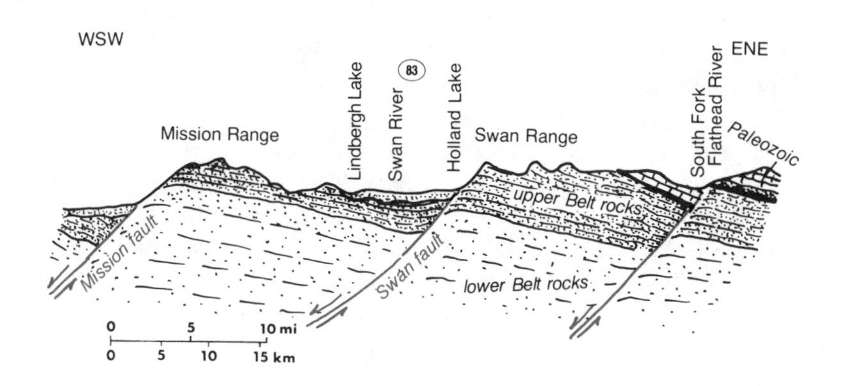

Section across the Swan Valley. Except for the Paleozoic formations near the Flathead River, all the rocks are Precambrian Belt formations.

Glaciated peaks of the Swan Range rise in the distance beyond Holland Lake.
—U.S. Forest Service photo by K. D. Swan

summer days. Since then, ice buried in the outwash melted and the surface collapsed to form Harper Lake and several small ponds near the road.

Salmon Lake, at the southern end of the Swan Valley, lies north of the Pinedale moraine, which formed a natural earth fill dam. Then the Clearwater River eroded its valley through the moraine right down to hard Precambrian bedrock. Had that bedrock been a little more deeply buried, the river would have drained the lake.

North of Salmon Lake, the road passes close to many large and small lakes that spangle the forested floor of the Swan Valley, all flood depressions in the glacial deposits. Picture the scene as those lakes formed at the end of the last ice age. The best evidence suggests that the last ice age ended with an abrupt change in climate approximately 10,000 years ago, and that the great glaciers melted within less than 3000 years. As the annual supply of snow that drove them forward dwindled, the glaciers stagnated, and then the ice melted where it stood. Warm summer days must have melted enormous amounts of water that collected in pools on the ice, and meltwater streams ran from pool to pool and eventually off the glacier. That must have been quite a scene.

Wherever the meltwater flowed, it carried sediment melted out of the ice, and wherever it ponded, that sediment accumulated on the ice as deposits of outwash. Meanwhile, the high areas of the stagnant glacier continued to melt, and to shed their content of sediment into the low areas. When the ice finally melted, the outwash sediments remained, and became the higher parts of the modern valley floor. The low areas many of the lakes now flood were the high parts of the ice surface, where no sediment collected. So the modern landscape of the valley floor is the inverse of the icescape that existed when the stagnant glacier was melting about 10,000 years ago.

After the ice finally melts, the gravel beds of the meltwater streams remain as long ridges of sand and gravel called eskers. One of those ridges winds for several miles parallel to the road in and beside Salmon Lake. A small public recreation area is on the southern end of the ridge, and its northern end forms several small islands, one nearly covered by a large house.

After the mass of ice that filled the Swan Valley was finally gone, large glaciers continued to creep down the valleys of the Mission and Swan ranges. Imagine those long tongues of ice thrusting out of the mountains onto the broad expanse of watery mud and sand that the floor of the Swan Valley then was. Those last lobes of ice left high moraines around their edges that now enclose Holland and Lindbergh lakes as well as several others that poke into the Swan Valley from the mouths of large canyons in the Mission and Swan ranges.

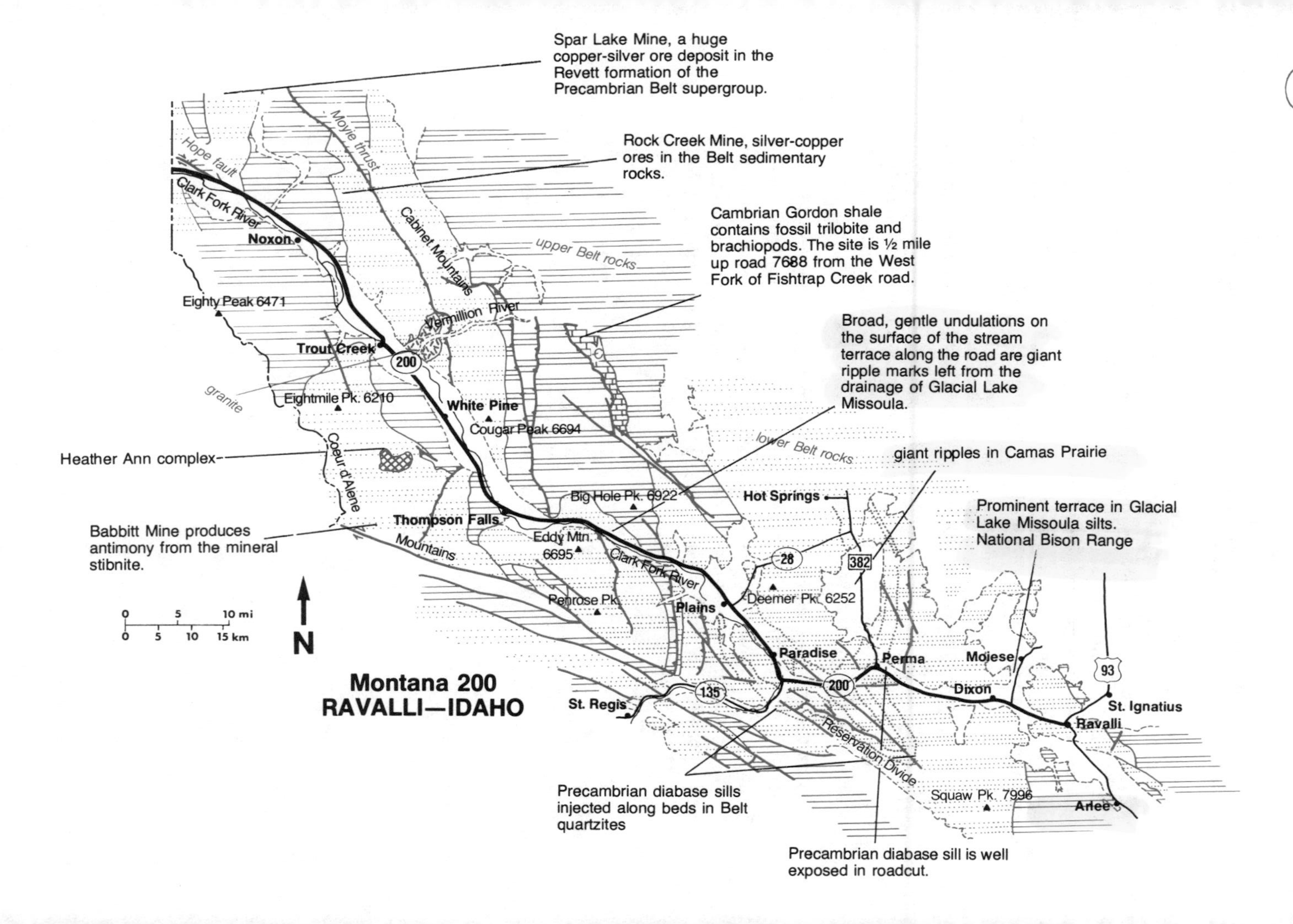
Spar Lake Mine, a huge copper-silver ore deposit in the Revett formation of the Precambrian Belt supergroup.
Rock Creek Mine, silver-copper ores in the Belt sedimentary rocks.
Cambrian Gordon shale contains fossil trilobite and brachiopods. The site is ½ mile up road 7688 from the West Fork of Fishtrap Creek road.
Broad, gentle undulations on the surface of the stream terrace along the road are giant ripple marks left from the drainage of Glacial Lake Missoula.
giant ripples in Camas Prairie
Prominent terrace in Glacial Lake Missoula silts. National Bison Range
Heather Ann complex
Babbitt Mine produces antimony from the mineral stibnite.
Precambrian diabase sills injected along beds in Belt quartzites
Precambrian diabase sill is well exposed in roadcut.
Hope fault
Clark Fork River
Noxon
Moyie thrust
Cabinet Mountains
upper Belt rocks
Eighty Peak 6471
Vermillion River
Trout Creek
200
granite
Eightmile Pk. 6210
White Pine
Cougar Peak 6694
lower Belt rocks
Coeur d'Alene
Mountains
Big Hole Pk. 6922
Hot Springs
Thompson Falls
Eddy Mtn. 6695
Clark Fork River
28
382
Penrose Pk.
Plains
Deemer Pk. 6252
Paradise
Perma
Mojese
93
Dixon
St. Ignatius
Ravalli
135
St. Regis
Reservation Divide
Squaw Pk. 7996
Arlee
0 5 10 mi
0 5 10 15 km
N
Montana 200
RAVALLI—IDAHO

The Clark Fork River Valley east of Thompson Falls

Montana 200: Ravalli—Idaho Border

115 mi./184 km.

Montana 200 follows the Jocko River between Ravalli and Moiese, the Flathead River between Moiese and Paradise, and the Clark Fork River between Paradise and the Idaho line. No part of the valley contained ice age glaciers. But this was the valley that carried all the drainage from Glacial Lake Missoula, so the ice ages did leave their emphatic mark.

All the bedrock along the road formed during Precambrian time. Most of it is sedimentary formations, Belt rock. Much of that rock is the Prichard formation, thousands of feet of very dark gray mudstone that weathers into reddish brown outcrops because it contains little crystals of iron pyrite that break down into reddish iron oxide. The road also passes several of the big diabase sills that sandwiched themselves between the sedimentary layers about a billion years ago, when the western part of the continent split off. The diabase is black, and in places it breaks into distinctive columns.

Cambrian rocks exist in a considerable area north of Thompson Falls. They contain a few fossil animals and numerous tracks and burrows, things the Precambrian formations lack. Several quarries have worked the Flathead sandstone, the oldest Cambrian formation, for flagstone and building material. The rock comes out of the quarries in colorful slabs beautifully patterned in shades of red, brown, and yellow.

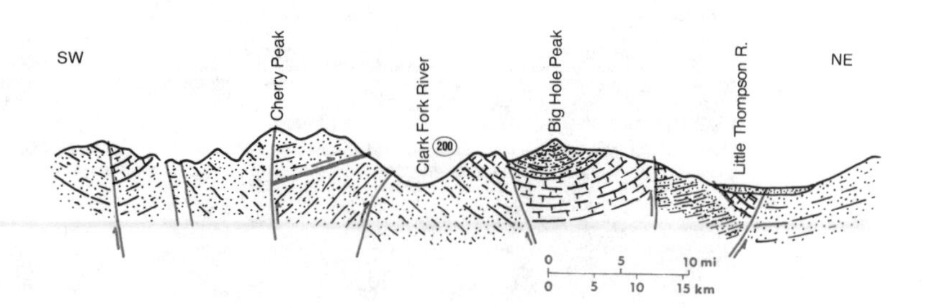

Section across Montana 200 about 10 miles east of Thompson Falls. All the rocks are Precambrian Belt formations. The situation near Cherry Peak, where steep faults that dip vertically offset older thrust faults that shoved the rocks east, is typical of the region.

The Precambrian sedimentary rocks in this part of Montana are the older formations that must have been deeply buried. They contain metamorphic minerals that formed as the rocks recrystallized at a temperature of several hundred degrees during Precambrian time, a billion or so years ago. About 65 million years ago, this part of Montana shed its thick cover of younger rocks eastward into the overthrust belt, uncovering the older rocks. It seems reasonable to speculate that the steep faults shown in the cross-section may have developed as the earth's crust floated upward after that heavy burden moved off it.

Glacial Lake Missoula

During the hours and days when Glacial Lake Missoula was draining, this part of the Clark Fork Valley carried more water than the combined flow of all the streams of the world. The passage of that incredible torrent left the wider stretches of the valley almost untouched except for occasional trains of giant ripple marks on the valley floor. Watch for gentle undulations that rise and fall rhythmically every several hundred feet on the stream terraces. Narrower parts of the valley show more obvious signs of flood scouring.

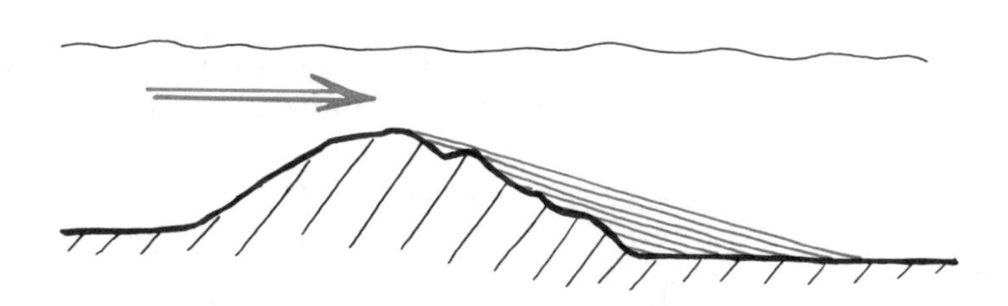

Deposits of sand and gravel accumulated where bedrock knobs broke the force of the flood current.

Flood flow through the narrower stretches of the valley was very fast and highly erosive, especially between Perma and Plains, and for several miles east of Thompson Falls. The torrents scrubbed off most of the soil, leaving ragged bedrock on the lower valley walls and on the valley floor. Many of the rough bedrock knobs on the valley floor have smooth, grassy slopes on their downstream sides. Those are flood deposits of sand and gravel swept into the lee side of the knob. The scoured lower part of the valley contrasts sharply with the softly upholstered slopes still covered with soil above the reach of the flood.

Water rushing through the narrow reaches of the Flathead River Valley eddied into the mouths of tributaries, and dumped sediment there to form deposits that look almost like small earthen dams. Watch for those gulch filling deposits between Perma and Paradise, and east of Thompson Falls.

A gulch filling partially dams a tributary valley north of Montana 200 between Perma and Paradise.

You can calculate the amount of water flowing through a channel by multiplying the area of its cross section by the speed of the flow. The gulch fillings between Perma and Paradise record the upper limit of the flood waters, and therefore make it possible to reconstruct the cross section of the flood. It is possible to approximate the speed of the flow by looking for the largest rocks it rolled along the bottom. In this case those are almost the size of basketballs, which suggests a flow speed of about 45 miles per hour. And that is a minimum figure because the size of the largest transported rocks may be limited by the size of the largest chunks that break free of the bedrock, instead of by the speed of the flow. The tendency of torrential floods to bury the largest boulders they carry under smaller debris as the water subsides further minimizes the figure.

Multiplying the estimated flow speed by the area of the channel cross section shows that something between 8 and 10 cubic miles of water per hour surged down the valley between Perma and Paradise while Glacial Lake Missoula was draining. To put that into better perspective, consider that the greatest flood discharge ever measured on the Mississippi River at Memphis was 0.02 cubic miles per hour.

Cold Air Springs

The lower part of Thompson Falls contains several cold air springs long used as natural refrigerators. The town of Thompson Falls stands on a hill that looks like the debris dump of a big rock fall of

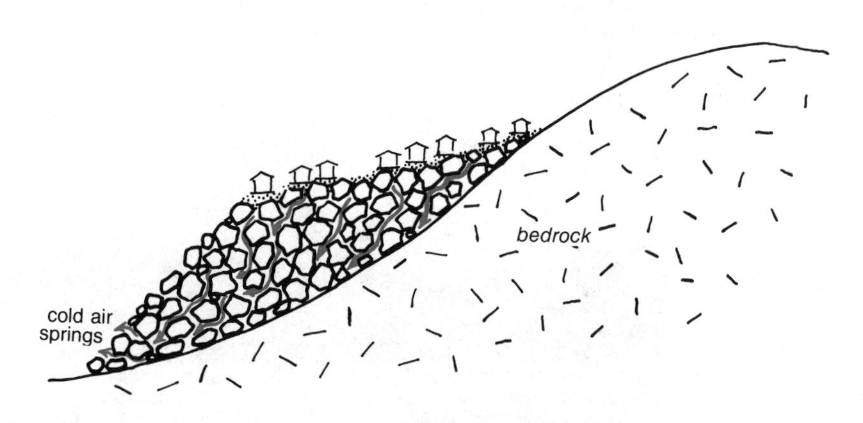

The underground plumbing of a cold air spring. Only air colder than that already there can sink into the open spaces underground, and it continues to sink until it emerges near the base of the debris pile.

many thousands of years ago. The ground is somewhat lumpy, and full of angular boulders that project at odd angles from the soil. Air can sink into open spaces between those rocks only on winter days when the air outside is colder and therefore denser than that underground. That, combined with the good insulating qualities of soil, keeps the interior of the hill well refrigerated.

Strange Rocks

Except for the diabase sills, igneous rocks are rare in this part of Montana. There are a few small masses of fairly ordinary granite such as the intrusion east of Trout Creek, and some very strange rocks. The Heather Ann complex west of Thompson Falls contains an assortment of pale igneous rocks that tend to contain excessive amounts of sodium. Some of the magmas lost part of their sodium into the neighboring sedimentary formations, converting them into unusual metamorphic rocks called fenites.

This isolated complex of odd rocks may be the top of an intrusion like the Rainy Creek complex near Libby, which also wears a cap of pale igneous rock rich in sodium. Perhaps in a few million years erosion will bite through the light rocks into a deeper mass of black igneous rock, pyroxenites, like those in the Rainy Creek complex. A magnetic survey over the Heather Ann complex revealed far more magnetism than the light rocks exposed at the surface can explain. That is good evidence that a large mass of dark igneous rocks may lie below the surface.

Old Mines, Possible New Mines

Prospectors discovered gold in the Vermillion River east of Trout Creek in 1887. A brief flurry of activity followed, but did not produce much. Some quartz veins along the Vermillion River still produce small samples of spectacular ore, but not in mineable quantity. The granite intrusion shown on the map undoubtedly explains the gold.

Drilling during the early 1980s around Goat Peak in the Cabinet Range north of Noxon turned up several impressive bodies of silver and copper ore. Although they lie within the Cabinet Wilderness Area, the deposits can be mined because they were discovered before the wilderness area was closed to mineral development. Whether or not they will be mined depends upon the markets for copper and silver, an uncertain prospect because both are victims of solid state electronics. Copper wires are no longer much used in communications, and photographic emulsions, the main industrial use for silver, are now being replaced by video tapes and laser disks.

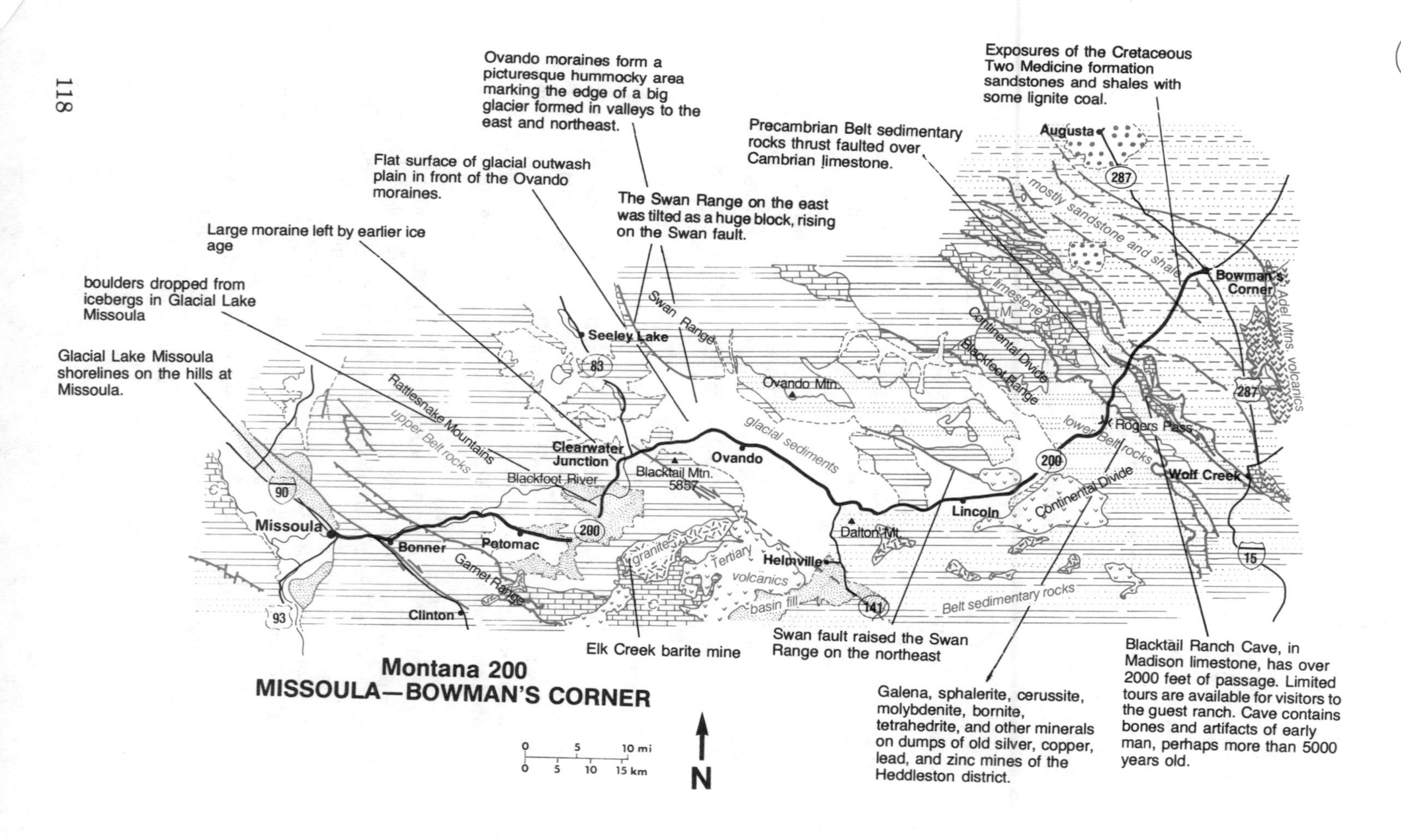

Exposures of the Cretaceous Two Medicine formation sandstones and shales with some lignite coal.
Ovando moraines form a picturesque hummocky area marking the edge of a big glacier formed in valleys to the east and northeast.
Precambrian Belt sedimentary rocks thrust faulted over Cambrian limestone.
Flat surface of glacial outwash plain in front of the Ovando moraines.
The Swan Range on the east was tilted as a huge block, rising on the Swan fault.
Large moraine left by earlier ice age
boulders dropped from icebergs in Glacial Lake Missoula
Glacial Lake Missoula shorelines on the hills at Missoula.
Augusta
287
mostly sandstone and shale
Bowman's Corner
Adel Mtns. volcanics
limestone
Continental Divide
Blackfoot Range
Swan Range
Seeley Lake
83
Ovando Mtn.
Rattlesnake Mountains
upper Belt rocks
glacial sediments
Rogers Pass
lower Belt rocks
Wolf Creek
15
Clearwater Junction
Blackfoot River
Blacktail Mtn. 5857
Ovando
200
Lincoln
Continental Divide
90
Missoula
Bonner
Potomac
Dalton Mt.
granite
Tertiary
volcanics
Helmville
basin fill
141
Garnet Range
Clinton
93
Belt sedimentary rocks
Elk Creek barite mine
Swan fault raised the Swan Range on the northeast
Galena, sphalerite, cerussite, molybdenite, bornite, tetrahedrite, and other minerals on dumps of old silver, copper, lead, and zinc mines of the Heddleston district.
Blacktail Ranch Cave, in Madison limestone, has over 2000 feet of passage. Limited tours are available for visitors to the guest ranch. Cave contains bones and artifacts of early man, perhaps more than 5000 years old.
0 5 10 mi
0 5 10 15 km
N

Montana 200
MISSOULA—BOWMAN'S CORNER

Montana 200: Missoula—Bowman's Corner
116 mi./186 km.

Most of the route between Missoula and Ovando skirts the western end of the Garnet Range, crossing the broad Potomac and Clearwater valleys along the way. Between Ovando and Rogers Pass, the road follows the Blackfoot River through the southern part of the overthrust belt to its headwaters on the continental divide. The route between Rogers Pass and Bowman's Corner crosses the disturbed belt.

The Sapphire Block

Between Missoula and Ovando, the road angles across the Garnet Range past steeply dipping sedimentary layers that mean the rocks are tightly folded. This is the north edge of the Sapphire tectonic block, the great mass of rock that moved east into Montana off the top

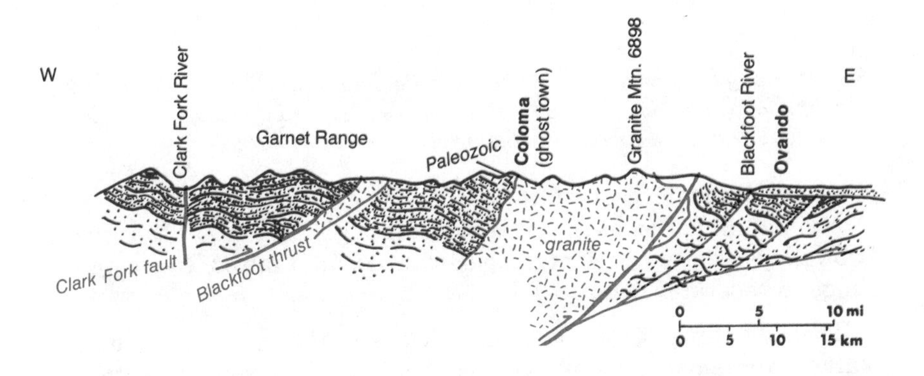

Section south of Montana 200 between Missoula and Ovando. Except for the granite and the small patch of Paleozoic rock, all the bedrock is Precambrian Belt formations.

of the Idaho batholith between about 75 and 70 million years ago. The folds are the crumpled leading edge of that block and the rocks bulldozed ahead of it.

Granite intrusions in the Garnet Range appear to have been emplaced along the big thrust faults that carried the Sapphire block east. We think the Sapphire block began to move before the Idaho batholith crystallized, and that it may have moved on a sole of molten granite magma, the granite we now see in the intrusions.

Belt Rocks

Mountains near the highway between Missoula and Rogers Pass, consist mostly of Precambrian sedimentary rocks, Belt formations, as do most of the rocks exposed in roadcuts. They began as beds of sand and mud laid down on land sometime around one or one and a half billion years ago, give or take a couple of hundred million years. Rocks near the road are mostly pink and gray sandstone, and colorful red and green mudstone. Exposures are especially good in the Blackfoot Canyon between Milltown and the Potomac Valley.

The Potomac and Clearwater Valleys

The Potomac Valley and the broad valley around Clearwater Junction are both structural basins deeply filled with sediment that accumulated during Tertiary time, when this region had a dry

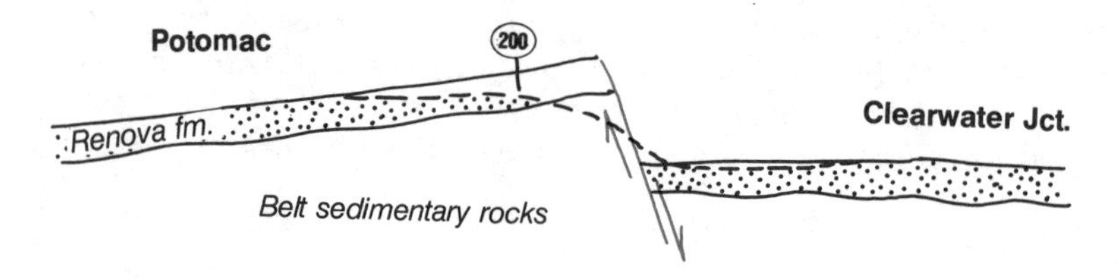

Section along the line of the road between Potomac and the Clearwater area showing how an active fault is raising a new ridge, and dividing one valley into two. The dashed line marks the present ground surface.

climate. Roadcuts east of the Potomac Valley to the Lubrecht Experimental Forest at the crest of Greenough Ridge show that the two valleys were one continuous basin until a fault raised Greenough Ridge to separate them.

Roadcuts north of the road where it passes through the Potomac Valley expose pale silts and clays of the Renova formation, with a thin veneer of coarse Pliocene gravel, the Six Mile Creek formation, on top. Roadcuts on both sides of the road on the southwest side of

Greenough Ridge expose precisely the same material—valley fill sediments to the top of the ridge. Obviously, Greenough Ridge was not there when those valley fill sediments were deposited, and has risen since then.

A northwest trending fault along its north side raised Greenough Ridge as though it were a giant cellar door hinged where it blends into the Potomac Valley. Occasional earthquakes, some sharp enough to rattle windows in Missoula, come from the Greenough area. They may mean that the fault is moving, that Greenough Ridge is rising, perhaps eventually to become a high mountain range.

The Overthrust Belt

Between Lincoln and Rogers Pass, Montana 200 follows a narrow valley cut through the high ridges of the overthrust belt. The ridges are hard Precambrian sedimentary rock that moved eastward in great slabs during formation of the Rocky Mountains. Masses of molten granite magma as much as several miles in diameter intruded the overthrust slabs, bringing small quantities of metallic minerals with them.

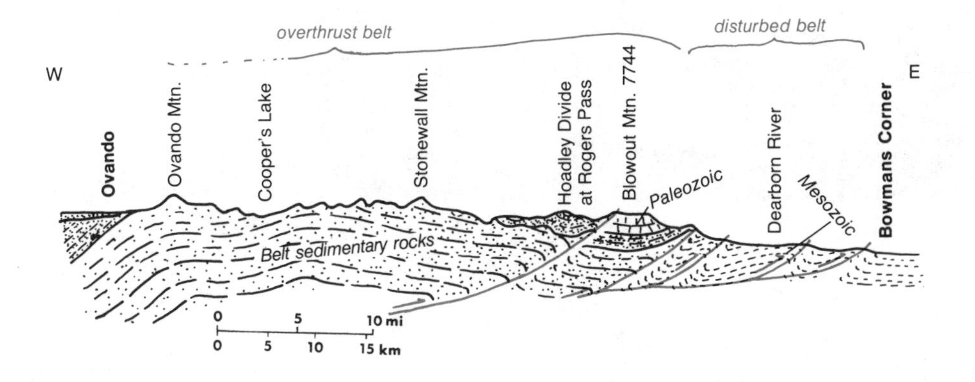

Section along a line north of the highway from Ovando to Bowman's Corner

The Disturbed Belt

Between Rogers Pass and Bowman's Corner, Montana 200 crosses the Montana disturbed belt. Rock formations beneath the High Plains are the same as those in the disturbed belt, and the landscapes of the two provinces are nearly the same. However, rock structures in

the disturbed belt closely resemble those in the overthrust belt.

Layers of brown sandstone and dark shales, relatively weak rocks, lie beneath this part of the western High Plains. Those rocks accumulated in shallow sea water that flooded this region during Cretaceous time, between about 145 and 65 million years ago. When the Rocky Mountains formed, great slabs of older sedimentary rocks moved eastward and piled up to make the overthrust belt, meanwhile rumpling the weak Cretaceous sedimentary rocks of the western High Plains into tight folds to make the disturbed belt. So the rocks in the disturbed belt are as severely deformed as those farther west in the mountains, but they do not resist erosion well enough to stand up as mountains.

Tilted layers of Cretaceous sedimentary rock in the disturbed belt east of Rogers Pass.

Mike Horse Mine

The Heddleston District between Lincoln and Rogers Pass includes several mines of which the Mike Horse, just south of the highway, was by far the largest. It produced gold, silver, and lead, probably nearly 10 million dollars worth, during the first half of this century. Continued exploration after the Mike Horse Mine closed revealed a large deposit of copper and molybdenum. There was much talk, and much controversy, during the late 1960s of plans to develop a large open pit mine to work that deposit. But the project was abandoned pending radical improvement in the economics of mining copper. The ore bodies are associated with igneous intrusions that invaded the overthrust belt, probably about 35 million years ago.

This hummocky topography near Ovando is a glacial moraine. The flat surface in the foreground is the outwash plain.

Souvenirs of the Ice Ages

Between the area a few miles south of Clearwater Junction and Rogers Pass, the highway crosses several areas of typical morainal topography. Recognize them as tracts of lumpy landscape full of humps and hollows. The humps are liberally strewn with erratic boulders, and the hollows commonly hold marshes, shallow ponds, and a few small lakes.

Several miles south of Clearwater Junction, the road crosses a tract of wooded hills, a big glacial moraine in which the humps and hollows are subdued enough to look fairly old. That moraine probably dates from the earlier Bull Lake glaciation. South of that moraine, the road follows the abandoned valley of an old meltwater stream onto a broad outwash plain where the marks of old stream channels are still faintly visible when the light is right.

Between Clearwater Junction and Lincoln, Montana 200 follows the Blackfoot River Valley, crossing a series of glacial moraines and their associated outwash plains. Ice poured down the big canyons in the mountains to the north into the Blackfoot Valley, where it spread out to form large ponds of nearly stagnant ice—piedmont glaciers. Most of those expanses of ponded ice were several miles across. Muddy meltwater pouring off them during the warm days of the ice age summers spread sand and gravel across the ice-free parts of the valley to create large outwash plains.

Ovando stands on a smooth outwash plain between two expanses of hummocky morainal topography that mark areas where ice draining down from the high mountains to the north ponded on the floor of the Blackfoot Valley. All along the way between Ovando and Lincoln, the road passes alternately across similar areas of smooth outwash and hummocky moraine. They are easy to recognize. Watch also for roadcuts in bouldery glacial till.

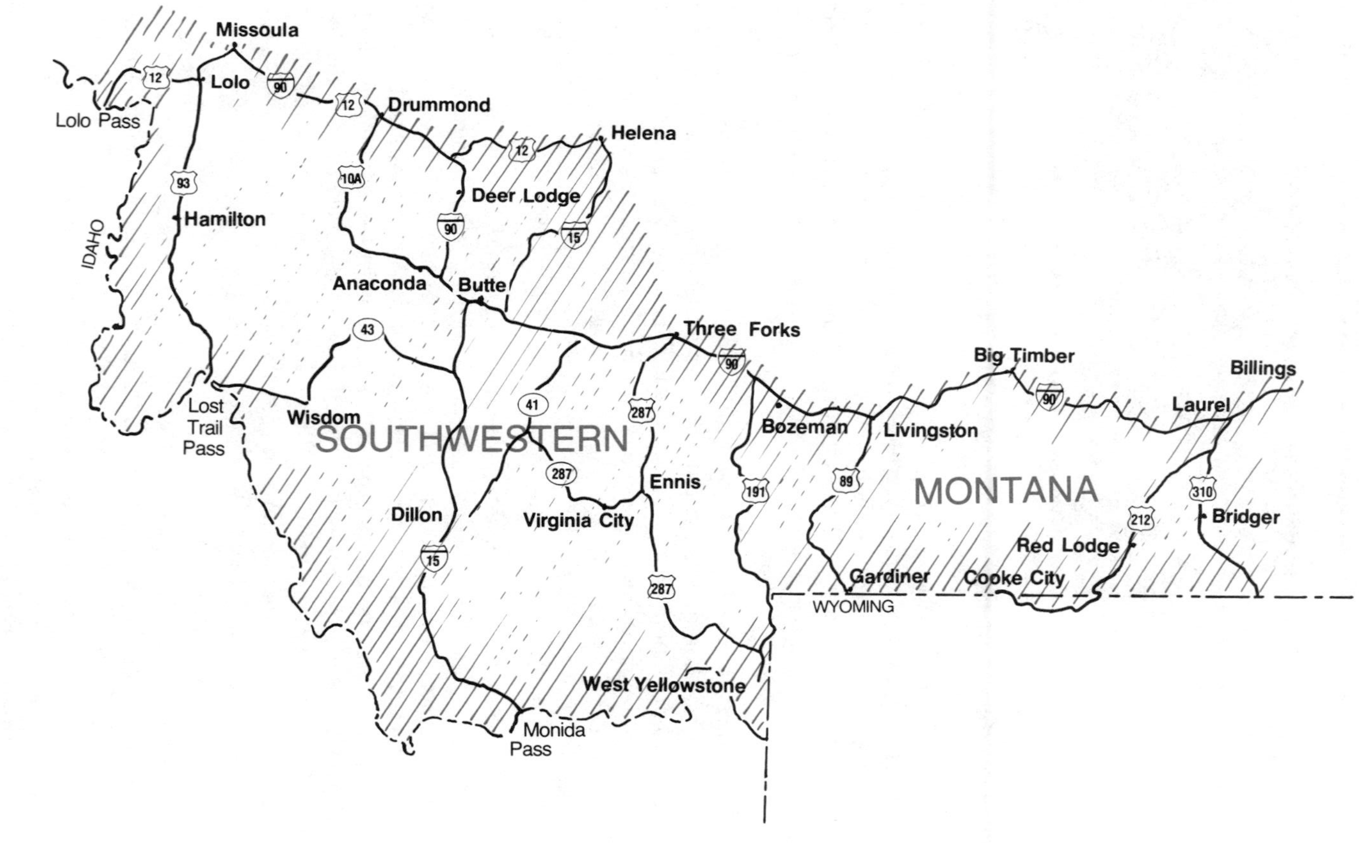

Roads covered in this section.

III
Southwestern Montana—Isolated Ranges, Spacious Valleys

BASEMENT ROCKS AND THEIR SEDIMENTARY COVER

Several of the mountain ranges in southwestern Montana contain large areas of continental crust without a cover of younger rocks, enough exposed basement rock to open an interesting window onto the most ancient part of the state's history. The great age of the rocks and the inability of geologists to read more than a fraction of the story they have to tell cloud our view through that window. Even more than the Precambrian Belt formations that cover them in much of western Montana, the basement rocks are messengers from a planet so remote in time from the one we know that it might as well be distant in space, too.

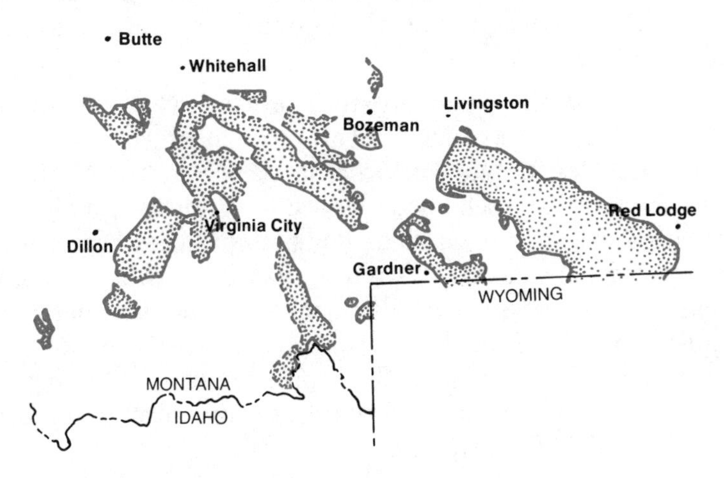

Areas of basement rock exposure in southwestern Montana

Like basement rocks everywhere, those of Montana are an extremely complex assortment of metamorphic rocks thoroughly scrambled into large masses of granite. The metamorphic rocks, mostly gneisses, schists, and marbles, surely formed through recrystallization of older sedimentary and volcanic rocks at extremely high temperature. In most of southwestern Montana, the temperature of metamorphic recrystallization was so extremely high that most of the gneisses and schists were at least on the verge of melting. Some actually did partially melt to form rocks composed of a mixture of metamorphic and igneous material; others melted completely to form large masses of granitic magma.

Throughout most of southwestern Montana, the basement rocks contain many large diabase dikes that trace dark, twisting paths through the metamorphic and igneous rocks. Dikes form as magma fills fractures, so they must be younger than the rocks that contain them. Dikes are normally straight, so we can safely assume that the curved dikes in the basement were folded along with the older metamorphic rocks they intrude.

Some of the basement rocks, including the dikes, contain minerals that form at relatively low metamorphic temperatures. Those minerals would have been obliterated during the recrystallization at high temperature, so must have formed later during a second metamorphism at a much lower temperature. One very important mineral that formed during the second metamorphism is the talc that now supports a large mining industry.

Age dates on the basement rock of southwestern Montana scatter so widely that it is difficult to be quite sure what they mean. The oldest dates come from the Beartooth Plateau, where some of the rocks are about 3.3 billion years old. Very few rocks anywhere on earth are older than those. Elsewhere in southwestern Montana the oldest dates cluster around 2.7 billion years. That must be about the time the basement rocks outside the Beartooth Plateau recrystallized at high temperature to essentially their present form. The original sedimentary and volcanic rocks certainly formed sometime before 2.7 billion years ago; no one knows how long before. Another tight cluster of age dates at about 1.6 billion years probably tells us when the low temperature metamorphic minerals formed. Some of the dikes are also about 1.6 billion years old. Others are younger, some as young as 800 million years, the time when the continent split during deposition of the Belt rocks.

The Sedimentary Veneer

In the eastern part of the area, the Willow Creek fault neatly separates the northern limit of exposed basement rock in southwestern Montana from the southern limit of the Precambrian Belt sedimentary formations. It seems certain that the area south of that fault rose, and supplied sediment to the accumulating Belt formations farther north. Farther west, the parallel Horse Prairie fault similarly separates areas of exposed basement rock from those in which Precambrian sedimentary rocks accumulated. Most of the Paleozoic and Mesozoic sedimentary formations that eventually covered the basement rock in southwestern Montana tend to be thinner there than elsewhere. Evidently, the area remained relatively high even though it must have been deeply eroded during Precambrian time.

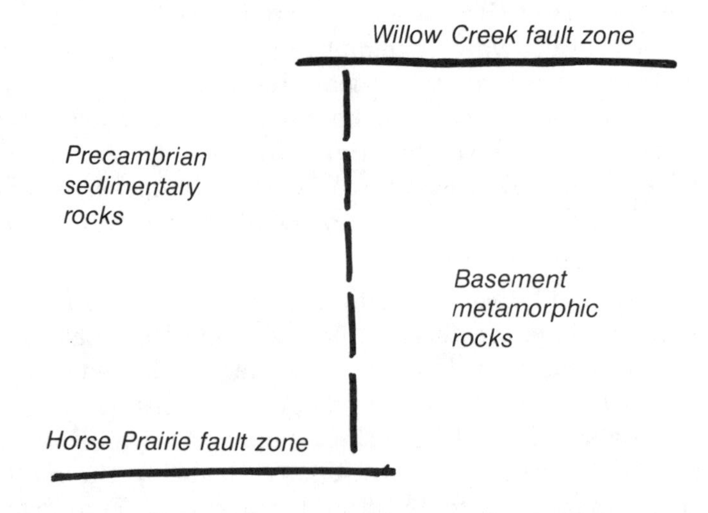

The basement, the Belt, and the Willow Creek and Horse Prairie faults

THE ACTION BEGINS

The geologic events that created most of the modern scene in southwestern Montana began as the Atlantic Ocean began to open and the floor of the Pacific Ocean started sinking beneath the old western margin of North America. As the sinking ocean floor reached a depth of about 60 or 70 miles, it lost steam and perhaps molten basalt magma, both at an incandescent white heat. Meanwhile, the push of the Pacific Ocean floor jammed the western margin of the continent into itself, raising a broad welt across Idaho and western Montana. After the glowing steam and basalt magma had risen from the sinking oceanic crust for some tens of millions of years, the lower continental crust began to melt. Expansion of the hot rocks further raised the welt along the western margin of the continent. That completed the setting of the stage.

The Idaho Batholith

Water is the only common substance that freezes more readily with a decrease in pressure, melts with an increase in pressure. Ice skates glide so slickly because the pressure of the skate melts a thin film of water that lubricates its passage across the ice. Like water, the melting point of granite magma that contains its full quota of steam increases as the pressure on it diminishes. So granite magmas heavily charged with steam rise through the earth's crust until the decreasing pressure raises their melting point to the actual temperature of the magma. Then they crystallize into granite at constant temperature, and the solid rock cools later.

Between about 90 and 70 million years ago, vast quantites of granite magma melted deep within the broad welt that formed along the old continental margin. The magma melted about 15 miles below the surface, rose in great masses until it was about 10 miles below the surface, then began to crystallize. The pressure at that depth was low enough to bring the melting point of the magma up to its actual temperature. That granite is now exposed throughout a large region of central Idaho and in the Bitterroot Range of western Montana. Geologists call it the Idaho batholith. It would still be far below the surface had the 10 miles or so of rock that covered the crystallizing magma not moved off in great slabs.

All that molten magma within the welt along the old western margin of the continent made it unstable. Big pieces began to detach and move east. The welt, as we vaguely call it, was a mountain range that must have been quite a bit higher than the modern Bitterroot Range; no one knows how much higher. The most visible results of that movement in southwestern Montana are the Sapphire and Pioneer blocks.

Magma is not nearly as strong as solid rock, so the continental crust was much weaker than it normally is while it was heavily engorged with the masses of granitic magma destined to crystallize to become the Idaho batholith. And the enormous masses of magma within the crust must have bulged the surface, further steepening the welt that had already developed as the collision with the floor of the Pacific Ocean crushed the western margin of North America.

Sometime between 75 and 70 million years ago, great slices of that bulge broke and moved eastward, probably gliding on a sole of weak granitic magma as though they were riding on a thick smear of grease. During a period that probably lasted something more than a million years, those detached pieces of the upper continental crust moved almost directly east a distance of about 50 miles, from Idaho into nearby western Montana. Geologists call those detached chunks of upper crust the Sapphire and Pioneer blocks. Both contain similar arrays of related geologic features.

The Sapphire Block

The Bitterroot mylonite was the first of those features to attract attention. It is a layer of extremely slabby gneiss a half mile or so thick that lies along the east front of the Bitterroot Range, facing the Bitterroot Valley. If you look closely at the surfaces of the slabs, you can see faint lines and grooves, a lineation, that invariably points slightly south of east. A hike into the mouth of any of the canyons in the range will pass good exposures.

Geologists interpret the Bitterroot mylonite as a fault zone that developed deep below the surface where the rocks are so hot and under such great pressure that they tend to flow like modeling clay instead of breaking and sliding. They think of movement in the slabby gneiss as resembling that in a sliding

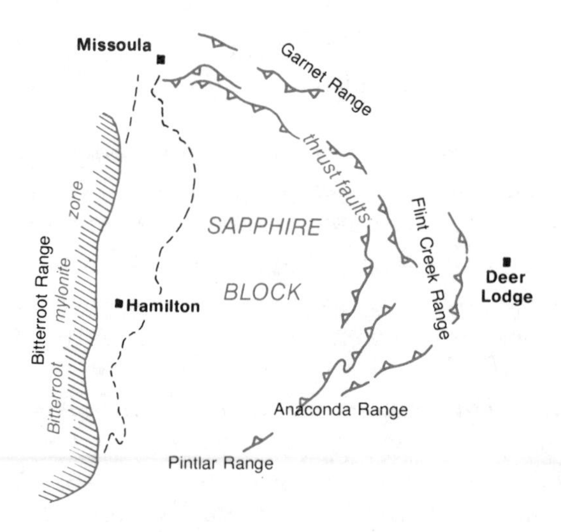

The Sapphire block lies east of the Bitterroot Valley within an arc of mountain ranges that define its leading edges.

deck of cards, each slab or card moving only a relatively small part of the total distance the zone or deck moves. The lineation points the way the rocks moved.

The Bitterroot Valley lies between the Bitterroot mylonite on the west, and the Sapphire Range on the east. Deep wells drilled to explore for uranium produced cores of platy gneiss from beneath the basin fill sediments near the northern end of the valley. So it seems that the mylonite continues beneath the Bitterroot Valley, indeed beneath the entire Sapphire block. It

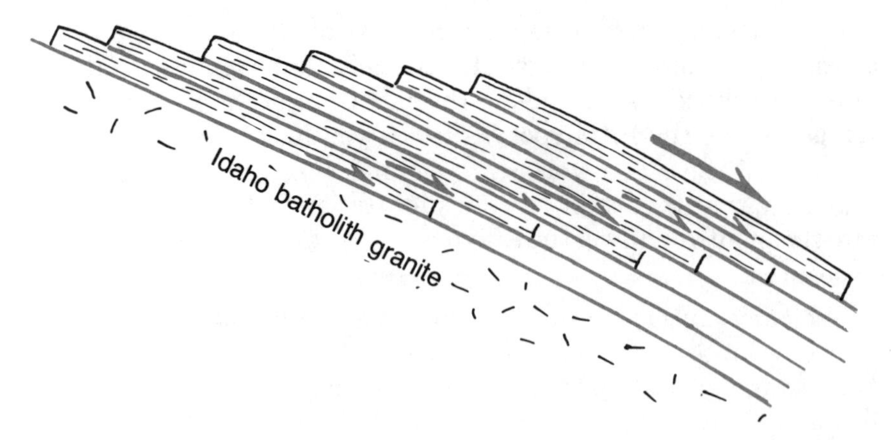

The Bitterroot mylonite probably moved like a sliding deck of cards.

makes sense then to think of the Bitterroot Valley as simply the gap behind the Sapphire block.

Low mountains in the interior of the Sapphire block, the Sapphire Range, consist mostly of Precambrian sedimentary rocks, Belt formations, that show relatively little folding and faulting compared to the intensely deformed rocks around the margins of the block. Those rocks contain scattered intrusions of granite, all apparently related to the Idaho batholith. If the Sapphire block did indeed ride east on a sole of granite magma, then it is easy to imagine some of that granite rising into the rocks above.

The Garnet, Flint Creek, and Anaconda-Pintlar ranges define the northern, eastern, and southern boundaries of the Sapphire block. All three ranges consist basically of tightly folded sedimentary formations of all ages from Precambrian through Cretaceous that were shoved east along big thrust faults. Large masses of granitic magma intruded all three ranges, and crystallized to make big intrusions. Almost invariably, the granite intrusions lie along the big thrust faults, but the fault does not break the granite. Evidently, the granite magma intruded along faults that moved while the magma was still melted, but did not move after it was crystallized. It seems very likely that the magma lubricated the faults.

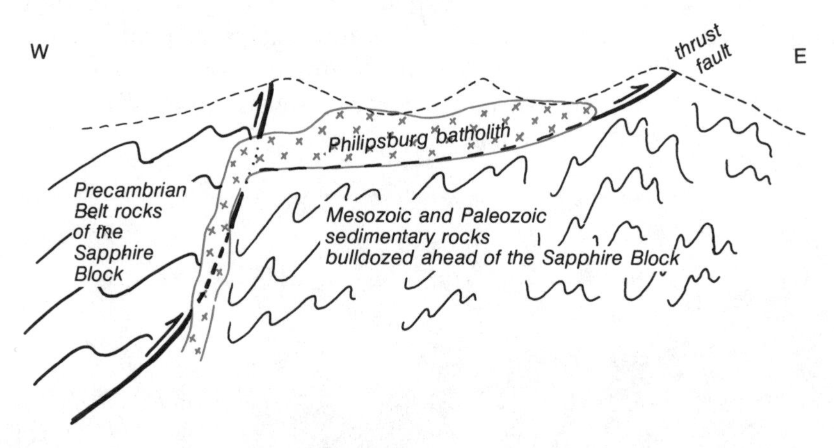

Schematic diagram through the Flint Creek Range that fringes the eastern edge of the Sapphire block showing its probable origin as sedimentary formations crushed in and ahead of the leading edge of the moving slab. A granite intrusion penetrates the folds.

It seems likely that all three of the mountain ranges that fringe the Sapphire block consist partly of the crumpled leading edge of the detachment block, partly of rock bulldozed ahead of it. It is difficult to distinguish between the two masses of rock because both are tightly folded and thrust faulted in much the same way. But in a general way, the younger Paleozoic formations tend to lie on the outboard sides of the fringing ranges and seem not to have moved as far as the Belt rocks, so they were probably bulldozed.

The Pioneer Block

The Pioneer Mountains appear to be another detachment block similar in most basic respects to the Sapphire block. However, a shear zone resembling the Bitterroot mylonite defining the east side of the Idaho batholith has not yet been found. The broad valley of the Big Hole may, like the Bitterroot Valley, have opened as a gap behind the detachment block. The Pioneer Range east of the Big Hole appears to be the detached block.

Rocks along the eastern front of the Pioneer Range include a wide variety of sedimentary formations, all rumpled into very tight folds and broken along numerous large thrust faults. The shape of the folds and the displacement of the faults both indicate that the rocks moved east. That kind of deformation is precisely what one would expect in the leading edge of a displaced detachment block. Vast quantities of granite, the enormous Pioneer batholith and many lesser intrusions, penetrate the folded sedimentary rocks, mostly along the faults that moved the rocks east.

Granite outcropping in an alpine meadow in the Pioneers.

The Big Hole, western Montana's largest valley, appears to be the gap behind the Pioneer block. Deep exploration wells drilled in an unsuccessful attempt to find oil and natural gas, revealed that the bedrock floor of the valley lies some 14,000 feet below the modern surface. If the early mountain men had known that, they might have called the valley the "Deep Hole." Volcanic rocks fill the lowermost part of that great depth, and a thick accumulation of Renova formation covers them. It is not clear whether the volcanic rocks erupted during Cretaceous time, as the Pioneer block moved, or about 50 million years ago when volcanic rocks erupted over much of southwestern Montana and nearby Idaho. Either time would make sense.

The Boulder Batholith

The Boulder batholith in the general area between Butte and Helena lies about 100 miles east of where a self-respecting batholith its size should be to fit into the plate tectonics scheme of things. So far, no one has explained why it is where it is. And the magma differed from the melts that became the Idaho batholith in coming all the way to the surface, instead of crystallizing at great depth. Evidently, the magma contained very little water.

Many laboratory experiments have shown that the melting point of granite magmas that contain very little steam drops with decreasing pressure—that is exactly opposite to the behavior of magmas rich in steam. Dry granite magmas remain melted as they rise through the crust, and erupt if they are still dry when they reach the surface. We can be sure that the magma that became the Boulder batholith contained very little steam because much of it erupted and the part that crystallized to become granite did so at very shallow depth. That magma must have melted somewhere near the base of the continental crust, where the rocks contain very little water.

Enormous volumes of the magma of the Boulder batholith erupted to build the volcanic pile that we call the Elkhorn Mountains volcanic rocks. At least some of the eruptions were explosive enough to spread volcanic rocks far beyond the volcanic center, overspraying a large area of older rocks. Meanwhile, the reservoir of molten magma beneath continued to rise, and invaded the lower part of its own volcanic pile before it finally crystallized into granite.

While it was in business, the big volcano that erupted the Elkhorn Mountains volcanic rocks over the Boulder batholith probably resembled the one now active in Yellowstone National Park, a type of volcano called a resurgent caldera. But the Elkhorn Mountains volcano erupted gray andesite in addition to white rhyolite. That distinction is important because andesite melts at a considerably higher temperature than rhyolite. Something, probably basalt magma rising from the mantle, must have added an enormous amount of heat to the lower part of the continental crust. Mixing of basalt magma with granitic magma from the crust could have formed andesite.

Andesite eruptions typically form broken and rubbly rocks that erode fairly rapidly. Within 20 million years, enough of the andesite cover had eroded off to expose large areas of granite in the Boulder batholith. As the Elkhorn Mountains volcanic pile eroded, streams carried most of the debris southeast, and dropped it along what was then the coast of a shallow inland sea. Now we see those deposits in the Livingston formation, an enormous dump of volcanic debris that covers a large part of the area that became the Crazy Mountain basin of central Montana. The next phase of volcanic activity in Montana will bury much of the erosion surface developed on the exposed granite of the Boulder batholith beneath a new generation of volcanic rocks.

A NEW GENERATION OF IGNEOUS ROCKS

During Eocene time, about 50 million years ago, igneous activity resumed along a broad zone on both sides of a northeast-trending axis that stretches from somewhere southwest of Boise through central Idaho and southwestern Montana, and into central Montana. In general, the igneous activity of 50 million years ago was most voluminous near the southwestern end of the trend, and became weaker northeastward. And the rocks change in composition along the trend, with andesite becoming less abundant northeastward. It makes sense for that change in volume and composition to go together as they do, because andesite melts at a higher temperature than rhyolite. Evidently, more heat was entering the crust in the more southwesterly parts of the trend, thus melting more rock, and forming melts at a higher temperature.

The best known 50 million-year-old igneous rocks in southwestern Montana are the Lowland Creek volcanic rocks that cover a large area in the general vicinity of the Boulder batholith. They consist mostly of rhyolite, along with minor amounts of andesite and basalt.

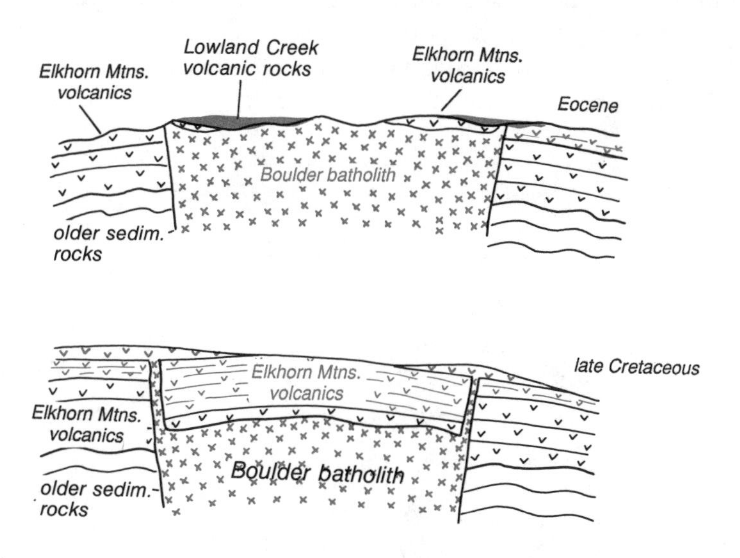

The Lowland Creek volcanic rocks lie on an erosion surface planed across the older Elkhorn Mountains volcanic rocks and the Boulder batholith beneath them.

Parts of extreme southwestern Montana contain large fields of volcanic rocks that have so far received very little attention. Those rocks lie between the Lowland Creek rocks of Montana and the Challis volcanic rocks of central Idaho, both 50 million years old, so it seems reasonable to assume that at least some of them are part of the same trend, and also 50 million years old. Unlike the Lowland Creek volcanic rocks, those in Beaverhead County include substantial quantities of andesite, as do the Challis volcanic rocks of central Idaho. At least some andesite probably erupted in southwestern Montana about 70 million years ago, when the Cretaceous granites were forming. The problem of distinguishing between the two generations of andesite will probably keep graduate students in geology busy for a long time.

Also 50 million years ago, numerous large masses of granitic magma crystallized without erupting to become masses of intrusive granite that exist here and there along the entire igneous trend, a swarm of new batholiths and stocks spread across northern Idaho and western and central Montana. All of those younger granite intrusions crystallized at extremely shallow depth, within a few thousand feet of the surface. Most of those granites contain blocky crystals of feldspar set in a matrix of fine-grained rock, a distinctive texture that makes them easy to recognize.

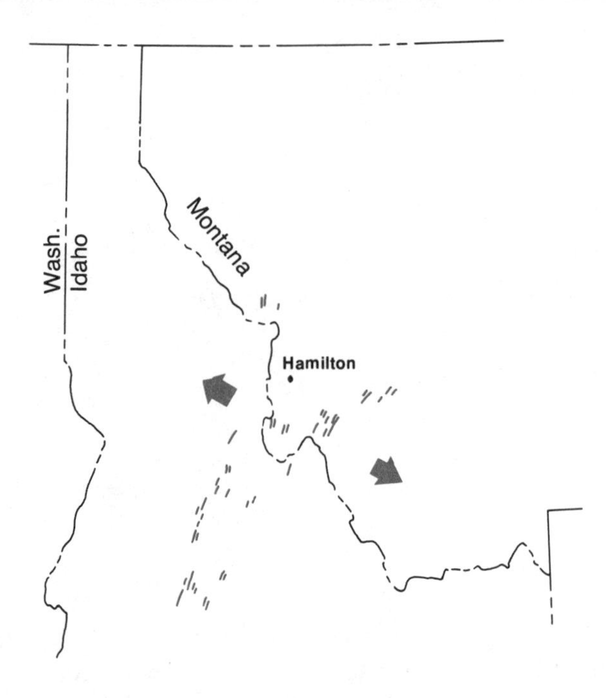

Known parts of the 50 million-year-old dike swarm. Arrows show the direction in which their emplacement must have stretched the earth's crust.

Swarms of large dikes, thousands of them, all trending northeast, follow much of the 50 million-year-old igneous trend. Emplacement of dikes must unavoidably involve stretching of the earth's crust because it is necessary to open a crack so magma can fill it to form the dike. Where the earth's crust stretches, it normally breaks along faults that separate the upper crust into blocks, which become mountain ranges and broad basins as they move away from each other. Although southwestern Montana does contain such basins and ranges, they trend in the wrong direction and are probably too young to fit in with the 50 million-year-old dike swarm. What happened?

We suggest, as bald conjecture, that the cause of the igneous activity of 50 million years ago was a long pool of basalt magma at the base of the continental crust. Perhaps it was trapped in the crest of a broad crustal arch that trended northeastward. Molten basalt exists at a temperature several hundred degrees higher than the melting temperature of andesite, so a reservoir of basalt magma could melt enough of the already very hot lower continental crust to form large quantities of andesite and rhyolite. If our hypothetical pool of basalt magma became larger southwestward, that could explain the increased volume, the higher temperature, and the increasingly basaltic composition of the igneous activity southwestward.

The Absaroka Volcanic Field

Also about 50 million years ago, the Absaroka volcanic center was erupting vast quantities of andesite in what is now the eastern part of Yellowstone National Park and the surrounding area. Some of those eruptions spread andesite north into the Gallatin and Absaroka ranges of Montana.

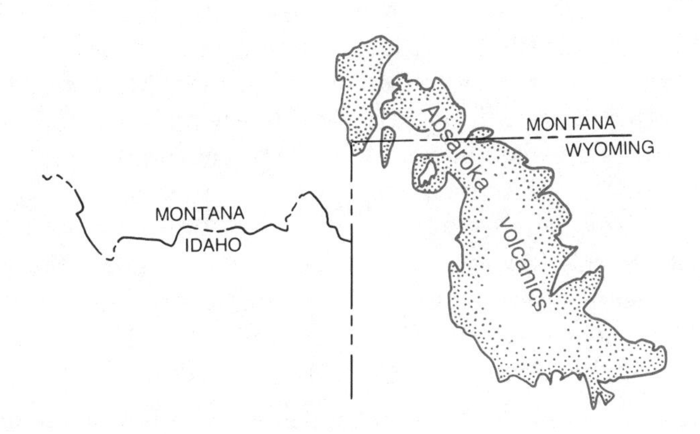

Only the northern fringe of the Absaroka volcanic field reaches into Montana.

The Absaroka volcanic rocks cover an approximately rectangular area about the size and shape of the Elkhorn Mountains volcanic field, except that the long axis of the rectangle points northwest instead of northeast. Like the Elkhorn Mountains volcanic rocks, those in the Absaroka volcanic pile consist almost entirely of andesite. The broad similarities between the Cretaceous Elkhorn Mountains and Eocene Absaroka volcanic piles show that they are two examples of the same thing, presumably of large resurgent calderas much like the one now active in Yellowstone National Park. Further erosion may well reveal a batholith, like the Boulder batholith, beneath the Absaroka volcanic pile.

The Eocene andesites of the Absaroka Mountains contain abundant petrified wood, entire forests carried down the mountain and buried in great volcanic mudflow deposits. The late Cretaceous Elkhorn Mountains andesites, on the other hand, do not contain plant fossils. Evidently the climate of Eocene time was wet enough to support dense forests, whereas that of late Cretaceous time was not. The generally brown color of the Absaroka volcanic rocks is an iron stain that may also reflect weathering in a wet climate.

The Latest Folds

To judge from the age dates, the spasm of igneous activity that peaked 50 million years ago wore itself out around 40 million years ago, about the time deposition of the Renova formation began. Nearly everywhere, the sedimentary layers of the valley-filling Renova formation are tilted on the flanks of folds and broken along faults. No one seems to have studied that deformation enough to find out whether it happened while the formation was accumulating, or afterward. If deformation accompanied deposition, then it would be reasonable to suggest that the mountain building crustal movements simply continued through Oligocene and early Miocene time while the sediments of the Renova formation accumulated. If deformation followed deposition, it would follow that crustal movements ceased during Oligocene and early Miocene time, then vigorously resumed in late Miocene time.

In either case, most of the crustal movements had certainly ended by the time the Six Mile Creek gravels began to accumulate during latest Miocene time. Although many faults break the Six Mile Creek gravels, that formation is generally only slightly tilted. So, most of the region has been relatively quiet for about 10 million years. But the northern Rocky Mountains aren't complete yet. Crustal movements continue.

Active Faults

Occasional earthquakes remind everyone that southwestern Montana contains most of Montana's active faults. The great majority trend northwest, a few trend northeast, and several have other orientations. Those active faults are now dissecting the older ranges and valleys of southwestern Montana into a pattern of new ranges and valleys overprinted on the old ones, another new page in the geologic record printed directly on top of the old ones.

Some of the newly rising mountain ranges, such as the Blacktail Range, have patches of Renova formation on them, showing that they were in the valley floor when those sediments were deposited. Some of the new valleys, including the Blacktail Deer Valley, contain no exposures of the Renova formation, little or none of the Six Mile Creek formation. The simplest interpretation is that the valleys did not exist in anything resembling their present form when those deposits were laid down.

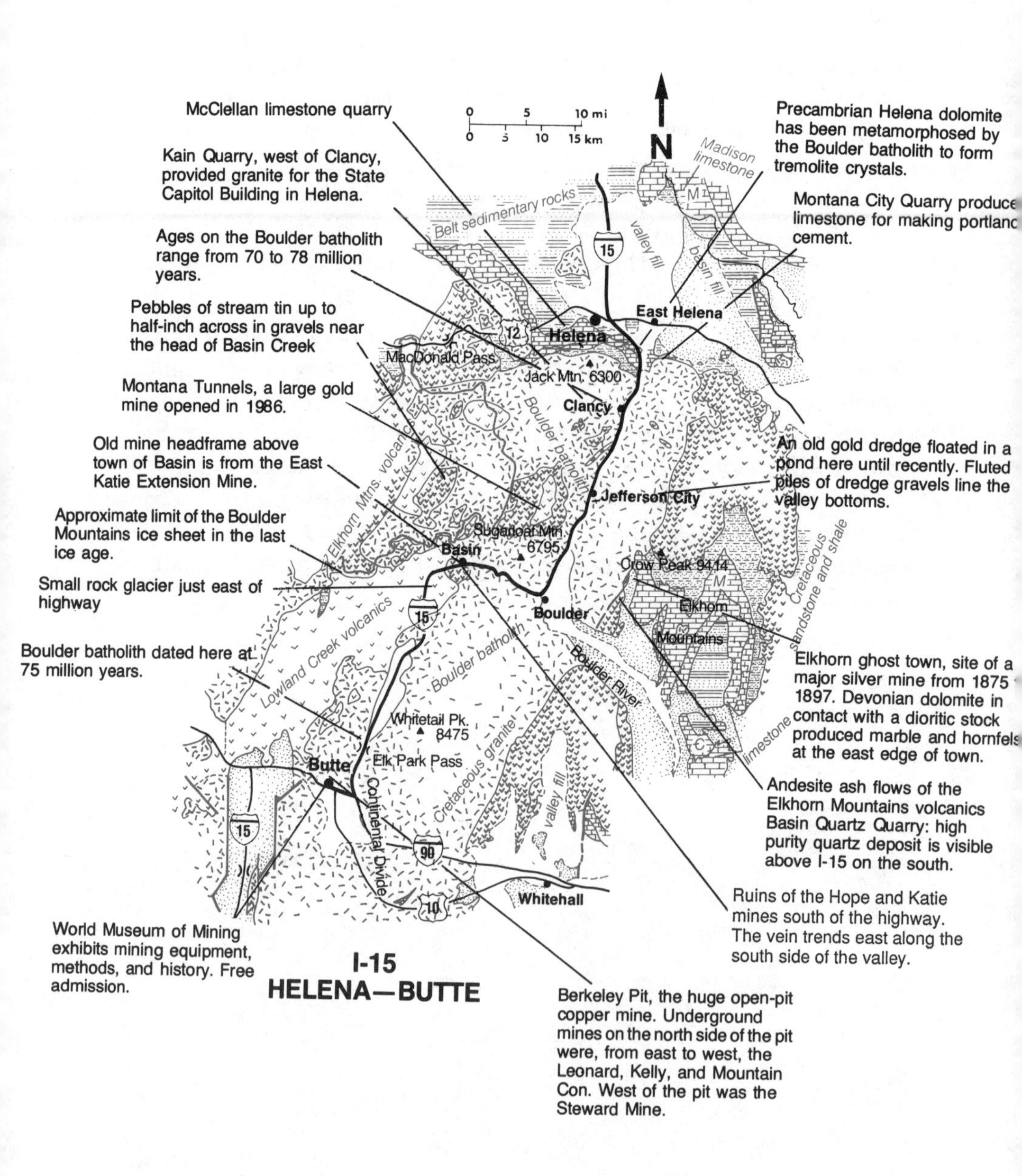
McClellan limestone quarry
Kain Quarry, west of Clancy, provided granite for the State Capitol Building in Helena.
Ages on the Boulder batholith range from 70 to 78 million years.
Pebbles of stream tin up to half-inch across in gravels near the head of Basin Creek
Montana Tunnels, a large gold mine opened in 1986.
Old mine headframe above town of Basin is from the East Katie Extension Mine.
Approximate limit of the Boulder Mountains ice sheet in the last ice age.
Small rock glacier just east of highway
Boulder batholith dated here at 75 million years.
World Museum of Mining exhibits mining equipment, methods, and history. Free admission.
Precambrian Helena dolomite has been metamorphosed by the Boulder batholith to form tremolite crystals.
Montana City Quarry produce limestone for making portland cement.
An old gold dredge floated in a pond here until recently. Fluted piles of dredge gravels line the valley bottoms.
Elkhorn ghost town, site of a major silver mine from 1875 1897. Devonian dolomite in contact with a dioritic stock produced marble and hornfels at the east edge of town.
Andesite ash flows of the Elkhorn Mountains volcanics Basin Quartz Quarry: high purity quartz deposit is visible above I-15 on the south.
Ruins of the Hope and Katie mines south of the highway. The vein trends east along the south side of the valley.
Berkeley Pit, the huge open-pit copper mine. Underground mines on the north side of the pit were, from east to west, the Leonard, Kelly, and Mountain Con. West of the pit was the Steward Mine.
0 5 10 mi
0 5 10 15 km
N
Madison limestone
Belt sedimentary rocks
valley fill
basin fill
Helena
East Helena
MacDonald Pass
Jack Mtn. 6300
Clancy
Boulder batholith
Elkhorn Mtns. volcanics
Jefferson City
Sugarloaf Mtn. 6795
Basin
Crow Peak 9414
Cretaceous sandstone and shale
Elkhorn Mountains
Boulder
Boulder River
Lowland Creek volcanics
Whitetail Pk. 8475
Elk Park Pass
Cretaceous granite
Butte
Continental Divide
limestone
Whitehall
15
12
90
10
I-15
HELENA—BUTTE

Interstate 15: Helena—Butte

64 mi./102 km.

Precambrian sedimentary rocks along the road between Helena and Clancy include conspicuous exposures of white marble, limestone bleached and baked by heat from the nearby Boulder batholith. About three miles north of Clancy watch for the change from ledges of white marble to bouldery outcrops of granite of the Boulder batholith. Although the fresh granite is pale gray, it weathers to a reddish brown surface. Between Clancy and Butte, the road passes through the Boulder Mountains, a picturesque tract of rugged hills that show little sign of ice age glaciation. All the rocks between Clancy and Butte are igneous: granite of the Boulder batholith and two generations of volcanic rock.

The Boulder Batholith

It is easy to recognize fresh granite exposed in roadcuts as a pale to medium gray rock that sparkles as flat crystal faces, tiny mirrors, reflect sunlight. A closer view reveals milky white or pink crystals of feldspar, along with shiny black crystals of hornblende, and a few dark gray grains of glassy looking quartz. Most of the mineral grains are about the size of a split pea. In natural exposures, the weathered granite generally forms pinkish brown boulders, or piles of boulders, heavily crusted with black lichens.

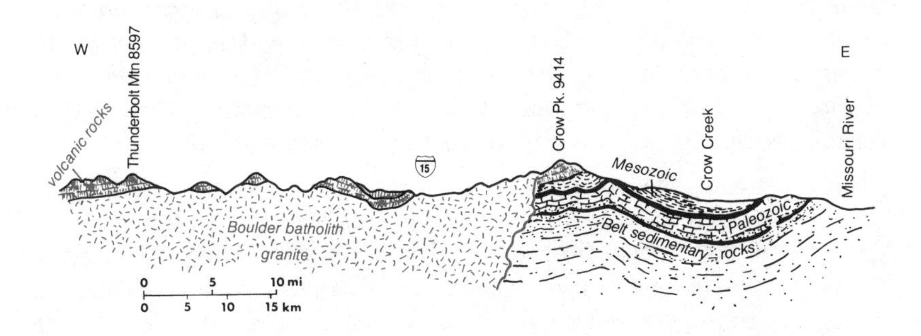

Section across the highway north of Boulder showing granite intruding older sedimentary rocks. Some of the volcanic rocks erupted 70 million years ago from the batholith; others are only 50 million years old.

Two Generations of Volcanic Rocks

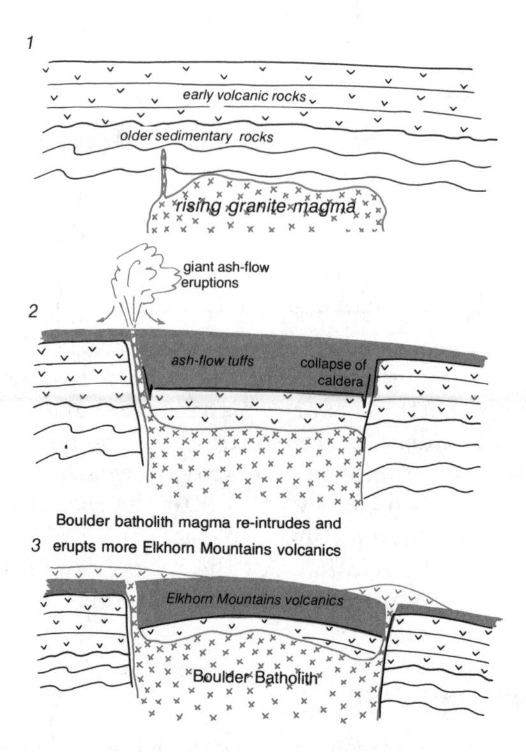

The Elkhorn Mountains volcanic rocks erupted from the same magma that formed the Boulder batholith, and were then intruded by that magma. The younger Lowland Creek volcanic rocks lie on an erosion surface that cuts across granite and the older volcanic rocks.

The older volcanic rocks are andesites named after the Elkhorn Mountains. They erupted between 70 and 80 million years ago from the same mass of magma that crystallized below the surface to become the Boulder batholith. Recognize them as fine-grained gray or greenish gray rocks that typically contain scattered needles of glossy black hornblende. The younger Lowland Creek volcanic rocks erupted about 50 million years ago. They are also fine-grained, and range in color from brown through red to almost white.

Both granite and volcanic rocks acquire a crust of black lichens that may thoroughly obscure the rock, but not the tendency of the volcanic rocks to weather into talus slopes instead of bouldery outcrops. Water seeps into closely spaced fractures in the volcanic rocks, expands as it freezes, and pries off blocks that tumble into the talus slopes.

Several miles west of Basin, on the opposite side of the river, the road passes a large talus slope that is turning itself into a rock glacier. Normal talus slopes have smooth surfaces and straight profiles; rock glaciers are wrinkled, and have the distinct look of something trying to move. Rock glaciers form where talus slopes fill with ice, which, like the ice in a ordinary glacier, begins to flow under its own weight. Although rock glaciers are common in the high mountains of western Montana, it is rare to see one at such low elevation.

The Elkhorn Mountains

Hills in the distance east of the highway between Helena and Boulder belong to the Elkhorn Mountains, a complex range beyond the margin of the Boulder batholith. The much faulted and tightly folded sedimentary rocks in the Elkhorn Mountains include formations that range in age from Precambrian to late Cretaceous.

Andesites, like those along the road, that erupted from the Boulder batholith during late Cretaceous time cover much of the older sedmentary rocks in the Elkhorn Mountains. Even though it looks quite different, the greenish Elkhorn Mountains andesite has the same chemical composition as the pale gray granites of the Boulder batholith. That is one excellent reason for believing that both crystallized from the same magma.

Elkhorn, about 10 miles southeast of Boulder, sprang up after gold was discovered in 1869, but the mines there produced far more silver than gold. Like most Montana silver mines, those at Elkhorn closed after the financial panic of 1893 took the bottom out of the price of silver. New owners pumped out the mine and put it back in production in 1901, then closed it again in 1912, this time because the vein was worked out. The mine and town have been dead since then despite occasional efforts to pluck a few last morsels of ore out of the vein.

Tin

Early prospectors found placer gold in the headwaters of Basin Creek in the 1860s, and miners worked the deposits until the 1890s, again during the Great Depression. They also found occasional small but heavy pebbles of cassiterite, a rare tin oxide mineral in a variety called "wood tin," because it looks almost like petrified wood. The tin probably weathered out of the Lowland Creek volcanic rocks, which include rhyolite that contains small amounts of topaz, a mineral commonly associated with tin.

Hydraulic mining with hoses at Jefferson Bar in 1890.
—Montana Historical Society photo by F. Jay Haynes

Gold

Between Clancy and Jefferson City, the highway passes several miles of old gold mine tailings, piles of gravel left behind the dredge that started chewing its way through the floodplain in 1939. Some of the dredged gravel found its way into the interstate highway, but plenty remains in neatly stacked ridges.

The Hope-Katie Mine across the river from Basin produced many fantastic museum specimens of fabulously rich ore—quartz literally butter yellow with disseminated flakes of gold. The mine worked more or less steadily from 1890 until 1925, then sputtered along off and on for another 40 years. The larger than necessary roadcut south of the highway marks surface mining operations that worked the vein almost down to river level after the underground ore bodies were gone. A few old buildings and foundations survive.

Despite its history of producing spectacular bonanzas of rich ore, most, if not all, of the many companies that operated the Hope-Katie Mine went broke. Their total production is unknown, but probably about 2 million dollars, not much to show for so many years of operation. As in many gold mines, the occasional bonanza ore shoots were not big enough to offset the vast expanses of barren and profitless rock that separated them.

Elk Park

For about 10 miles north of Elk Park Pass, the highway follows the floor of Elk Park, an obvious stream valley complete with tributaries, but without a stream. To judge from the way the dry tributaries angle

into Elk Park, the stream that eroded the valley must have flowed south, opposite to the way its floor now slopes. Another stretch of road a few miles north of Boulder follows a similar dry valley.

Elk Park and the valley north of Boulder are probably just what they appear to be, remnants of an old stream valley eroded millions of years ago that survive in the modern landscape. Both valleys are wide enough to contain a large stream of the kind that must have flowed in western Montana during the latter part of Miocene time, when the region enjoyed a wet, tropical climate. We think they are probably segments of one of those Miocene valleys abandoned when the climate again became very dry during latest Miocene time.

Bison Creek and its tributaries have dissected the north end of Elk Park into a hilly landscape.

The Elk Park fault, a branch of the Continental fault that bounds part of the Butte Valley, abruptly chops off the south end of Elk Park. Some geologists think the abandoned valley continues southwest from the Butte Valley, but that is hard to see.

Butte

Between Elk Park and Butte, the road follows the eastern part of the area where copper was mined from big open pits between 1955 and 1983. The pit near the road was the last to operate before the mines finally closed in June, 1983, the end, in all probability, of the main period of activity of one of the world's greatest mining districts.

Butte started as a placer gold camp in 1864, but gold was almost as

The derelict head frame of the East Katie Extension Mine stands watch over the town of Basin. The mine worked a part of the Hope-Katie vein that was offset north of the river along a fault.

scarce as water to wash it out of the ground. The first miners hauled gravel down the hill to Silver Bow Creek in ox carts, only to find that it contained very few nuggets too much diluted with silver to bring the best price. Quartz veins stained black with manganese and full of silver minerals were discovered at about the same time as the placer gold, mainly because they were too big and obvious to overlook.

The red lines show veins containing silver; the black ones copper bearing veins as they were mapped in Butte about the turn of the century. Mineralized veins trend generally east throughout the Boulder batholith.

In fact, there were two important sets of veins: silver bearing quartz veins stained black with manganese minerals, and veins full of copper minerals. The copper veins congregated in a tight bullseye in the area that later became the Berkeley Pit and the silver veins swarmed like a halo around them. That pattern is typical: geologists commonly find copper veins passing into silver mineralization with increasing distance from the source of the metals. Some of the many mines that started working during the 1880s produced mostly silver, others mostly copper, depending upon the kind of vein they were following.

Many of the rich veins were completely worked out by the 1950s, and the end was in sight for the others. Then Butte changed its style by switching to production of enormous tonnages of low grade ore from the Berkeley Pit. Most of the ore from that pit contained less than one half percent copper, so little that it was difficult to see ore minerals in a specimen of the rock. The pit operated on an extremely narrow profit margin offset by its ability to produce large amounts of ore with a fairly uniform metals content.

Although the high grade veins were worked out years ago, the supply of pit grade ore is by no means exhausted. The mines finally closed because the price of copper persistently failed to keep pace with inflation. If renewed demand for copper were to restore the price to a level equivalent to that of the mining years, Butte could resume its former identity as a mining camp.

Meanderville area before excavation of the Berkeley Pit. —World Museum of Mining photo

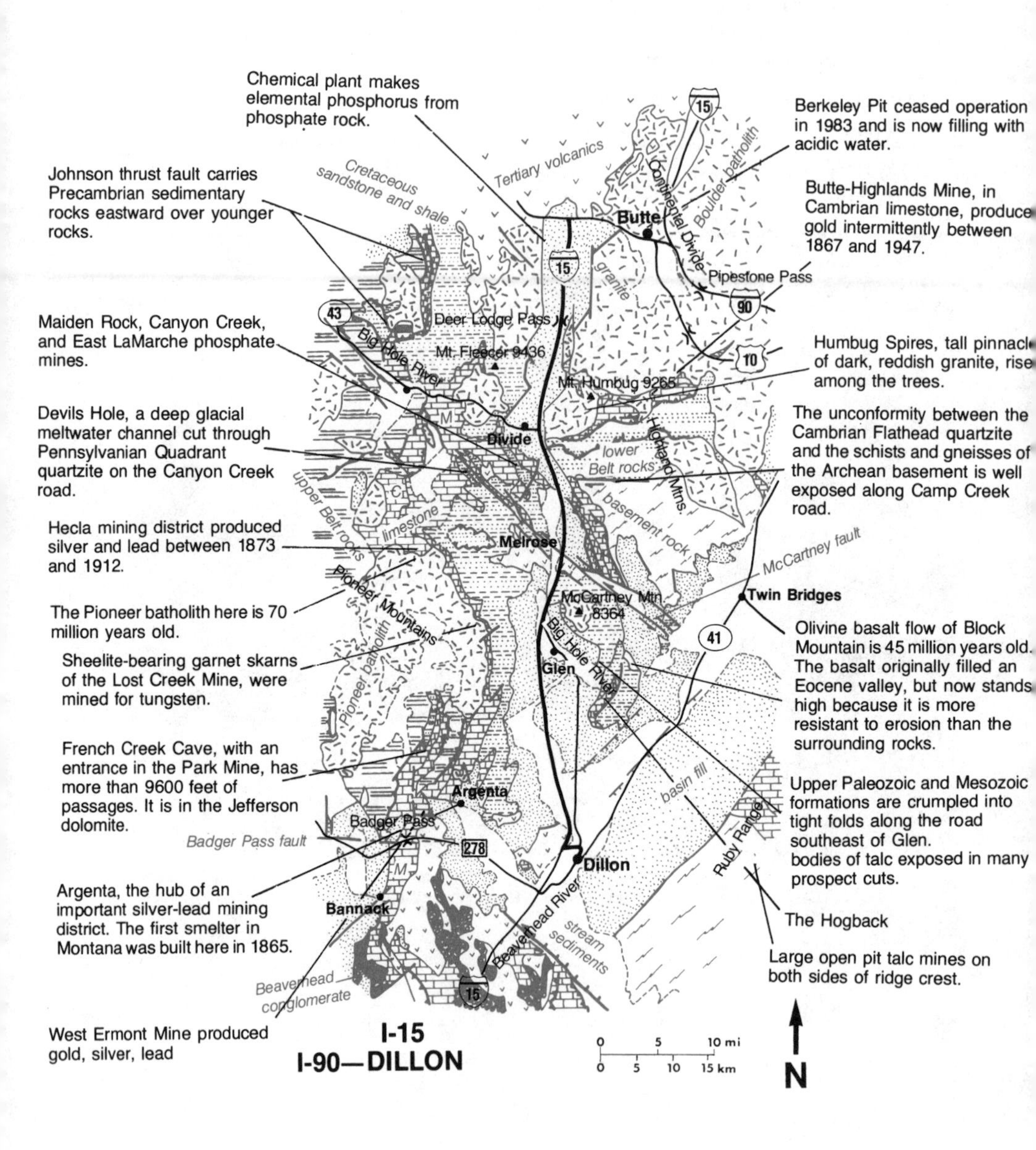
Chemical plant makes elemental phosphorus from phosphate rock.
Johnson thrust fault carries Precambrian sedimentary rocks eastward over younger rocks.
Maiden Rock, Canyon Creek, and East LaMarche phosphate mines.
Devils Hole, a deep glacial meltwater channel cut through Pennsylvanian Quadrant quartzite on the Canyon Creek road.
Hecla mining district produced silver and lead between 1873 and 1912.
The Pioneer batholith here is 70 million years old.
Sheelite-bearing garnet skarns of the Lost Creek Mine, were mined for tungsten.
French Creek Cave, with an entrance in the Park Mine, has more than 9600 feet of passages. It is in the Jefferson dolomite.
Badger Pass fault
Argenta, the hub of an important silver-lead mining district. The first smelter in Montana was built here in 1865.
West Ermont Mine produced gold, silver, lead
Berkeley Pit ceased operation in 1983 and is now filling with acidic water.
Butte-Highlands Mine, in Cambrian limestone, produce gold intermittently between 1867 and 1947.
Humbug Spires, tall pinnacl of dark, reddish granite, rise among the trees.
The unconformity between the Cambrian Flathead quartzite and the schists and gneisses of the Archean basement is well exposed along Camp Creek road.
Olivine basalt flow of Block Mountain is 45 million years old. The basalt originally filled an Eocene valley, but now stands high because it is more resistant to erosion than the surrounding rocks.
Upper Paleozoic and Mesozoic formations are crumpled into tight folds along the road southeast of Glen.
bodies of talc exposed in many prospect cuts.
The Hogback
Large open pit talc mines on both sides of ridge crest.
Cretaceous sandstone and shale
Tertiary volcanics
Butte
Continental Divide
Boulder batholith
granite
Pipestone Pass
Deer Lodge Pass
Mt. Fleecer 9436
Mt. Humbug 9265
Big Hole River
Divide
lower Belt rocks
Highland Mtns.
upper Belt rocks
limestone
basement rock
Melrose
McCartney fault
Twin Bridges
Pioneer Mountains
McCartney Mtn. 8364
Pioneer batholith
Glen
basin fill
Argenta
Badger Pass
Ruby Range
Dillon
Bannack
Beaverhead River
stream sediments
Beaverhead conglomerate
15
90
10
43
41
278
I-15
I-90—DILLON
0 5 10 mi
0 5 10 15 km
N

Granite split and weathered along fracture surfaces forms pinnacles in the Humbug Spires.

Interstate 15: Interstate 90—Dillon
58 mi./93 km.

Between the junction with Interstate 90 and Divide, the highway passes between the Highland Range to the east, and the long granite rise of Starlight Mountain, part of the Boulder batholith, to the west. The route between Divide and the area south of Melrose passes through the narrow gap between the craggy skyline of the Pioneer Range on the west and McCartney Mountain. Between McCartney Mountain and Dillon, the road continues past the Pioneer Range, and imposing row of craggy peaks, as it follows the west side of the Beaverhead Valley. Much lower and generally subdued mountains on the skyline east of Dillon are the Ruby Range, which was never glaciated.

The Highland Range

East of Interstate 15, north of Divide, the north end of McCartney Mountain blends into the southern Highland Range through a long

series of hills that conspicuously expose pale gray Madison limestone, a thick sedimentary formation deposited in shallow sea water about 350 million years ago. Now those rocks are crumpled into tight folds that give the hills an intricately twisted look.

Precambrian basement rock about 2.7 billion years old forms the core of the southern Highland Range, well east of the highway. Farther north, the entire range becomes 70 million-year-old granite, the southern end of the Boulder batholith. The Humbug Spires Primitive Area contains a separate mass of granite that is probably a satellite of the Boulder batholith. The name derives from the towering outcrops of granite that do indeed suggest steeples rising among the pine trees.

Pioneer Range

Rocks along the eastern flank of the Pioneer Range are tightly folded sedimentary formations sliced into big flat slabs along faults, and full of granite intrusions. Several great slabs of sedimentary rocks moved eastward and came to rest as a series of overlapping overthrust sheets. Masses of granite magma penetrated the wrinkled sedimentary rocks along the same faults. The situation is like that in the ranges around the edges of the Sapphire detachment block, so the Pioneer Range appears to be another big detachment block that moved east from the Idaho batholith while the granite there was still partly molten. High peaks on the skyline west of the highway between Melrose and Dillon are part of the Pioneer batholith, 70 million-year-old granite in the core of the Pioneer Range.

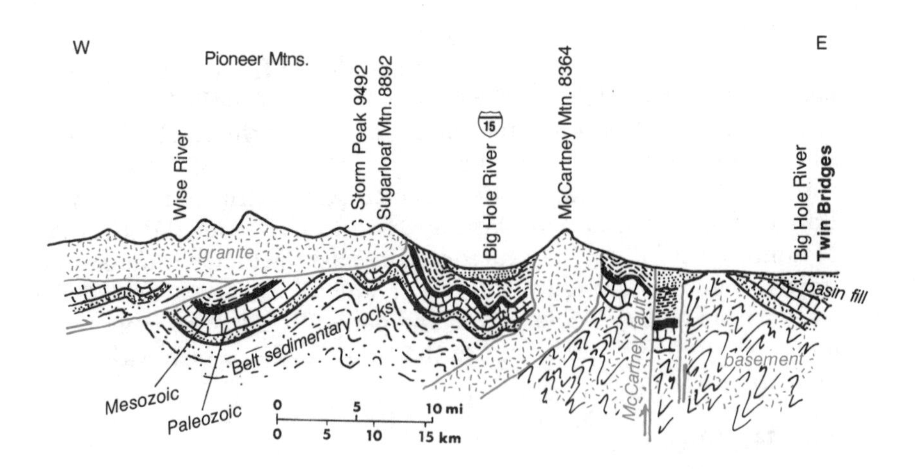

Section across the line of the highway near Melrose, through the Pioneer Range and McCartney Mountain. Granite magma moved east from the Idaho batholith along the large thrust faults.

Also about 70 million years ago, several masses of granitic magma, satellites of the Pioneer batholith, invaded limestone formations in the hills west of Melrose. Reaction between granitic magma and limestone produced zones of spectacular rock called skarn where the two meet, wild rocks typically full of oversized cystals of garnet and other minerals, a rockhound's dream. Some of the skarns west of Melrose contain the tungsten mineral scheelite. Mines there have produced modest quantities of scheelite and may produce more if the price of tungsten rises.

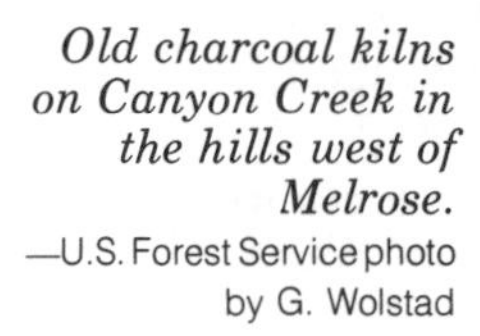

Old charcoal kilns on Canyon Creek in the hills west of Melrose.
—U.S. Forest Service photo by G. Wolstad

Prospectors discovered vein deposits of silver and lead ore in the Bryant Mining District west of Melrose in 1872. Several mines and small towns were in prosperous full blast within a few years. The pace of activity slackened after 1890, and most of the mines had closed by the turn of the century. A number of old buildings and charcoal kilns still stand back in the hills. The remains of the old town of Farlin mark another old mining district in Birch Creek between Dillon and Melrose. Mines there produced copper from skarns in the contact zone where granite magma reacted with the Madison limestone.

Volcanic Rocks

Low hills just west of Dillon, like those along the front of the Pioneer Range near Melrose, consist mostly of volcanic rocks. Exposures are especially good in the brown cliffs along the Big Hole

Brown cliffs of andesite rise above the Big Hole River south of Melrose.

River west of the highway. The age of those rocks is a bit uncertain, but we think they probably erupted about 70 million years ago from the masses of magma that became the big batholiths in the Pioneer Range. Other volcanic rocks in the same area may well have erupted during the episode of widespread igneous activity of 50 million years ago.

McCartney Mountain

McCartney Mountain is a small mountain range that closely resembles the much larger Tobacco Root Range to the east. A large mass of granite in its core intruded a variety of tightly folded Paleozoic and Mesozoic sedimentary rocks about 70 million years ago. Most of the granite intrusion is on the west side of the mountain, in the high forested area visible from the highway. The county road that follows the Big Hole River along the south side of McCartney Mountain between Melrose and Montana 41 passes spectacular exposures of the folded sedimentary rocks, as well as large areas of basement rock.

Tightly folded sedimentary layers on the south side of McCartney Mountain.

The Hogback

A conspicuous dark ridge, The Hogback, extends several miles south from McCartney Mountain. The name is slightly unfortunate because the term "hogback" refers technically to a symmetrical ridge eroded on a steeply tilted layer of resistant sedimentary rock. The Hogback south of McCartney Mountain is a tight anticlinal fold arched in the Quadrant quartzite. It is unusual to see an anticline standing up as a ridge. Such folds more commonly erode into valleys because the rocks crack as they stretch over the crest of the arch, and thus become easily vulnerable to erosion.

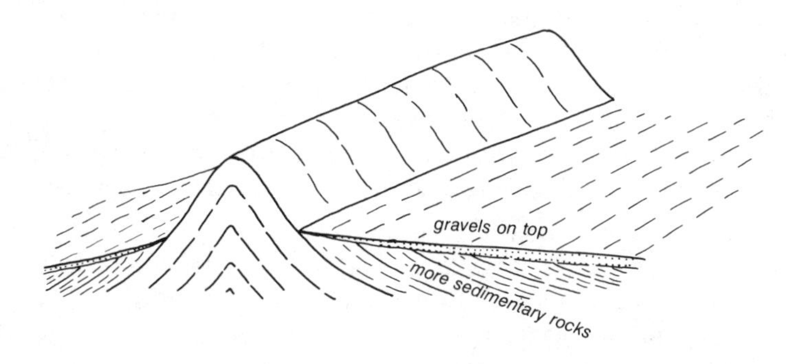

Section through The Hogback

The Ruby Range

An unpaved road quite passable in all but the worst weather crosses the Ruby Range between Dillon and Alder, a beautiful drive through high meadows where fresh spring flowers bloom into late July. This is one of the few good roads in southwestern Montana that cross large areas of Precambrian basement rock.

West of the crest of the range, watch for conspicuous ridges of gleaming white marble, a metamorphic rock that forms as limestone, a sedimentary rock, recrystallizes at high temperatures deep within the earth's crust. The same area contains many outcrops of schist and gneiss, metamorphic rocks that form through recrystallization of mudstones at high temperature. Almost all the rocks west of the range crest are pink granites and granitic gneisses that greatly resemble the granite except in having a streaky look. Age dates show that the main metamorphism happened about 2.7 billion years ago. That is impressively ancient even by the elongated standards of geology.

The Ruby Mountains owe their name to bright red gem garnets, which some early prospectors mistook for rubies—the two stones have nothing in common except color. Bright red garnets are common in many kinds of metamorphic and igneous rocks, but most are too fractured to withstand cutting, too full of impurities to look well polished, and too intensely colored to sparkle as gems should. Some Ruby Range garnets are clear and pale enough to make lovely gems. Most of the gem garnet localities are on the east side of the range. Look in the soil and in the creek beds for red garnets about the size of split peas that look clear and show good color against the light.

Wherever geologists find large volumes of Precambrian sedimentary rocks more than about 2 billion years old, they generally find banded iron formation among them. As a sedimentary rock, banded iron formation consists of alternating thin layers of white quartz and red hematite, which make a striking pattern of stripes. Metamorphism converts the rock into alternating thin layers of white quartz and black magnetite, quartz-magnetite gneiss, good iron ore.

The Ruby Range contains numerous small bodies of quartz-magnetite gneiss. One fairly large body of that strange rock forms a hill on the west flank of the Ruby Mountains almost within sight of Dillon. Estimates vary, but that hill probably contains at least 70 million tons of iron ore. That is a modest iron reserve, but enough nevertheless to supply a small smelter for quite a few years. Dillon may have an iron mine in its future.

Montana Talc

Several large mines in the Ruby and Gravelly ranges produce enough talc to make this the world's largest talc district. The area began producing talc during the early years of this century, but high shipping costs retarded development of the industry until the 1970s. Then Montana talc became extremely valuable and competitive on the world market because it contains no asbestos, a common impurity in talc from most other districts. Asbestos causes cancer.

Talc is a greenish-white magnesium silicate mineral so soft that you can scratch it with your fingernail or whittle it with a knife. Many people actually do carve talc into pretty little figurines. Talc is most familiar as talcum powder, simply powdered talc sweetened with a touch of perfume. The mineral is also used as a filler in products ranging from pills to chocolate candy. The most important uses are industrial: many paints contain talc, because it makes them brush more easily, and talc is useful in the manufacture of paper, precision ceramic products, and an amazing variety of other things.

Interstate 15: Dillon—Monida Pass
64 mi./102 km.

Interstate 15 follows the Beaverhead River between Dillon and Clark Canyon Reservoir, the Red Rock River between the reservoir and Lima. South of Lima, the road climbs to Monida Pass. An exceptional variety of rocks appear near and along the highway: Precambrian basement in the Ruby Range, folded Paleozoic and Mesozoic sedimentary formations in the Blacktail and Tendoy ranges, the Beaverhead gravel, and volcanic rocks.

Volcanic Rocks

Low hills directly west of Dillon consist mostly of volcanic rocks, as do many along the road south of town. The most conspicuous volcanic rocks are sparkling white rhyolite ash and reddish brown andesite. There is some uncertainty about their age.

The volcanic rocks near Dillon closely resemble the Lowland Creek volcanics in the area of the Boulder batholith and the Challis volcanics of central Idaho, both known to have erupted about 50 million years ago. So it seems likely that the similar rocks of southwestern Montana could be the same age. On the other hand, the granite intrusions in the Pioneer Range were emplaced at such shallow depth that they could easily have erupted, and there are a few age dates of 70 million years for volcanic rocks in this area. So it seems likely that two generations of volcanic rocks may exist here. If so, no one has yet figured out how to tell them apart.

The reddish brown hill south of the talc mill about 10 miles south of Dillon, is the one William Clark identified in his journal as Beaverhead Rock, the landmark that helped Sacajawea guide the Lewis and Clark expedition. Seen from the north, the direction Lewis and Clark came, its outline does look remarkably like the profile of a beaver headed west. Whether or not it is the authentic Beaverhead Rock, the hill is unquestionably andesite, of unknown age.

Argenta and Bannack

Montana 278, the route to Wisdom, passes near Bannack and

silver-lead mining town from 1865
Blue Wing Silver Mining District
In the Bannack area are garnet, specular hematite, coarsely crystalline calcite, epidote, pyrite, sulfides and carbonates.
Gray sandstone of the Pennsylvanian Quadrant formation is quarried and crushed as high-purity silica.
Crown-Smith-Dillon talc mine. Open cut.
metamorphism occurred about 1600 million years ago
Keystone Talc Mine in marble on Sweetwater road.
Crystal Graphite Mine produced single crystals of graphite several inches long.
Sillimanite-rich schists and gneisses. Metamorphism of these rocks has been dated as 2810 million years ago.
Dating of gneisses here indicated that their metamorphism occurred about 1600 million years ago.
Red River fault raised the Tendoy range.
Belt sedimentary rocks
Medicine Lodge thrust fault
Gypsum deposits in shale, siltstone, and limestone of the Mississippian Big Snowy Group. Gypsum is used in wallboard, as a setting retarder in cement, and as a soil conditioner.
0 5 10 mi
0 5 10 15 km
N
upper Belt rocks
Argenta
Dillon
Bannack
15
41
278
324
basin fill
Ruby Range
basement rock
Beaverhead River
Blacktail Deer Creek
Blacktail Range
stream sediments
Gallagher Mtn.
8477
Tertiary volcanics
Clark Canyon Res.
Tertiary basalts
Tertiary granite
Red Rock River
Tendoy Mountains
basin fill
Snowcrest Range
Beaverhead gravel
Dell
Dixon Mtn.
9665
Lima
MONTANA
IDAHO
Madison limestone
Centennial Valley
Centennial fault
Centennial Mtns.
Monida Pass
Garfield Mtn.
10935
Snake River Plain
rhyolite and basalt
Spencer

I-15
DILLON—MONIDA PASS

Argenta, two of Montana's earliest mining camps. Mining began rather feebly at Argenta with the discovery of mediocre deposits of placer gold in Rattlesnake Creek in 1862. The mining business picked up two years later when excellent vein deposits of lead and silver turned up. Most of the ore was in a contact zone where granite magma reacted with the Madison limestone, but some was in veins more than a mile from the granite. Montana's first smelter started operating in Argenta in 1866, using charcoal burned in kilns several miles up the creek. Although most of the action at Argenta ended before 1890, occasional revivals continue when the price of silver is high.

Charcoal kilns like this one made fuel to smelt the ores at Argenta.
Mark A. Sholes photo

Small scale mining began almost immediately at Bannack after prospectors found gold in Grasshopper Creek in the late summer of 1862. Operations expanded in 1866 when a hand dug flume brought a better supply of water from Horse Prairie to the dry gravels above the creek. Several other ditches soon followed, and Bannack thrived so mightily that it became the first capital of Montana, for several years the most important town in the state. Production from the rich placers declined so rapidly that those mines were almost idle by 1890. Then, in 1895, the first gold dredge in the country began working the stream gravels. Three more followed within the year, and Bannack prospered for a few years until the dredges had washed all the gravel they could reach. They left piles of gravel tailings and dredge ponds as souvenirs of their passage.

The bedrock source of the gold was in the contact zones around two small granite intrusions on opposite sides of the creek in the high ridge of Madison limestone that looms above Bannack. Bedrock mining began in 1863 after a stamp mill was built to pound the rock to a powder, and then wash the small particles of metallic gold free. By 1870, four stamp mills were operating at Bannack, and by the end of the century they had worked so much of the ore that the last stamp mill, built in 1920, operated only two years. So little remained that a cyanide mill built in 1914 to dissolve the gold out of the rock was idle within a few years.

Total production from the Bannack district is difficult to estimate because so much came before anyone kept formal records. Some estimates put the figure at about 12 million dollars. Even though most of those were the fat dollars of the last century, it is hard to escape concluding that very few of the thousands of people who mined in Bannack during 60 years of activity could have made much money.

Blacktail Range

Blacktail Deer Valley extends southeast from Dillon, a northwest trending slot that separates the southern end of the Ruby Range from the northern end of the Blacktail Range. The valley does not appear to contain any fill of Renova formation, so it must have formed since those deposits accumulated. Mountain fronts that confine the Blacktail Deer Valley between straight walls are the faults that dropped the valley block between the Ruby and Blacktail ranges. The highway crosses that dropped block for about 10 miles south of Dillon on stream sediments probably deposited since the last, Pinedale, ice age.

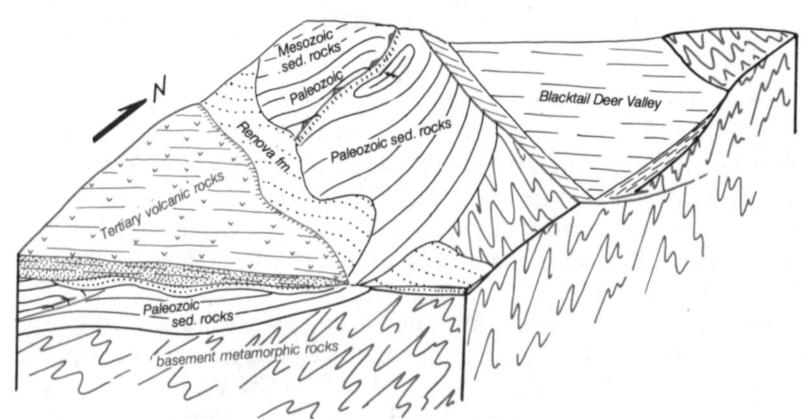

Structure of the Tendoy and northern Blacktail ranges seen looking generally northward. A slab of sedimentary rocks moving from the west slid over similar rocks, then volcanic rocks partially covered them.

View southeast across the Blacktail Deer Valley to the north edge of the Blacktail Range, an active fault scarp.

Bedrock roadcuts about 10 miles south of Dillon mark where the highway crosses the fault that bounds the north side of the Blacktail Range. Notice that the high end of the range is at the north, near the fault, and that it gets steadily lower southward. The fault is raising the range as though it were an old fashioned cellar door hinged along the south side. Although they are not visible from the road, outcrops of Renova formation cap parts of the Blacktail Range almost to its summit. Obviously, the area was valley floor when those valley fill sediments accumulated, and has risen since.

Older bedrock in the Blacktail Range is mostly folded and faulted Paleozoic and Mesozoic sedimentary formations. Much of the western part of the range is a slab of sedimentary formations that moved eastward onto the same formations. Thick deposits of volcanic rocks, mostly white rhyolite ash, that blanket parts of the Blacktail Range are much less deformed than the older sedimentary formations they cover. So it seems that the first stage of faulting and folding must have almost ended by the time the volcanic rocks erupted, probably about 50 million years ago.

Tendoy Range

The Tendoy Range lies west of the highway between Clark Canyon Reservoir and Lima. Like the Blacktail Range, it consists of folded and faulted Paleozoic and Mesozoic formations, along with Beaverhead gravel, and some younger volcanic rocks. The Red Rock fault defines the straight face of the Tendoy Range visible from the road. Notice that many of the mountain spurs end in triangular facets where movement along the Red Rock fault cut them off. The fault also

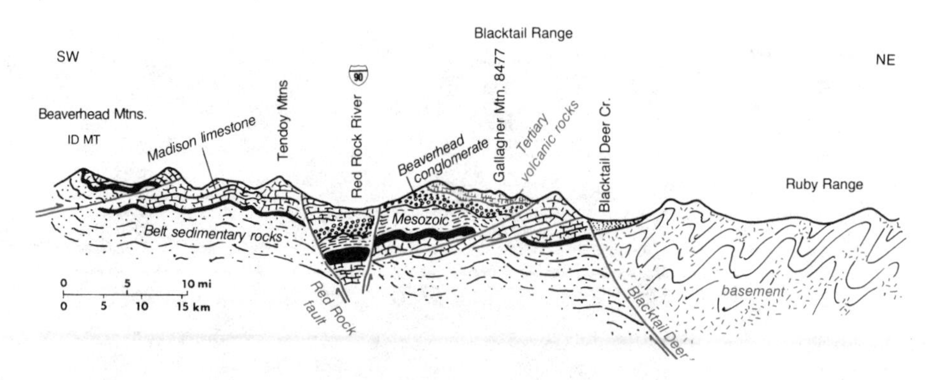

Section across the Blacktail and Tendoy ranges south of Clark Canyon Reservoir

offsets fairly young alluvial fans along the front of the range, certain evidence that it has moved recently, and must be active.

Beaverhead Conglomerate

Long stretches of the highway, especially in the area south of Lima, pass through expanses of the Beaverhead conglomerate, one of the most peculiar formations in southwestern Montana. It also appears here and there in the hills within a few miles north of the Clark Canyon Dam as conspicuous orange splashes in the landscape. Between Clark Canyon Dam and Monida Pass, the Beaverhead gravel forms softly rounded hills covered with short grass. A short walk in those grassy hills reveals that they consist of large, stream rounded pebbles and cobbles.

The Beaverhead gravel must be some thousands of feet thick, but no one knows how many. Fossil pollen grains indicate that the formation was probably deposited during latest Cretaceous time. Geologists find the Beaverhead formation wrapped around folds, or lying beneath great overthrust sheets of older rock, so we can be sure that the formation already existed when crustal movements were deforming the rocks of southwestern Montana. Most of the pebbles in the Beaverhead formation are cracked, many show deep dents where other pebbles squashed them, and a few are elaborately sheared. Evidently, the formation was deformed before the pebbles stuck together to become solid rock. Therefore, individual pebbles took up the strain of deformation where they touched, and mashed into each other as though they were a basketful of hard boiled eggs slowly crushed beneath a heavy load.

A crushed pebble from the Beaverhead conglomerate

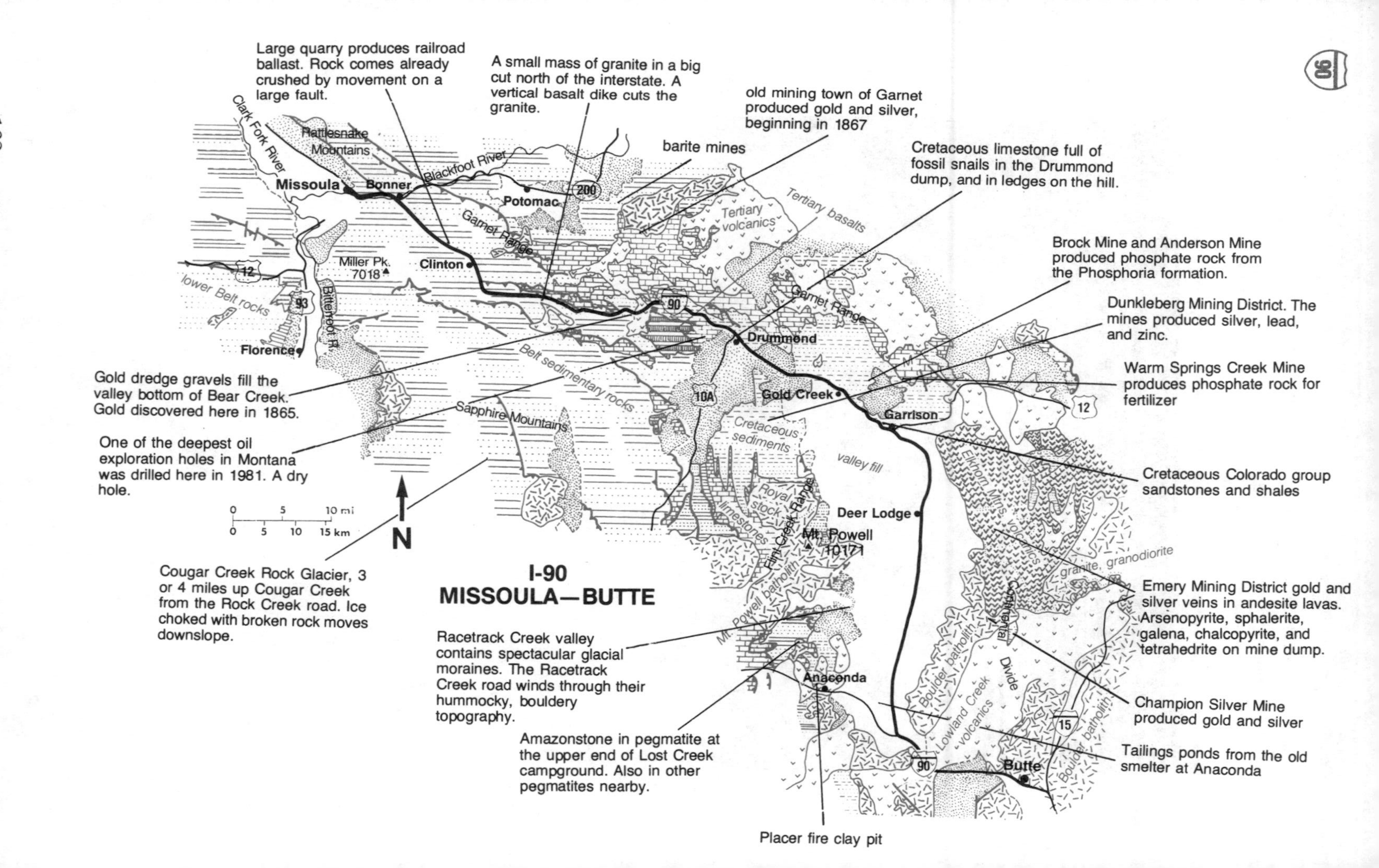
90
I-90
MISSOULA—BUTTE
Large quarry produces railroad ballast. Rock comes already crushed by movement on a large fault.
A small mass of granite in a big cut north of the interstate. A vertical basalt dike cuts the granite.
old mining town of Garnet produced gold and silver, beginning in 1867
barite mines
Cretaceous limestone full of fossil snails in the Drummond dump, and in ledges on the hill.
Brock Mine and Anderson Mine produced phosphate rock from the Phosphoria formation.
Dunkleberg Mining District. The mines produced silver, lead, and zinc.
Warm Springs Creek Mine produces phosphate rock for fertilizer
Cretaceous Colorado group sandstones and shales
Emery Mining District gold and silver veins in andesite lavas. Arsenopyrite, sphalerite, galena, chalcopyrite, and tetrahedrite on mine dump.
Champion Silver Mine produced gold and silver
Tailings ponds from the old smelter at Anaconda
Placer fire clay pit
Amazonstone in pegmatite at the upper end of Lost Creek campground. Also in other pegmatites nearby.
Racetrack Creek valley contains spectacular glacial moraines. The Racetrack Creek road winds through their hummocky, bouldery topography.
Cougar Creek Rock Glacier, 3 or 4 miles up Cougar Creek from the Rock Creek road. Ice choked with broken rock moves downslope.
One of the deepest oil exploration holes in Montana was drilled here in 1981. A dry hole.
Gold dredge gravels fill the valley bottom of Bear Creek. Gold discovered here in 1865.
Clark Fork River
Rattlesnake Mountains
Missoula
Bonner
Blackfoot River
Potomac
200
Garnet Range
Clinton
Miller Pk. 7018
12
93
lower Belt rocks
Bitterroot R.
Florence
Tertiary volcanics
Tertiary basalts
90
Drummond
Belt sedimentary rocks
Sapphire Mountains
10A
Gold Creek
Garrison
12
Cretaceous sediments
valley fill
Royal stock
limestones
Flint Creek Range
Deer Lodge
Mt. Powell 10171
Mt. Powell batholith
Elkhorn Mtns. volcanics
granite, granodiorite
Continental Divide
Boulder batholith
Lowland Creek volcanics
Anaconda
15
Butte
0 5 10 mi
0 5 10 15 km
N

Pinnacles of Madison limestone overlook the Clark Fork River near Bearmouth. The rocks were exposed as the slope lost its soil cover to erosion.

Interstate 90: Missoula—Butte
119 mi./190 km.

The great horseshoe curve of the interstate highway route between Missoula and the south end of the Deer Lodge Valley traces the northern and eastern borders of the Sapphire block, the enormous chunk of the earth's crust that moved east about 50 miles. Between the south end of the Deer Lodge Valley and Butte, the highway crosses part of the Boulder batholith.

The Garnet Range

Between Missoula and Drummond, the road follows the Clark Fork River along the south flank of the Garnet Range, the northern boundary of the Sapphire block. Rocks along this stretch of highway include an assortment of tightly deformed Precambrian, Paleozoic, and Mesozoic sedimentary formations, as well as patches of volcanic rocks erupted about 50 million years ago. Most of the deformation happened while the Sapphire block was bulldozing its way into Montana.

Roadcuts and cliffs near the western part of the route are in Precambrian rocks, mostly red and green mudstones. A big quarry north of the road near Clinton produces railroad ballast from one of

those Precambrian formations, the pink Bonner quartzite. It comes already crushed out of a large fault zone. Many large outcrops about midway between Missoula and Drummond expose the thick Madison limestone, easy to recognize as a pale gray rock splashed red in places with iron oxide stains. Most rocks exposed within about 10 miles on either side of Drummond are Cretaceous sedimentary formations.

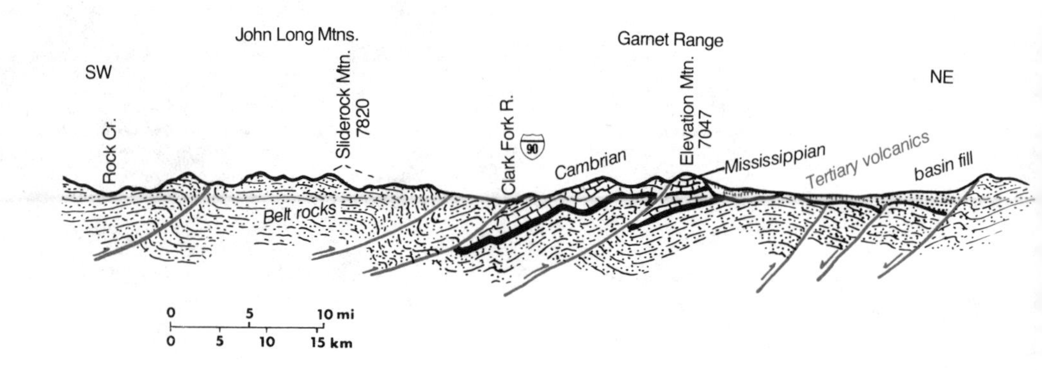

Section through the Garnet Range in the Bear Creek area

Bear Creek empties into the Clark Fork near the Bearmouth exit. Prospectors discovered placer gold in Bear Gulch in 1863, precipitating the usual stampede that soon brought thousands of miners to work the gravels along nearly the entire length of Bear Creek. There must have been a large population of generally unemployed people in those days. The early miners recovered an unknown amount of gold generally estimated at around 30 million old-fashioned dollars, an impressive haul for the relatively few people who worked there during the glory years.

Many of the early miners in Bear Gulch sank shafts to bedrock at depths as great as 70 feet, then shoveled out the gravel along the bedrock surface and hoisted it to the surface to wash out the gold. That is an extremely unusual and extraordinarily dangerous way to mine a placer deposit. But in those early days there was no alternative because placer gold sinks to the bottom of the gravel fill in the stream channel and concentrates on the bedrock surface. A large dredge turned the lower part of the floodplain inside out during the depression, leaving it a wasteland of gravel ridges.

Soon after the placer discovery, prospectors found bedrock lode deposits around the margins of granite intrusions high in the mountains near the head of Bear Creek. As elsewhere around the edges of the Sapphire block, the granite magma intruded along thrust faults in the older sedimentary rocks about 70 million years ago. Where magma penetrated Madison limestone, the two reacted to form skarns that consist mostly of garnet, no doubt explaining how the town of Garnet got its name. This skarn also contains gold. Most of the underground mining was done between 1880 and 1900, with some continuing until 1920.

Big stockpiles of brown phosphate rock usually stand beside the railroad at the Phosphate exit between Drummond and Garrison. Underground mines several miles north of the highway in the valley of Warm Springs Creek produce the rock from the sedimentary Phosphoria formation, which was deposited in shallow sea water during Permian time, a little more than 200 million years ago. This phosphate rock goes to Trail, British Columbia, where sulfuric acid made from the sulfur dioxide scrubbed out of smelter stack gas is used to make it into fertilizer.

There really is a warm spring in Warm Springs Creek. The water trickles in small waterfalls over the edge of a big mossy deposit of travertine, a kind of limestone that forms in caves and springs. The water probably picks up its heat as it circulates to a depth of several thousand feet along fractures.

The Flint Creek Range

At Drummond, the highway crosses the north edge of the Flint Creek Valley. The road passes north of the Flint Creek Range between Drummond and Garrison, past exposures of sedimentary formations, mostly sandstones, deposited during Cretaceous time. Watch for tilted sedimentary layers in folds wrinkled into the leading edge of the Sapphire block as it moved east.

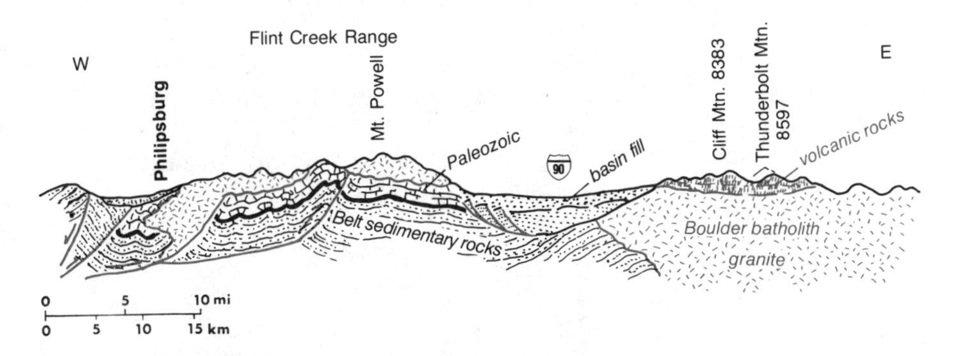

Section across Interstate 90 about 5 miles south of Deer Lodge. Granite magma moving east from the Idaho batholith along thrust faults penetrated the Flint Creek Range.

These piles of gravel among the trees near Pioneer are old hand-worked tailings piles, relics of the frantic search for gold in the last century.

Gold Creek, at the northern tip of the Flint Creek Range, was the site of the first gold discovery in Montana in 1852. Miners swarming into the area a few years later found that most of the gold was several miles south of town near the site of another town called Pioneer, now long abandoned. The richest and most accessible stream placers were thoroughly worked over before 1890 with production of an unknown quantity of gold. Then two large dredges, one of which was shipped in on 42 rail cars, worked the gravels in several thousand acres near Pioneer between 1933 and 1939, recovering another 1,400,000 dollars worth of gold. Long windrows of gravel tailings still cover the floodplain.

Between Garrison and Anaconda, a long parade of jagged peaks rises along the western horizon. They are the eastern front of the Flint Creek Range. Rocks on the east side of the Flint Creek Range were bulldozed into tight folds and sheared along big thrust faults as the Sapphire block moved east off the Idaho batholith between 70 and 75 million years ago. Those rocks form the low flanks of the range. Granite that intruded them along the thrust faults forms the high peaks. Canyon walls in Lost Creek State Park, west of Warm Springs, show spectacular views of a granite intrusion following thrust faults, stepping upward from one fault to the next.

One of the dredges that worked the gravel south of Gold Creek. —Wm. Napton collection, Mansfield Library, University of Montana

Glacial moraine at the mouth of Racetrack Creek, south of Deer Lodge

Ice age glaciers that gouged the high valleys of the Flint Creek Range descended to the elevation of the valley floor. They left their signature in the deeply carved high peaks, sharp ridges, and valleys shaped like deep troughs. Hummocky moraines littered with erratic boulders spread around the mouths of the canyons.

The Deer Lodge Valley

The rounded hills that overlook the Deer Lodge Valley from the east are eroded into granite of the Boulder batholith and its cover of darker volcanic andesite erupted from the same mass of magma. Large roadcuts and ledges on the hills in the northern part of the Deer Lodge Valley expose Cretaceous sandstone, tilted because it was buckled up ahead of the moving Sapphire block. Farther south, the highway crosses stream deposits, and no bedrock appears near the road.

Old mine car of the Granite Bimetallic Mine rests in hills behind Philipsburg.

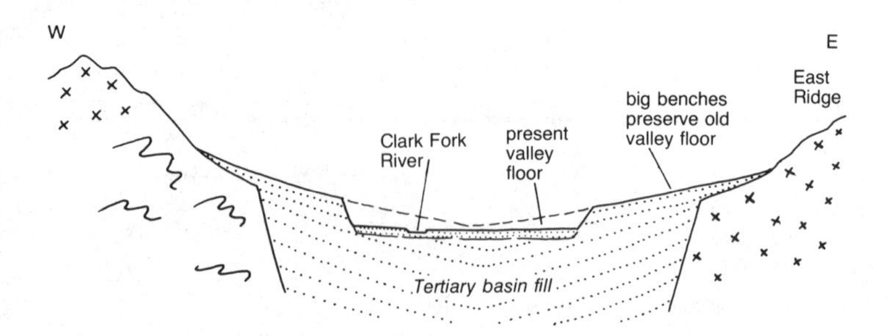

A schematic section across the Deer Lodge Valley to show how the big benches preserve parts of the old valley floor

Magnificent high benches in the southern part of the Deer Lodge Valley are basin fill deposits, mostly Renova formation deposited during the millions of desert years of Oligocene and early Miocene time. Connect those benches across the valley with your eye to mentally reconstruct the valley floor as it was before the modern streams began to erode their valleys into it sometime between two and three million years ago.

Five wildcat wells were drilled in the Flint Creek Valley during the early 1980s in a determined attempt to find oil in those valley fill sediments. The first well produced some very encouraging oil when it was tested, but that little bit turned out to be all there was, so the well never went into production. All the others were dry.

A wildcat well drilling below Mt. Powell in the Deer Lodge Valley in 1982. —Montana Bureau of Mines and Geology photo by H. L. James

The Emery Mining District overlooks the valley from its perch high on the ridge about 9 miles southeast of Deer Lodge. Meager placer gold deposits were discovered there in 1872, much better bedrock vein deposits of silver, gold and lead in 1887. Mining continued at Emery off and on until about 1950. The ore veins are in andesite lava flows, part of the Elkhorn Mountains volcanic series that erupted from the magmas of the Boulder batholith.

When the mines and mills at Butte were in operation, ore concentrate was shipped to Ananconda for smelting. The large ponds on both sides of the road in the southern end of the Deer Lodge Valley were used to purify waste water from the smelter before it was discharged into the Clark Fork River. The water was treated with limestone to make it slightly alkaline, and cause most of the dissolved metals to precipitate out of solution. Although the treatment was fairly effective, it didn't catch everything, and didn't even begin until after the smelter had been in operation for many years. Large volumes of dissolved metals did go downstream. Sediments in the bottom of the Milltown Reservoir near East Missoula contain large concentrations of arsenic and other heavy metals that almost certainly came down the river from the area around Butte and Anaconda.

The Boulder Batholith

Between the area near Opportunity and Butte, the highway crosses the western margin of the Boulder batholith. Lowland Creek volcanic rocks, erupted about 50 million years ago, locally cover the 70 million-year-old granite. Recognize the granite through its tendency to weather into rounded boulders. It makes a picturesque landscape in which small juniper trees grow in a garden of pinkish brown boulders. The younger volcanic rocks come in shades of red and brown, and do not weather into bouldery outcrops.

The Richest Hill on Earth

The last mine closed during the summer of 1983, so Butte, one of the world's great mining camps, ceased to produce ore for the first time in considerably more than a century. Nevertheless, Butte still lays claim to being the "Richest Hill on Earth," and that is no idle boast. Butte mines may well have produced more mineral wealth than any others.

During the early decades in Butte, underground miners followed the richer veins of ore, leaving behind vast tonnages of rock that contained small amounts of fairly evenly disseminated ore minerals.

Open pit mines thrive on such ores. The Berkeley Pit opened in 1955, and underground mining gradually phased out in the next 20 years. At Butte, the average ore grade in the Berkeley Pit was a bit less than one half of one percent copper. It is difficult to find copper minerals in it, even with the help of a strong hand lens. Nevertheless, the pit produced incredible quantities of metal.

According to estimates made after mining ceased in Butte, the district produced a total of more than 20 billion pounds of copper—all these figures refer to the entire district, not just the Berkeley Pit. Zinc, the next most abundant metal, amounted to just under 5 billion pounds.

The earliest prospectors in Butte commented on the big quartz veins that made bold outcrops stained black as coal with the manganese mineral pyrolusite. Butte went on to become one of the few districts in North America that ever produced much manganese, a total of more than 3.7 billion pounds. Manganese is an absolutely vital metal used mostly for making steel alloys. No industrial economy can function without manganese.

Butte production of lead and silver, two metals that commonly occur together, amounted to a total of 855 million pounds of lead, 704 million ounces of silver. Those metals came mostly from the veins around the margins of the mineralized area, the part of the district farthest from the copper core.

Even though Butte never made it as a gold mining camp, total production of gold did amount to some 2.9 million ounces, far more than any gold mining camp in Montana. But all that gold, like several of the other metals, was a by-product that could never have been mined if the ore had not also contained much larger amounts of copper. The district lived on copper, and the other metals sweetened the ore.

The Berkeley Pit looking northwest from the visitors' viewing stand.

Interstate 90: Butte—Bozeman

82 mi./131 km.

Between Butte and Bozeman, Interstate 90 threads a path through an astonishing variety of geologic sights: Butte with the Berkeley Pit, the Boulder batholith, the Tobacco Root Range, The Bull Mountains, and the Bridger Range.

The Richest Hill on Earth

From some places along the highway, people coming into Butte from the east see the Berkeley Pit yawning ahead of them as though it were about to swallow the highway. It is a big hole, more than a mile across, and about 1800 feet deep. All that in only 28 years of digging. Altogether, the Berkeley Pit produced some 316 million tons of rather lean ore. Three tons of waste rock were mined for every ton of ore.

Big quartz veins loaded with silver were obvious to the first gold miners in Butte, but the silver was so awash in copper that there seemed no hope of recovering it. In those days, copper was not especially valuable. Then, the rapid rise of big electrical industries during the final two decades of the last century created an enormous market for copper and a place in the sun for Butte—as a copper producer, with silver and gold as by-products. The copper mines of

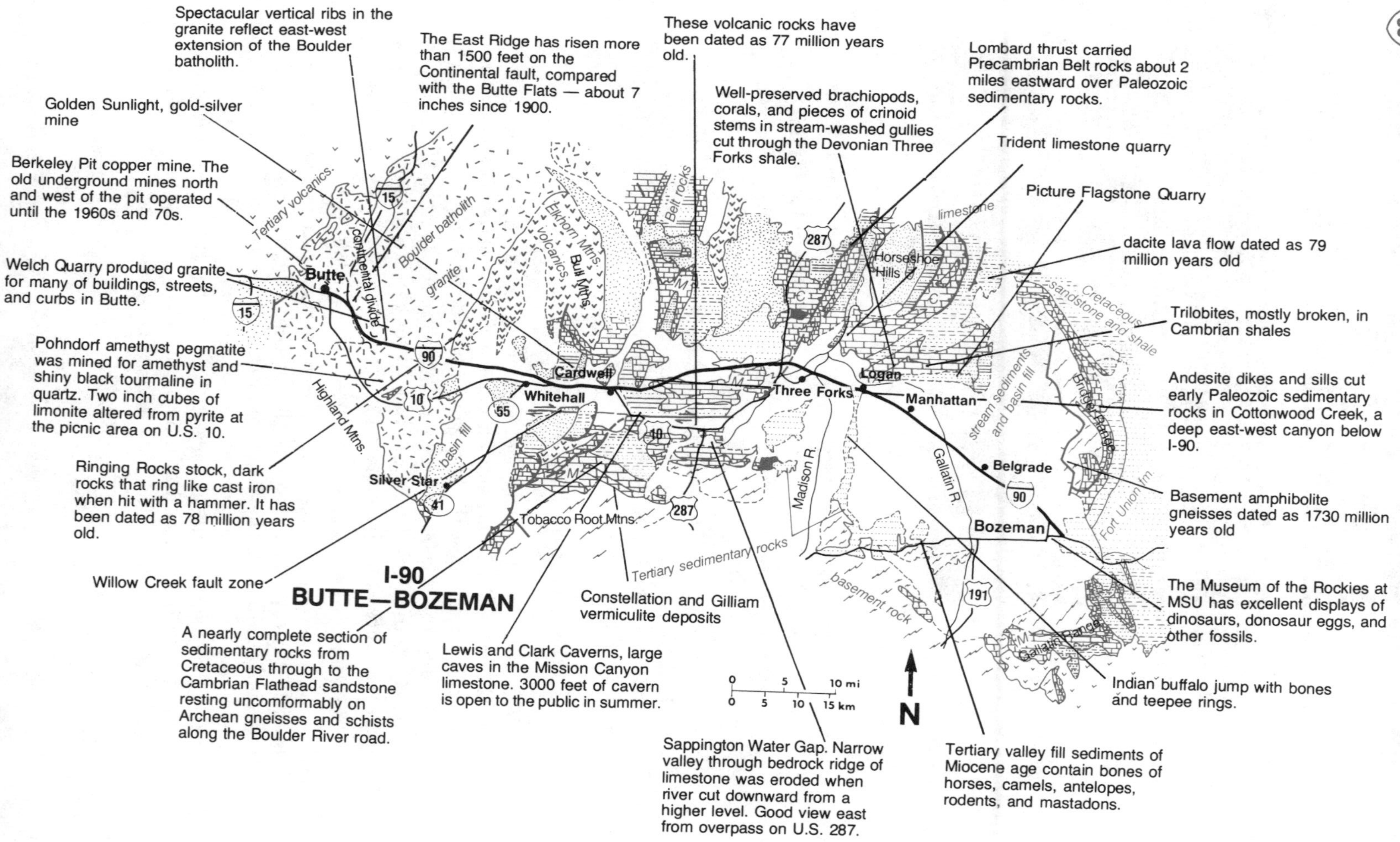
I-90
BUTTE—BOZEMAN
Spectacular vertical ribs in the granite reflect east-west extension of the Boulder batholith.
Golden Sunlight, gold-silver mine
Berkeley Pit copper mine. The old underground mines north and west of the pit operated until the 1960s and 70s.
Welch Quarry produced granite for many of buildings, streets, and curbs in Butte.
Pohndorf amethyst pegmatite was mined for amethyst and shiny black tourmaline in quartz. Two inch cubes of limonite altered from pyrite at the picnic area on U.S. 10.
Ringing Rocks stock, dark rocks that ring like cast iron when hit with a hammer. It has been dated as 78 million years old.
Willow Creek fault zone
A nearly complete section of sedimentary rocks from Cretaceous through to the Cambrian Flathead sandstone resting uncomformably on Archean gneisses and schists along the Boulder River road.
The East Ridge has risen more than 1500 feet on the Continental fault, compared with the Butte Flats — about 7 inches since 1900.
These volcanic rocks have been dated as 77 million years old.
Well-preserved brachiopods, corals, and pieces of crinoid stems in stream-washed gullies cut through the Devonian Three Forks shale.
Lombard thrust carried Precambrian Belt rocks about 2 miles eastward over Paleozoic sedimentary rocks.
Trident limestone quarry
Picture Flagstone Quarry
dacite lava flow dated as 79 million years old
Trilobites, mostly broken, in Cambrian shales
Andesite dikes and sills cut early Paleozoic sedimentary rocks in Cottonwood Creek, a deep east-west canyon below I-90.
Basement amphibolite gneisses dated as 1730 million years old
The Museum of the Rockies at MSU has excellent displays of dinosaurs, donosaur eggs, and other fossils.
Indian buffalo jump with bones and teepee rings.
Tertiary valley fill sediments of Miocene age contain bones of horses, camels, antelopes, rodents, and mastadons.
Sappington Water Gap. Narrow valley through bedrock ridge of limestone was eroded when river cut downward from a higher level. Good view east from overpass on U.S. 287.
Lewis and Clark Caverns, large caves in the Mission Canyon limestone. 3000 feet of cavern is open to the public in summer.
Constellation and Gilliam vermiculite deposits
Tertiary volcanics
Butte
15
continental divide
Boulder batholith
granite
Elkhorn Mtns. volcanics
Bull Mtns.
Belt rocks
Highland Mtns.
90
10
Cardwell
Whitehall
55
basin fill
Silver Star
41
Tobacco Root Mtns.
287
Tertiary sedimentary rocks
Three Forks
Madison R.
Logan
Horseshoe Hills
limestone
Manhattan
Gallatin R.
stream sediments and basin fill
Cretaceous sandstone and shale
Bridger Range
Belgrade
Bozeman
Fort Union fm
191
basement rock
Gallatin Range
0 5 10 mi
0 5 10 15 km
N

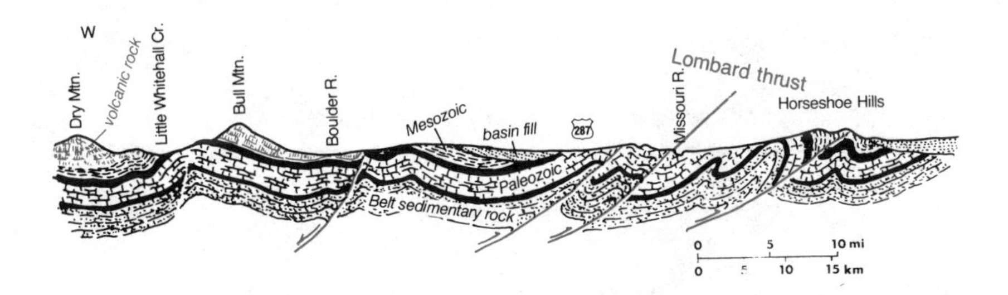

Section from the area north of Whitehall to that north of Manhattan

Butte thrived as people built big transformers and motors, wired their houses, and wove a tight web of telegraph, telephone, and transmission lines across the continent. By 1898, the United States was producing about 60 percent of the world's copper, and Butte was producing 40 percent of that.

Aluminum wires and modern electronic equipment killed the copper boom. Many modern buildings and transmission lines use aluminum wires. And as satellite and microwave relay systems grow, the old telephone and telegraph wires come down for scrap. The combination of less demand for new metal and recycling of old has caused the value of copper to lag behind the rate of inflation ever since the 1950s. Now Butte, like many other large copper mining camps, has fallen on hard times.

Rising Water and Moving Faults

Old maps show marshland along Silverbow Creek and in the flats south of the interstate highway. Those wetlands dried out as the mines pumped the ground water level down, and buildings now cover much of the formerly marshy ground. Now that the last pumps are silent, the Berkeley Pit is filling with water, and the regional water table is slowly returning to its natural level. If that continues, the rising water level will again flood the old marshland, along with the buildings that now stand there. Continuing crustal movements compound the water problem.

A member of the U.S. Geological Survey concluded in 1906 that Butte is sinking relative to the East Ridge. When surveyed benchmarks established then to test that conclusion were resurveyed during the 1970s, it became absolutely clear that Butte is indeed sinking relative to the East Ridge at an annual rate between 1 and 3 millimeters. In other words, parts of Butte flats have sunk almost a

Abandoned head frames of the old underground mines still punctuate the skyline of Butte

foot since the early years of the century. The movement appears to follow the Continental fault.

On first thought, it might seem that the Continental fault is simply slipping, a type of movement that should cause little trouble except continued subsidence of Butte. But there are no signs of surface slippage where the highway and other roads cross the fault. Therefore, it seems more likely that the fault is stuck, and the earth's crust is gradually bending, accumulating strain energy like a slowly drawn bow. If that is the case, then the fault will eventually slip suddenly, and release the accumulated energy in an earthquake, possibly a large earthquake.

Boulder Batholith

Roadcuts along the highway across Homestake Pass east of Butte provide spectacular exposures of the Boulder batholith. Watch for pale gray granite, featureless except for numerous round patches of darker rock that differs from the granite in containing a greater proportion of dark minerals. Weathered granite surfaces acquire a

These granite hills east of Homestake Pass lost their soil cover.

dark mask compounded of a reddish brown stain of iron oxide and a crust of dark lichens.

A close look at a specimen of the granite reveals that it consists mostly of milky white feldspar. Glossy black flakes of biotite and needles of hornblende pepper the white background. Scattered grains of quartz appear dark gray, partly because you look through the glassy mineral into the shadowed interior of its nest in the rock.

Roadcuts near Homestake Pass illustrate the origin of the granite boulders that litter the hills around Butte. Notice that the rock has weathered into softly incoherent debris along fractures while

Granite weathering into boulders within the soil

remaining fresh and sparkling between them. That pattern reflects the tendency of water to penetrate fractures, and react with the rock along them to break it down into soil. So the soil cover on the slope fingers down into the fractures in the rock beneath. If something such as a catastrophic forest fire or a dry climatic period destroys much of the plant cover, that leaves the soil without its protective umbrella of leaves. Then, splashing raindrops can erode the soil, leaving the rounded cores of unweathered rock between fractures as boulders on the surface.

The Jefferson Valley

Between Pipestone and Cardwell, the highway crosses the north end of the Jefferson Valley. The high ridge along the western edge of the valley is the Boulder batholith, and the towering Tobacco Root Mountains loom snowcapped in the southern horizon. Unconsolidated material exposed in roadcuts is basin fill sediment, mostly the Renova formation, that accumulated during Oligocene and early Miocene time. Seismic profiles show that it probably fills this part of the valley to a depth between 3 and 4 thousand feet.

Interstate 90 leaves the eastern edge of the Jefferson Valley by crossing between Doherty Mountain, north of the road, and the London Hills. The prominent white cliffs in Doherty Mountain are Cambrian limestone formations wrapped around tight folds. Roadcuts south of Doherty Mountain expose those same formations, as well as yellowish Precambrian sandstones, which appear to grade rapidly southward into the LaHood formation exposed in nearby Jefferson Canyon. Occasional greenish stripes in the roadcuts are deeply weathered sills of andesite sandwiched between the sedimentary layers.

Bull Mountains

The darkly forested mass of the Bull Mountains overlooks most of the Jefferson Valley from the north. The rocks there are mostly the Elkhorn Mountains andesite that erupted from the magma of the Boulder batholith before it crystallized. The volcanic rocks lie on a base of tightly folded Belt and Paleozoic sedimentary rocks also exposed in a few small roadcuts near Cardwell.

People approaching Cardwell from the east see the Golden Sunlight Mine as great scars high on the east flank of the Bull Mountains, about a mile north of the highway. People going the other way must crane their necks a bit. The gold mine, discovered in 1892, produced off and on for many years until it opened a large open pit operation in 1982. Now it is one of the more productive gold mines in Montana, one

of the few to produce large quantities of ore since the Great Depression. It probably has a long future of productivity whenever the price of gold is favorable.

The Tobacco Root Mountains

Glaciated high peaks of the Tobacco Root Range rise directly south of Whitehall and Cardwell. A large granite batholith, punched up through Precambrian basement rock about 70 million years ago, forms the core of the range, Precambrian basement rocks surround it, and Paleozoic and folded Paleozoic and Mesozoic formations drape the flanks. An unpaved but generally passable road goes south from Cardwell to the headwaters of the South Boulder River in alpine

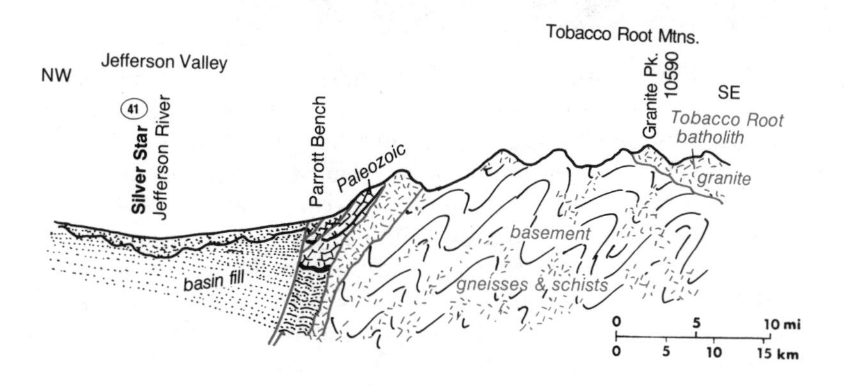

Cross section of the Tobacco Root Range

meadows high in the Tobacco Root Range, to the little that remains of the old gold mining town of Mammoth. The road passes southward through a nearly complete section of successively older sedimentary formations until it enters Precambrian basement rock, and finally gets onto the granite batholith—a nice cross section of the range. The highest, most southern, parts of the route get into spectacular landscapes of mountain glaciation.

The LaHood Conglomerate

At the west end of Jefferson Canyon, just east of LaHood Park, old US 10 passes high cliffs in somber gray rock, the Precambrian LaHood formation. A close look reveals large and small chunks of many kinds of igneous and metamorphic rocks eroded from the Precambrian basement set in a matrix of dark greenish mudstone. The rock appears to have been deposited as a series of mudflows.

Dark conglomerates of the LaHood formation exposed beside old US 10 in Jefferson Canyon. Most of the chunks are several inches across.

The peculiar mudflow conglomerates of the LaHood formation mark the southern margin of the Precambrian Belt sedimentary rocks in this area. Geologists call that margin the Willow Creek fault, and trace it east and west of Jefferson Canyon through much of western Montana. The Willow Creek fault moved during Precambrian time to separate a high southern area from lowlands to the north where the Belt sedimentary rocks accumulated. That billion year old fault line still separates the region to the north, where thick Belt formations deeply bury the basement, from the region of widespread basement exposure in southwestern Montana.

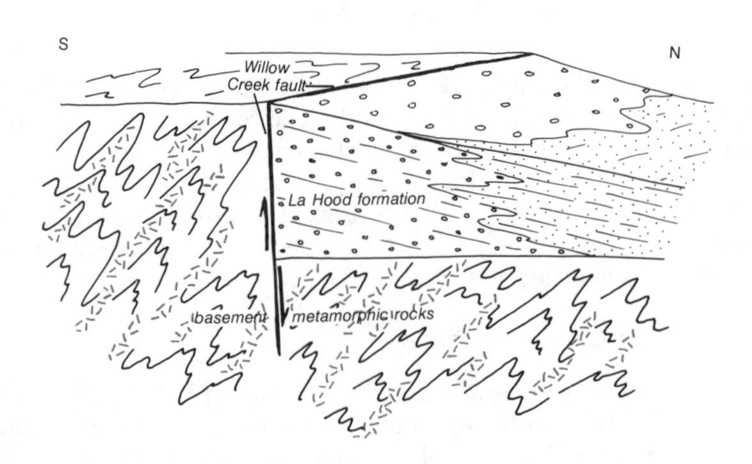

Section showing how the debris that became the LaHood formation was shed in Precambrian time off the scarp of the Willow Creek fault

Lewis and Clark Cavern

Near the eastern end of Jefferson Canyon, a side road winds north up the side of Cave Mountain to Lewis and Clark Cavern State Park, the only developed cave in Montana. The tour starts near the top of the cave, then winds downward through narrow passages that connect large rooms full of spectacular cave formations, and finally exits through a large tunnel.

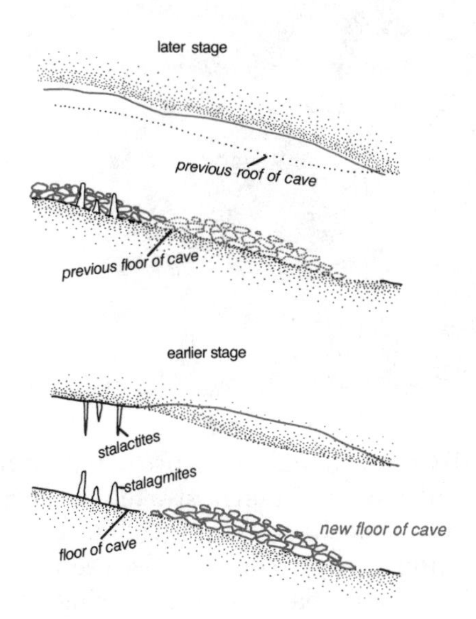

How a cave works its way toward the surface as the collapsing roof buries the original floor under rubble

Lewis and Clark Cavern began as slightly acid ground water seeping through steeply tilted layers of Madison limestone dissolved the initial cave openings. Open holes are inherently unstable, so the roof collapsed again and again, burying the original cave floor under broken debris as the opening migrated up. Notice that in many of the cave rooms a steeply arching roof rises above a roughly conical pile of angular rubble on the floor.

Ornamental dripstone formations grow as groundwater seeps into a cave and redeposits limestone that it had dissolved above the roof. Evaporation of dripping groundwater leaves a hanging icicle of dripstone called a stalactite. A corresponding pedestal called a stalagmite grows below as water dripping from the stalactite splashes onto the floor. Water trickling along the wall of a cave deposits its mineral load as thin curtains of dripstone that may eventually grow into impressive draperies.

Two stalagmite sentinels, dripstone formations in Lewis and Clark Cavern

Caves typically contain little evidence of their age, so no one knows when Lewis and Clark Cavern started to form. But it seems reasonable to guess that the cavern is geologically young, probably no more than a million or so years old. It also seems reasonable to guess that most of the cave probably formed during the ice ages, when a wetter climate provided a better supply of water to erode caverns and deposit dripstone.

The Gallatin Valley

Between Manhattan and Bozeman, the highway crosses the lush Gallatin Valley on a thin veneer of stream gravels exposed in occasional borrow pits. Although it is not visible from the road, exposures elsewhere leave no doubt that pale gray desert valley fill sediments of the Renova formation lie beneath that gravel, probably to a depth of several thousand feet. The Horseshoe Hills defines the western edges of the valley; the high Bridger Range sharply bounds it on the east, from Bozeman north. The Gallatin Range, and beyond it the Madison Range, both tumbled seas of craggy peaks, stretch into the southern distance.

The Stillwater River near Nye at the north edge of the Beartooth Plateau.
—Montana Bureau of Mines and Geology photo by H. L. James

Interstate 90: Bozeman—Billings
141 mi./226 km.

Between Bozeman and Livingston the highway passes spectacular rock outcrops in Bozeman Pass in the low southern end of the Bridger Range. The long route between Livingston and Billings follows the Yellowstone River along the southern edge of the Crazy Mountain basin. High peaks, snowcapped most of the year, rise north and south of the road: the jaggedly glaciated Crazy Mountains in the north between Livingston and Big Timber, the high Beartooth Plateau all along the southern horizon.

The Bridger Range

The Bridger Range is a steep fold that trends slightly west of north. It formed as sedimentary rocks buckled on top of a fault that shoved a big slice of continental crust and its cover of sedimentary formations an unknown but modest distance up and to the east. Several small faults cross the range, chopping it into segments. The Willow Creek fault divides the range into a southern part where the Paleozoic

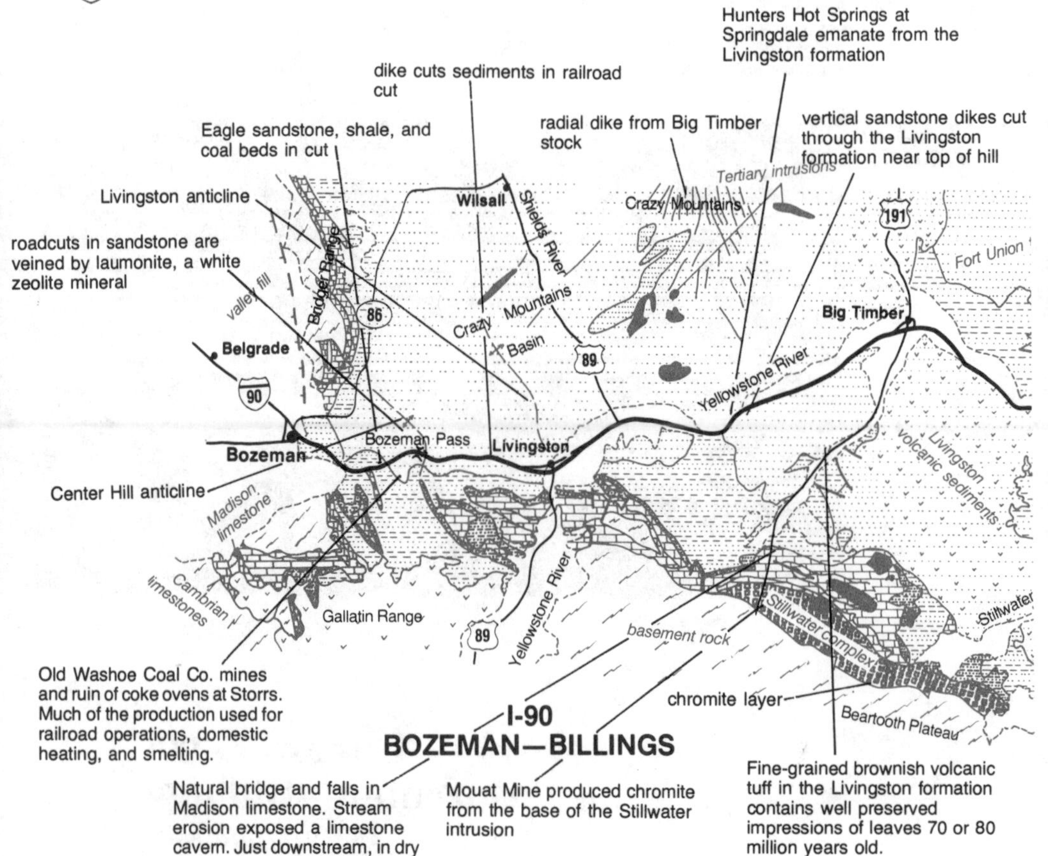
Hunters Hot Springs at Springdale emanate from the Livingston formation
dike cuts sediments in railroad cut
radial dike from Big Timber stock
vertical sandstone dikes cut through the Livingston formation near top of hill
Eagle sandstone, shale, and coal beds in cut
Livingston anticline
roadcuts in sandstone are veined by laumonite, a white zeolite mineral
Tertiary intrusions
Wilsall
Shields River
Crazy Mountains
Crazy Mountains Basin
Bridger Range
valley fill
Belgrade
Fort Union
Big Timber
Yellowstone River
Bozeman Pass
Livingston
Bozeman
Center Hill anticline
Madison limestone
Cambrian limestones
Gallatin Range
Yellowstone River
Livingston volcanic sediments
basement rock
Stillwater complex
Stillwater
chromite layer
Beartooth Plateau
90
86
89
191
Old Washoe Coal Co. mines and ruin of coke ovens at Storrs. Much of the production used for railroad operations, domestic heating, and smelting.
I-90
BOZEMAN—BILLINGS
Natural bridge and falls in Madison limestone. Stream erosion exposed a limestone cavern. Just downstream, in dry season, the river again goes underground for a half mile.
Mouat Mine produced chromite from the base of the Stillwater intrusion
Fine-grained brownish volcanic tuff in the Livingston formation contains well preserved impressions of leaves 70 or 80 million years old.

Big outcrops of pale Madison limestone full of small caverns rise above the trees in Bozeman Pass.

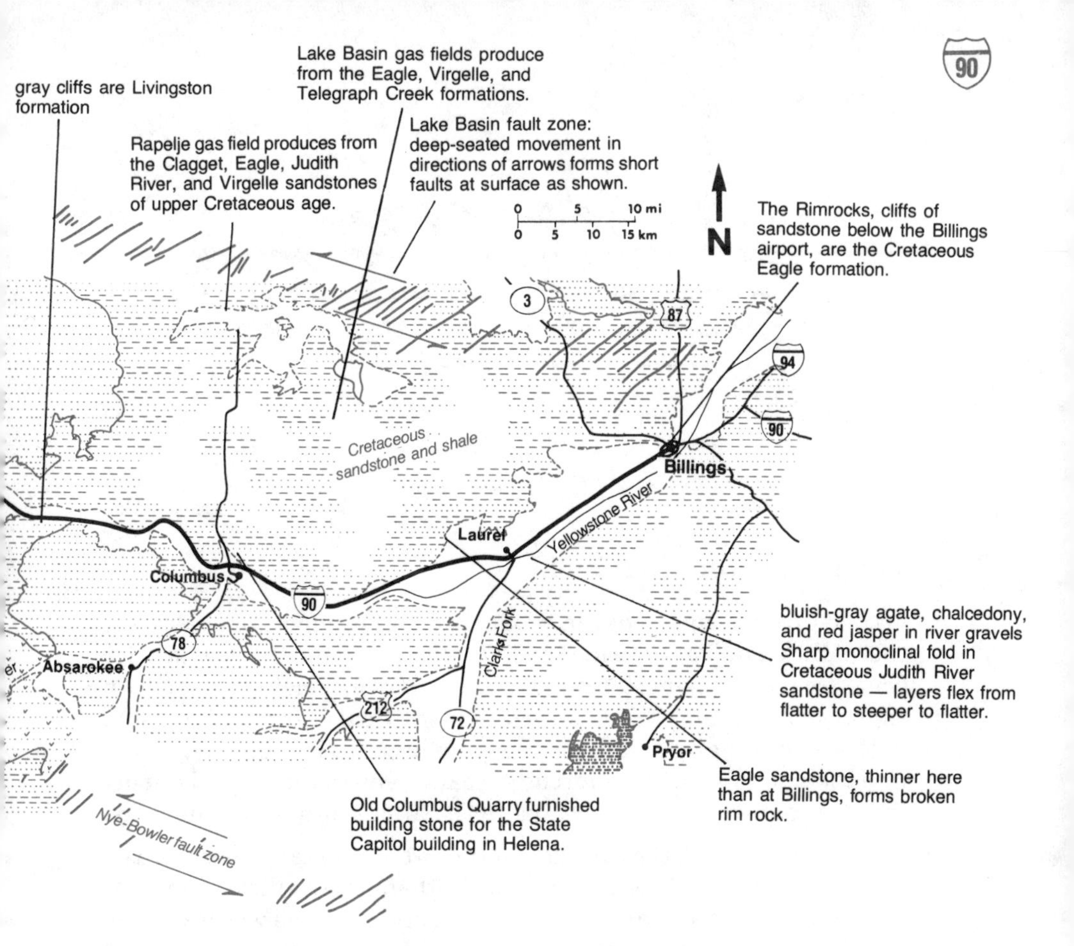
gray cliffs are Livingston formation
Lake Basin gas fields produce from the Eagle, Virgelle, and Telegraph Creek formations.
Rapelje gas field produces from the Clagget, Eagle, Judith River, and Virgelle sandstones of upper Cretaceous age.
Lake Basin fault zone: deep-seated movement in directions of arrows forms short faults at surface as shown.
0 5 10 mi
0 5 10 15 km
N
The Rimrocks, cliffs of sandstone below the Billings airport, are the Cretaceous Eagle formation.
90
3
87
94
90
Cretaceous sandstone and shale
Billings
Yellowstone River
Laurel
Columbus
90
78
Absarokee
Clarks Fork
212
72
Pryor
bluish-gray agate, chalcedony, and red jasper in river gravels
Sharp monoclinal fold in Cretaceous Judith River sandstone — layers flex from flatter to steeper to flatter.
Eagle sandstone, thinner here than at Billings, forms broken rim rock.
Old Columbus Quarry furnished building stone for the State Capitol building in Helena.
Nye-Bowler fault zone

formations lie on basement rocks, and a northern part in which they rest on Precambrian sedimentary formations, Belt rocks. Mesozoic sedimentary formations cap the succession of rock layers in both parts of the range.

The Fort Union formation, no less than 55 million years old, lies steeply tilted on the eastern flank of the range, but the Renova formation, which might be as much as 40 million years old, is considerably less disturbed in the Gallatin Valley. All that makes it seem likely that the Bridger Range probably rose sometime in that time window between 55 and 45 million years ago. About 50 million years ago, when there was so much igneous activity in Montana, seems a reasonable guess.

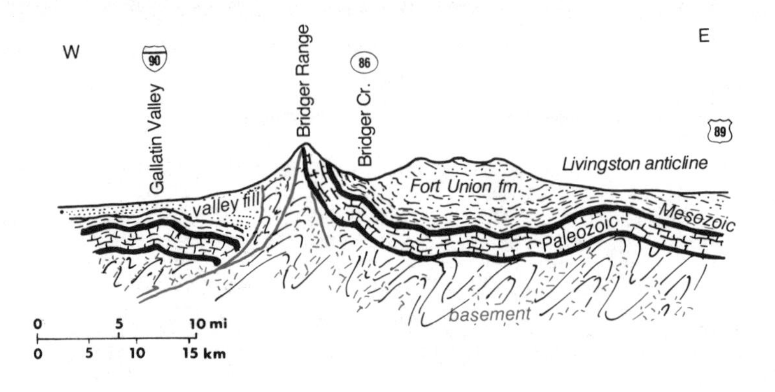

Section across the Bridger Range north of Bozeman

The Livingston Coal Field

Underground mines began producing coal around Cokedale, just east of Bozeman Pass, during the 1870s. Early production was sold for domestic use in Livingston and Bozeman, but ore smelters and steam engines soon became the largest consumers. Coal from the Livingston field was especially suited for smelting because it could be roasted into coke. Many of the hundreds of coke ovens that operated during the last century survive as ruins in the area south of the highway.

After about 1910, the Cokedale mines went into rapid decline. The most easily mineable coal was gone by then, more efficient smelter designs diminished the demand for coke, and cheap oil was becoming readily available. Production diminished further after the first World War, and finally ceased about 1942. Far more coal remains in the ground than the mines produced.

The Cokedale mines worked several seams near the top of the Eagle sandstone, a late Cretaceous formation. During late Cretaceous time, about 70 million years ago, the shore of the shallow sea that still flooded much of inland North America lay in central Montana. Most of the Eagle sandstone accumulated along the beaches. The coal probably began as thick deposits of peat laid down in broad tidewater marshes that may have resembled those along the modern west coast of Florida. The weight of younger sedimentary rocks deposited on top of the peat compressed it into coal.

The Crazy Mountain Basin

Between Livingston and Columbus, Interstate 90 crosses the southern part of the Crazy Mountain basin. Most of the rocks more conspicuously exposed in roadcuts and in stream banks near the road

An exposure of agglomerate in the Livingston formation. The rock, angular chunks of dark andesite suspended in volcanic ash, probably formed as a mudflow deposit.

belong to the Livingston formation, a thick accumulation of late Cretaceous sediments, mostly volcanic debris. Watch for generally greenish or gray roadcuts full of angular chunks of dark andesite. Geologists call such sedimentary rocks composed of volcanic rubble "agglomerate."

Volcanic sedimentary rocks must come from a source in a volcanic pile. The largest area of late Cretaceous volcanic activity in Montana was that of the modern Boulder batholith, where large volcanoes were then erupting the Elkhorn Mountains andesites. Sediments eroded from the Elkhorn Mountains volcanic pile went down the late Cretaceous rivers to accumulate as the Livingston formation in the southern part of the Crazy Mountain basin. The steep tilt of the Livingston formation along the edge of the Bridger Range shows that it is older than the big downfold of the Crazy Mountain basin.

Layered volcanic sediments of the Livingston formation exposed in a hill west of Big Timber

The Stillwater Complex

The road south from Big Timber follows the Boulder River almost to its source in the northern edge of the Beartooth Plateau, a block of basement rock that rose along faults about 50 million years ago. The basement rock in the headwaters of the Boulder River is the Stillwater Complex, an enormous mass of dark gabbro and related rocks, all of which must have come from somewhere in the earth's mantle. Gabbro, a fairly rare igneous rock, is a coarsely crystalline version of basalt. As in similar masses elsewhere, the gabbro of the Stillwater Complex is layered. It looks from a distance like sedimentary rock.

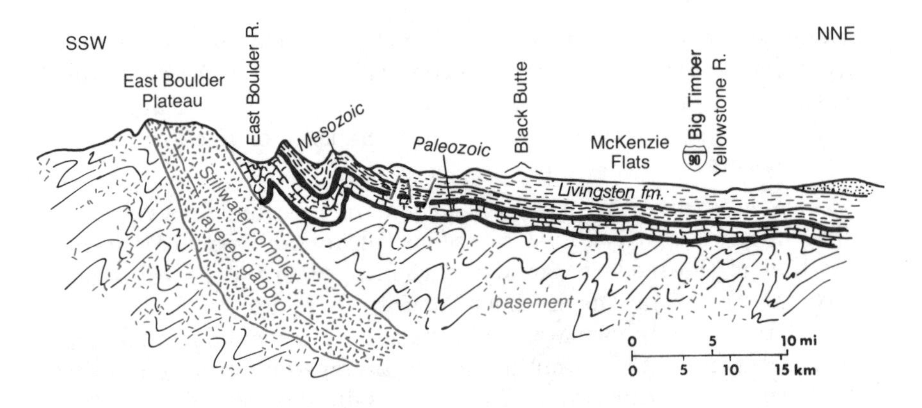

Section from the Stillwater Complex in the north edge of the Beartooth Plateau to Big Timber

Natural Bridge Falls on the Boulder River.
—U.S. Forest Service photo by E. M. Welton

Many, but by no means all, geologists believe that masses of gabbro like the Stillwater Complex form in huge craters opened by the impacts of large meteorites that plunge through the atmosphere and strike the earth so hard they explode with extreme violence. The crater reduces pressure on the extremely hot rocks beneath the crust enough to permit them to partially melt. Basaltic magma wells up into the crater to form a lava lake that very slowly crystallizes into gabbro, the coarse-grained form of basalt. The layers form as growing crystals of denser minerals sink within the magma.

However layered gabbro complexes may form there is no controversy over their tendency to contain valuable ore bodies. Many years of mining exploration have shown that certain layers in the Stillwater Complex contain large concentrations of platinum, chromite, nickel, and other metals. There is little doubt that those dark rocks at the head of the Boulder and Stillwater rivers will someday support large mines like those that work in the similar rocks of the Bushveld Complex in South Africa.

Sandstone

Much of the route between the area a few miles west of Columbus and Billings crosses forested hills eroded in late Cretaceous sandstone. Watch for bouldery outcrops of pale gray or tan rock among the pine trees that flourish on sandstone everywhere in central and eastern Montana. About the turn of the century, a quarry just north of the Columbus exit supplied handsome pale gray stone for the older part of the State Capitol in Helena, as well as for many other buildings in several Montana towns.

High cliffs of late Cretaceous Eagle sandstone, the Rimrock, rise above Billings. That massive sandstone appears to be the remains of a barrier island, perhaps about like modern Galveston Island, Texas, that stood between a coastal lagoon west of Billings and the shallow inland sea that still flooded much of North America during late Cretaceous time.

Rimrocks of Cretaceous Eagle sandstone at Billings.

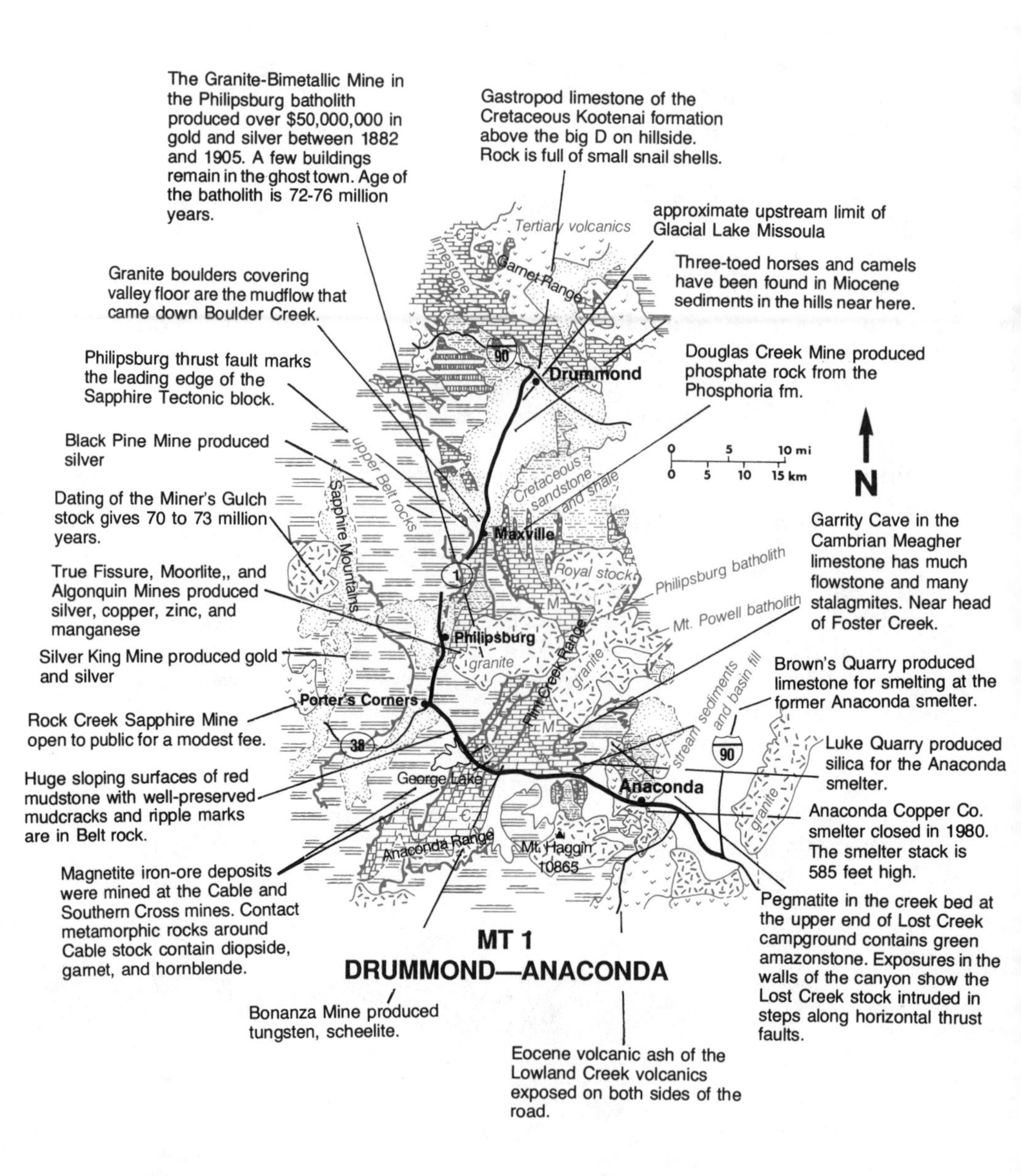
1
The Granite-Bimetallic Mine in the Philipsburg batholith produced over $50,000,000 in gold and silver between 1882 and 1905. A few buildings remain in the ghost town. Age of the batholith is 72-76 million years.
Gastropod limestone of the Cretaceous Kootenai formation above the big D on hillside. Rock is full of small snail shells.
approximate upstream limit of Glacial Lake Missoula
Three-toed horses and camels have been found in Miocene sediments in the hills near here.
Granite boulders covering valley floor are the mudflow that came down Boulder Creek.
Philipsburg thrust fault marks the leading edge of the Sapphire Tectonic block.
Douglas Creek Mine produced phosphate rock from the Phosphoria fm.
Black Pine Mine produced silver
Dating of the Miner's Gulch stock gives 70 to 73 million years.
True Fissure, Moorlite,, and Algonquin Mines produced silver, copper, zinc, and manganese
Silver King Mine produced gold and silver
Rock Creek Sapphire Mine open to public for a modest fee.
Huge sloping surfaces of red mudstone with well-preserved mudcracks and ripple marks are in Belt rock.
Magnetite iron-ore deposits were mined at the Cable and Southern Cross mines. Contact metamorphic rocks around Cable stock contain diopside, garnet, and hornblende.
Bonanza Mine produced tungsten, scheelite.
Eocene volcanic ash of the Lowland Creek volcanics exposed on both sides of the road.
Garrity Cave in the Cambrian Meagher limestone has much flowstone and many stalagmites. Near head of Foster Creek.
Brown's Quarry produced limestone for smelting at the former Anaconda smelter.
Luke Quarry produced silica for the Anaconda smelter.
Anaconda Copper Co. smelter closed in 1980. The smelter stack is 585 feet high.
Pegmatite in the creek bed at the upper end of Lost Creek campground contains green amazonstone. Exposures in the walls of the canyon show the Lost Creek stock intruded in steps along horizontal thrust faults.
Tertiary volcanics
limestone
Garnet Range
90
Drummond
upper Belt rocks
Sapphire Mountains
Cretaceous sandstone and shale
Maxville
1
Royal stock
Philipsburg batholith
Mt. Powell batholith
Philipsburg
granite
Flint Creek Range
granite
Porter's Corners
38
stream sediments and basin fill
90
George Lake
Anaconda
granite
Anaconda Range
Mt. Haggin
10865
0 5 10 mi
0 5 10 15 km
N

MT 1
DRUMMOND—ANACONDA

MT 1: Drummond—Anaconda
49 mi./78 km.

The road between Drummond and Philipsburg traverses the connecting Flint Creek and Philipsburg valleys. The high Flint Creek Range, snow capped much of the year, rises in the east along the entire route; the much lower Sapphire Mountains lie in the west. The valley floor contains a deep fill of basin sediments, nowhere well exposed near the road.

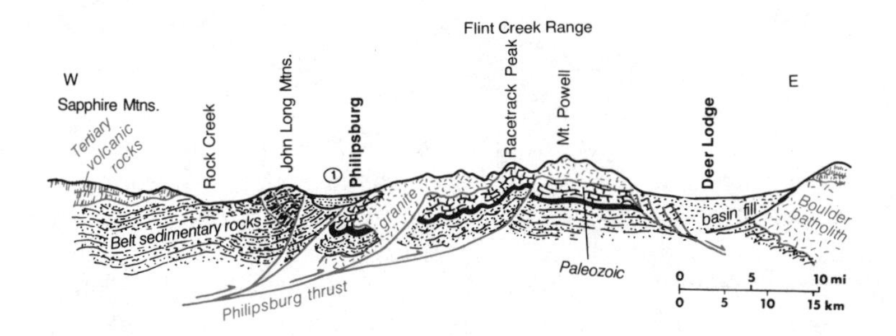

Section across the line of road at Philipsburg. Thrust faults in the Flint Creek Range define the leading edge of the Sapphire block.

The Leading Edge of the Sapphire Block

The Sapphire Mountains west of the road are part of the block that moved east out of Idaho about 70 million years ago. The Flint Creek Range at the leading edge of that block contains rocks crumpled and broken into tight folds and faults—the leading edge of the Sapphire block jammed into and bulldozed the rocks ahead of it. Large masses of magma spread west from the Idaho batholith along thrust faults to invade the folds, making structures of staggering complexity. As often happens, ore deposits developed around the margins of some of the granite intrusions.

Mining in the Flint Creek Range

A prospector found bedrock outcrops rich in silver and gold near Philipsburg in 1867. Mining got off to a vigorous start a few years

Machinery at the Bimetallic Mine in Granite before the big crash.
—Mansfield Library, University of Montana

later, flourished mightily during the 1880s, and has continued on and off ever since. Much of the ore is in quartz vein deposits in the granites and the rocks that enclose them; some is in zones of skarn, metamorphic rocks that formed where granite magma reacted with limestones.

The Philipsburg District is one of the few in North America that contain much manganese, an essential ingredient in some kinds of tool steel. But the deposits are lean by international standards, so Philipsburg mines manganese only when circumstances cut off imports. Nevertheless, the manganese oxide mineral pyrolusite is abundant enough to stain many of the mine dumps coal black. It formed through alteration of rhodochrosite, a shocking-pink manganese carbonate mineral in the skarns. Rhodochrosite also appears on some of the dumps, although not abundantly.

The sad history of Granite, high on the mountain four miles from Philipsburg, shows that events on Wall Street may be more important to a mining district than rocks. Mines there began to produce silver and some gold from deep vein deposits in late 1882, nearly 30 million dollars worth of bullion within a few years. Some authorities claim that Granite was the richest silver camp in the world during the 1880s. Then the terrible economic panic of 1893 cut the price of silver, and abruptly closed the mines on the morning of August 1. Within 36 hours, the entire population of Granite, some 3000 people, came streaming down the mountain carrying their belongings with them in wagons and wheelbarrows. Most of the refugees stopped in Philipsburg just long enough to withdraw their savings from the bank and buy a train ticket.

That happened just when the mines were running out of enriched high grade ore at the base of the weathered zone, ore that contained the original metals content of the rock plus metals that had seeped down from the weathered rock above. Although limited mining resumed in 1898, and continued sporadically for several decades, the mines never did well in the leaner primary ore below the enriched zone. Granite never really revived. Most of the wooden buildings burned in 1958.

The old town of Cable, four miles north of the road between Georgetown Lake and Anaconda, started slowly after rather poor ore was discovered in 1867. Then the mine broke into a bonanza of incredibly rich gold quartz ore in 1880, and Cable blossomed. So did many miners, who got rich by sneaking ore out of the mine in their pockets and lunchboxes. The fortunes of Cable rose and fell for the next 60 years as the mines alternately broke into new bonanzas, and worked through long stretches of barren rock. Nothing much has happened there since about 1940.

Sapphires

Sapphires were discovered in the West Fork of Rock Creek about 1892, and mined off and on between 1899 and 1943. The deposit is beside Montana 38 about 16 miles southwest of Philipsburg. Total production was probably more than three quarters of a million ounces, most of it for use in bearings and abrasives. Synthetic sapphires began to dominate the market during the 1930s. Gem quality sapphires still come from the Rock Creek deposit in the full rainbow of colors from almost red to dark blue, including shades of yellow, pink, orange, and lavender. The workings open, for a fee, in the summer months to people who want to collect their own gems.

Although the rocks that contain the sapphires are not exposed, it seems reasonable to assume that they probably resemble those at other sapphire deposits, black igneous rocks, of peculiar composition related to the diatremes of central Montana. The magma must have melted somewhere deep within the mantle, far below the continental crust. Miners wash sapphires from placer deposits in stream gravel and from the soil, where they remain after all the other minerals in the rock have weathered into clay.

Belt Rock

A long series of magnificent roadcuts on Flint Creek Hill between Georgetown Lake and the Philipsburg Valley provide spectacular exposures of Precambrian sedimentary rocks. Most of the rocks are yellow and red mudstones in which every bedding surface shows intricate patterns of ripple marks and suncracks, acres of them. It is hard to say how the original sediments accumulated in the vanished

This mud, now mudstone exposed in a roadcut on Georgetown Hill, dried and cracked about a billion years ago.

world of a billion or more years ago. But the rocks look very much like muds in modern desert lakes and tidal flats, places that alternately flood and bake in the sun.

Glaciation

Craggy granite peaks high in the Flint Creek Range drop off into cirques hollowed in the mountain, and those descend into deeply gouged valleys—all the mark of the ice. The softly rounded Sapphire Mountains show no sign of glaciation. Evidently, the Flint Creek Range snatched enough snow out of the ice age storms to nourish glaciers, while the much lower Sapphire Mountains did not.

Granite boulders that moved in a mudflow litter the floor of the Flint Creek Valley.

Thousands of rounded granite boulders, which could have come only from the Flint Creek Range, cover several square miles of the valley floor just north of Maxville. Although the deposit looks at first glance like a moraine, there is no evidence that glaciers ever reached the valley floor. Instead, it appears that a large mudflow poured down the narrow valley of Boulder Creek southeast of Maxville, then spread across the broad valley floor. That happened sometime since the last ice age, when a moraine damming a glacial lake in the canyon of Boulder Creek washed out, and mixed with the water to make a thick porridge. Mudflows carry such large boulders because they are much denser than clear water, and therefore exert a greater buoyant effect.

Between Georgetown Lake and Anaconda, the highway follows valleys that held large glaciers during the ice ages. Watch for the moraines, tracts of lumpy topography rather densely littered with large boulders. Roadcuts through moraines expose glacial till, easy to recognize as an unsorted mess of rocks of all sizes embedded in a matrix of mud.

The old stack at Anaconda seen across heaps of black slag, all that remains of the smelter that for more than 80 years reduced the ores mined at Butte.

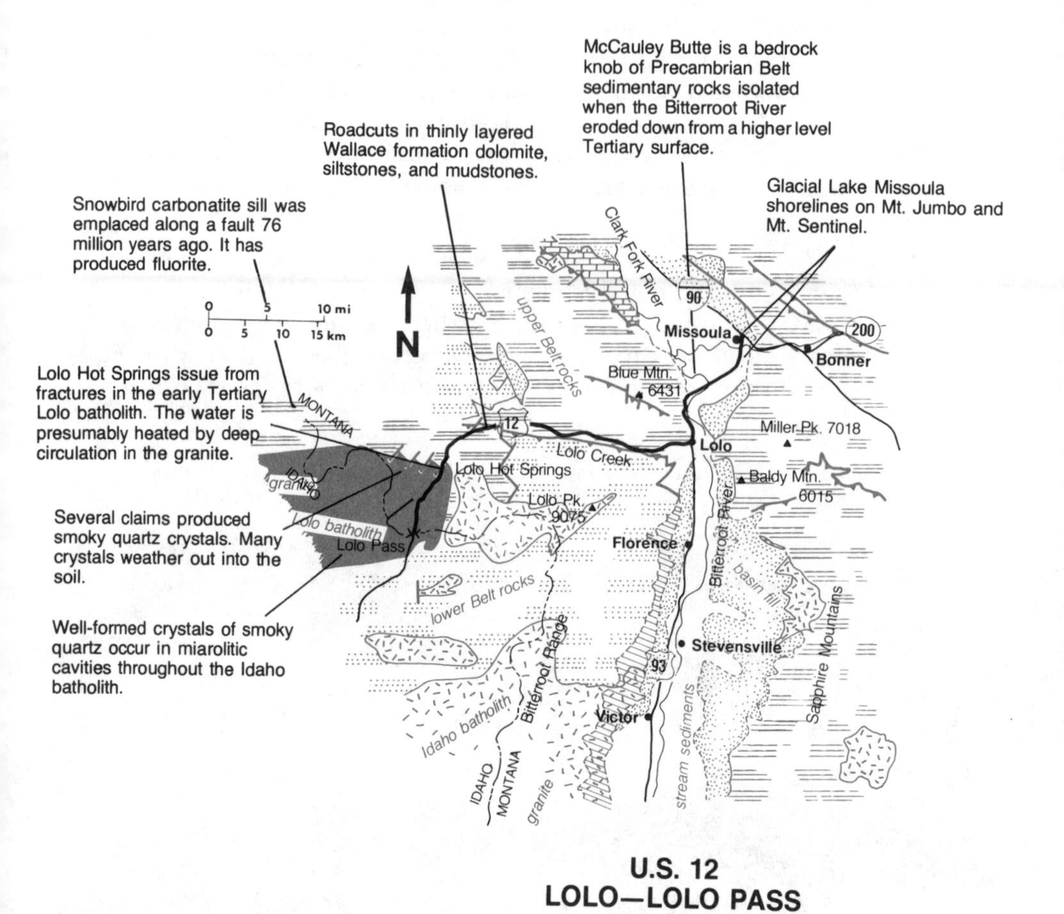
McCauley Butte is a bedrock knob of Precambrian Belt sedimentary rocks isolated when the Bitterroot River eroded down from a higher level Tertiary surface.
Roadcuts in thinly layered Wallace formation dolomite, siltstones, and mudstones.
Glacial Lake Missoula shorelines on Mt. Jumbo and Mt. Sentinel.
Snowbird carbonatite sill was emplaced along a fault 76 million years ago. It has produced fluorite.
0 5 10 mi
0 5 10 15 km
N
Lolo Hot Springs issue from fractures in the early Tertiary Lolo batholith. The water is presumably heated by deep circulation in the granite.
Several claims produced smoky quartz crystals. Many crystals weather out into the soil.
Well-formed crystals of smoky quartz occur in miarolitic cavities throughout the Idaho batholith.
Clark Fork River
upper Belt rocks
90
Missoula
200
Bonner
Blue Mtn. 6431
12
Lolo
Miller Pk. 7018
Lolo Creek
Lolo Hot Springs
Baldy Mtn. 6015
MONTANA
IDAHO
granite
Lolo batholith
Lolo Pass
Lolo Pk. 9075
Florence
Bitterroot River
basin fill
lower Belt rocks
Bitterroot Range
Stevensville
Sapphire Mountains
93
Idaho batholith
Victor
stream sediments
IDAHO
MONTANA
granite

U.S. 12
LOLO—LOLO PASS

US 12: Lolo—Lolo Pass
33 mi./53 km.

Rocks between Lolo and Lolo Hot Springs belong to the Precambrian Ravalli and Wallace formations, Belt rock. Most are sedimentary rocks in which the steeply tilted layers are hard to see. Lolo Hot Springs are almost at the contact between the ancient sedimentary rocks and the Lolo granite batholith, which the road crosses between Lolo Hot Springs and the pass.

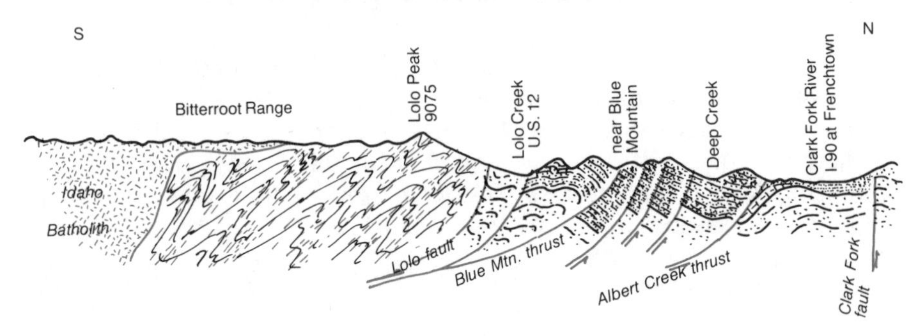

Section drawn across the line of the road several miles west of Lolo

The cross section shows big blocks of Precambrian rocks that moved north, away from the Idaho batholith, by sliding on large faults. That movement happened between 70 and 80 million years ago, when the large masses of molten magma that were to become the granite of the Idaho batholith were rising into the earth's crust. Had the section been drawn along a line a few miles farther west, it would have shown the much younger granite of the Lolo batholith cutting through those fault blocks.

Lolo Granite

Granite magma invaded the ancient Belt rocks about 50 million years ago. Most of it crystallized a few thousand feet below the surface to become the Lolo batholith; the rest erupted violently to form sheets of rhyolite ash, none visible from the road. The volcanic neck is west of the pass, also out of sight.

Boulders of granite among the trees near Lolo Hot Springs. They formed as the rock weathered within the soil, and was then exposed by soil erosion.

Like most granites emplaced at such shallow depth, the Lolo batholith is full of gas cavities lined with crystals. Most of the cavities are too minute to see easily without a lens, but a few are much larger. Those are lined with large crystals that include smoky quartz clear and flawless enough to facet into gem stones. Collectors find the quartz by walking dirt roads after a hard rain.

Lolo Hot Springs

Lewis and Clark and their men enjoyed a hot bath in Lolo Hot Springs on their way west, again on their way back. Thousands of people have followed their example since. The water issues from fractures in the granite at a temperature of about 100 degrees Fahrenheit, and flows through pipes into a pool.

Granite 50 million years old has had so long to cool that it can hardly be the source of the heat. It seems far more likely that rain water sinks to depths of at least several thousand feet through fractures along the contact between granite and Belt rocks. The temperature of the earth's crust increases at an average rate of a little less than 2 degrees Fahrenheit for every 100 feet of increasing depth. Therefore, if surface water sinks along fractures to a depth of approximately 5000 feet, it would be heated to the temperature of Lolo Hot Springs.

US 12: Garrison—Helena
47 mi./75 km.

The route between Garrison and Helena crosses sedimentary rocks in the overthrust belt, granite that intrudes them, and two generations of volcanic rocks that partly cover them. The magmas that became granite and volcanic rocks brought with them the gold and silver that made Helena a roaring mining camp in the early days of Montana.

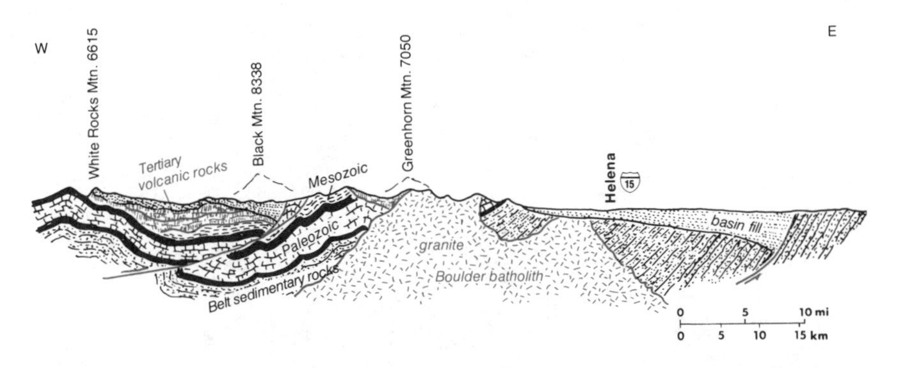

Section drawn just north of the line of US 12 between Garrison and Helena

Big slabs of sedimentary rock west of McDonald Pass must have moved east on overthrust faults sometime before the magma that became the Boulder batholith granite first invaded them about 75 million years ago, but probably not long before. Cretaceous brown sandstones and black shales are conspicuous within about 10 miles of Garrison. Sedimentary rocks exposed along the road near Helena, east of the Boulder batholith, are Precambrian formations.

Some of the magma of the Boulder batholith erupted to become andesites, the Elkhorn Mountains volcanic rocks, that still cover parts of the batholith. Watch for dark greenish rocks very poorly exposed along the western approach to McDonald Pass. About 50 million years ago, a second generation of volcanoes erupted the Lowland Creek volcanic rocks. Those volcanic rocks appear in

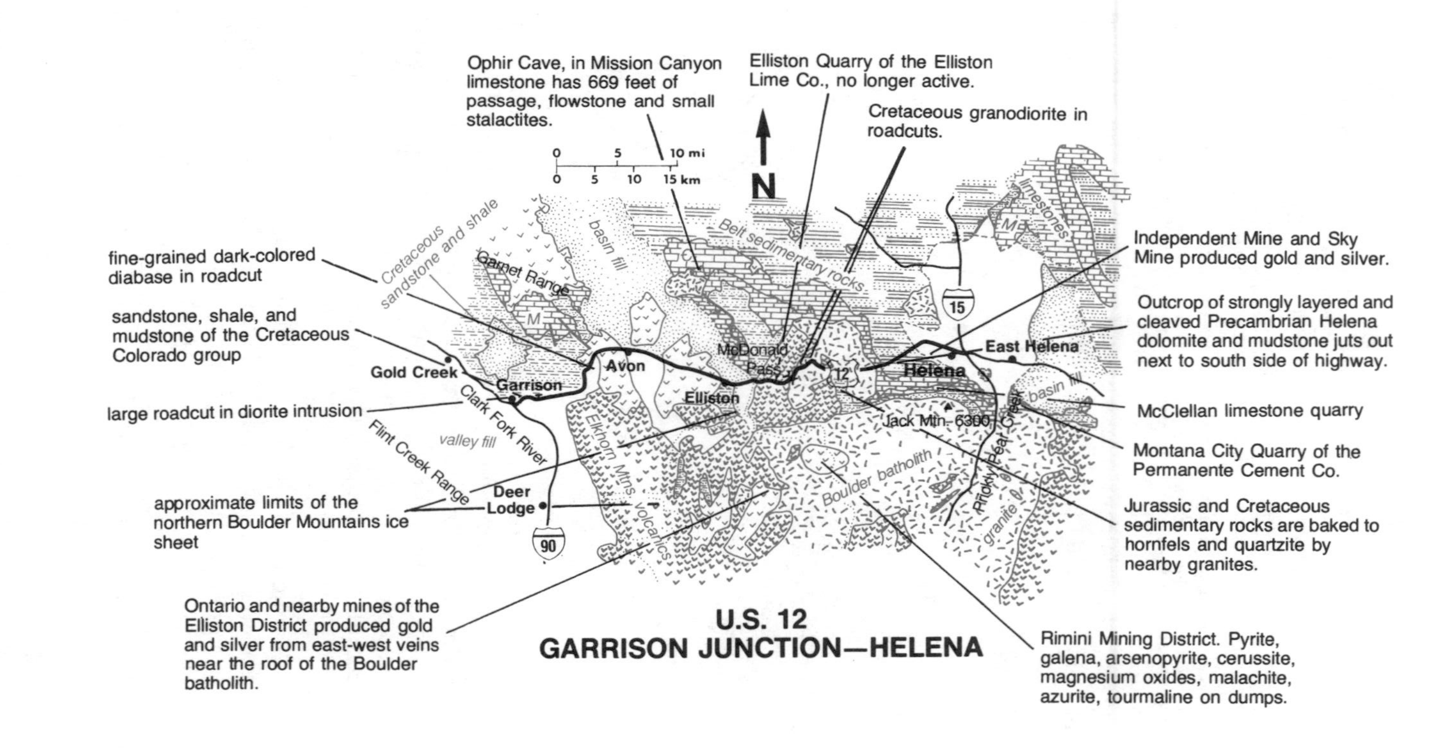
U.S. 12
GARRISON JUNCTION—HELENA
Ophir Cave, in Mission Canyon limestone has 669 feet of passage, flowstone and small stalactites.
Elliston Quarry of the Elliston Lime Co., no longer active.
Cretaceous granodiorite in roadcuts.
0 5 10 mi
0 5 10 15 km
N
fine-grained dark-colored diabase in roadcut
sandstone, shale, and mudstone of the Cretaceous Colorado group
large roadcut in diorite intrusion
approximate limits of the northern Boulder Mountains ice sheet
Ontario and nearby mines of the Elliston District produced gold and silver from east-west veins near the roof of the Boulder batholith.
Independent Mine and Sky Mine produced gold and silver.
Outcrop of strongly layered and cleaved Precambrian Helena dolomite and mudstone juts out next to south side of highway.
McClellan limestone quarry
Montana City Quarry of the Permanente Cement Co.
Jurassic and Cretaceous sedimentary rocks are baked to hornfels and quartzite by nearby granites.
Rimini Mining District. Pyrite, galena, arsenopyrite, cerussite, magnesium oxides, malachite, azurite, tourmaline on dumps.
Cretaceous sandstone and shale
Garnet Range
basin fill
Belt sedimentary rocks
limestones
Gold Creek
Garrison
Avon
Elliston
McDonald Pass
12
15
90
Helena
East Helena
basin fill
Jack Mtn. 6300
Prickly Pear Creek
Boulder batholith
granite
Elkhorn Mtns. volcanics
Clark Fork River
valley fill
Flint Creek Range
Deer Lodge

A regular pattern of fractures breaks the granite into neat blocks; a roadcut near McDonald Pass.

outcrops and roadcuts along about ten miles of road on each side of Avon. Watch for rather nondescript greenish or brown roadcuts and for natural outcrops that rise in rough brown cliffs above the trees on both sides of the river.

Helena Gold

Helena started as a gold mining camp after rich gravels were discovered in Last Chance Gulch in 1864. The early miners kept poor records, at best, so it is extremely difficult to estimate how much gold they produced. Perhaps we can believe an early geologist who estimated that the Last Chance Gulch placers produced about 16 million old-fashioned dollars worth of gold before 1910, most of that before 1868. A dredge working the downstream gravels north of town recovered some 2.5 million dollars worth of gold between 1935 and 1948. Although rumors persist of fortunes in placer gold sealed under the pavement of downtown Helena, it seems unlikely that the early miners left much behind.

The only known bedrock sources for the placer gold in Last Chance Gulch are the Spring Hill and Whitlatch-Union vein deposits in Orofino and Grizzly gulches several miles south of Helena, at the old settlements of Springtown and Unionville. Lode mining began in that area in 1864, and the big production came early, but the Whitlatch-Union Mine continued to produce until 1948. Both mines worked deposits in rocks adjacent to the Boulder batholith.

At Marysville, about 15 miles northwest of Helena, gold was mined in quantity from deposits around the margin of a granite intrusion

An exposure of Precambrian Helena dolomite in a roadcut nine miles west of Helena. Heat from the granite magma of the Boulder batholith bleached the rock, making the sedimentary layering easy to see.

with an age date of 78 million years, clearly a satellite of the Boulder batholith. Fairly sober estimates have the Marysville mines producing at least 30 million dollars in gold, more than half from the Drumlummon Mine. The last mines closed about 1948.

Rimini, on a side road eight miles south of US 12, produced lead and zinc, along with about eight million dollars worth of gold. The ore is in granite and volcanic rocks. Mining began in 1870 and continued for several decades. Mines in the Elliston District, just west of Rimini, produced sightly more than three million dollars worth of gold and silver between 1890 and 1909. The ore is in veins in the Elkhorn Mountains volcanic rocks within 100 feet of their contact with granite of the Boulder batholith.

Earthquakes

Indians warned the first settlers in the Helena Valley of frequent earthquakes, and their advice was good. Old newspapers report a moderate tremor in 1869, more than 60 others between 1903 and 1935, when the shaking got serious. In three months after October 3, 1935, Helena felt more than 1200 earthquakes, a few violent enough to cause damage amounting to several million dollars. Many of the older stone and brick buildings in Helena still show obvious earthquake damage. Look for long cracks in their walls, shifted lintels over doors and windows, and chimmeys with new tops. Masonry does not withstand much shaking.

A fault in the Prickly Pear Valley southeast of town appears to have caused the commotion. Helena is near the northern end of the intermountain seismic belt, a zone of frequent earthquake activity that extends straight south through Yellowstone National Park to the Wasatch Front at Salt Lake City.

Helena is especially vulnerable to earthquakes because most of the town stands on soft valley fill sediments that shake far more violently than hard bedrock. Think of a large bowl full of soft pudding; the bowl simulating bedrock and the pudding the soft sediments. Then imagine yourself sharply tapping the edge of the bowl to simulate an earthquake. The pudding will move far more visibly than the bowl, and so will the valley fill sediments move more violently than the bedrock during an earthquake.

Marysville was a busy place in its day. —Montana Historical Society photo

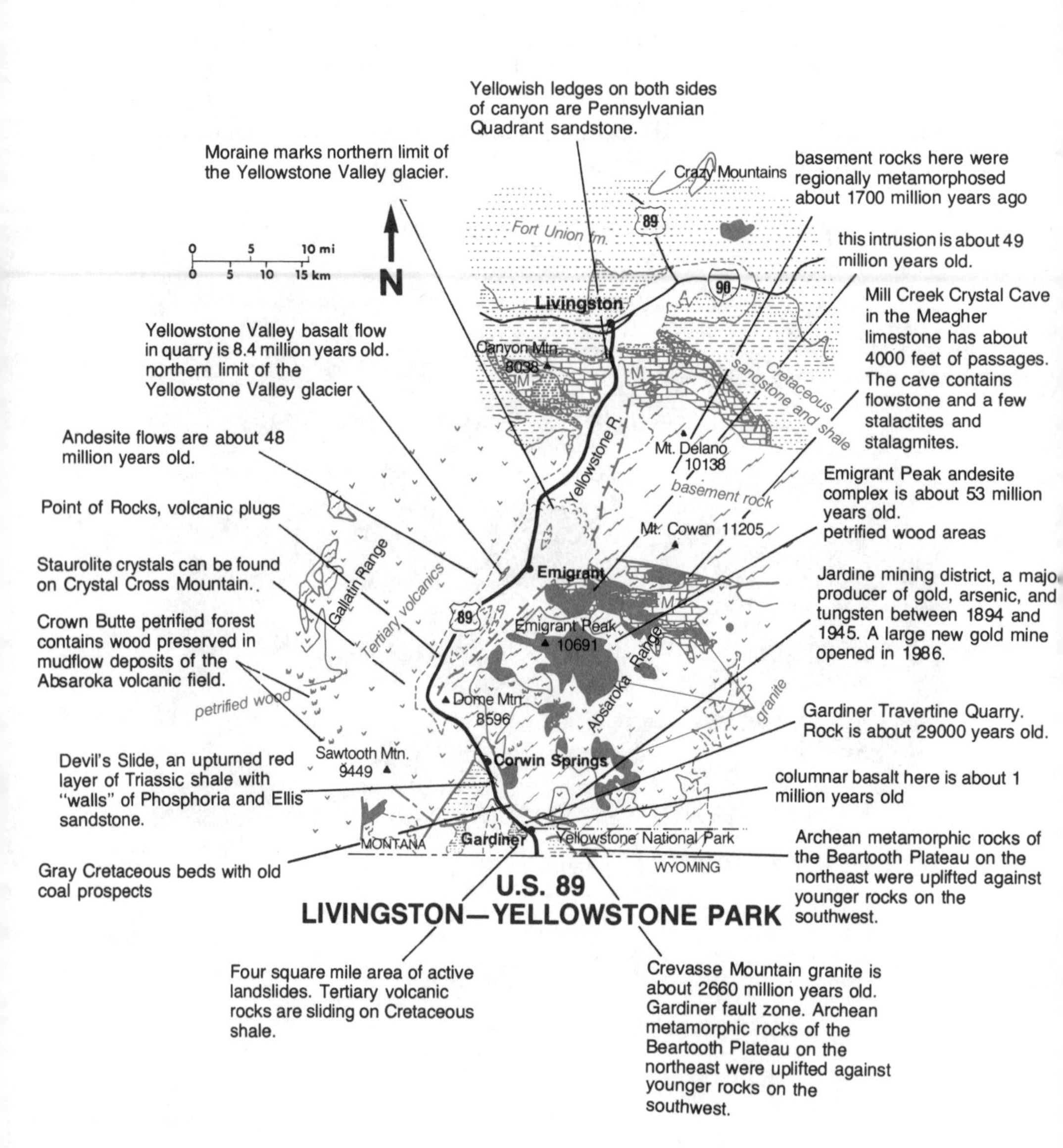
Yellowish ledges on both sides of canyon are Pennsylvanian Quadrant sandstone.
Moraine marks northern limit of the Yellowstone Valley glacier.
basement rocks here were regionally metamorphosed about 1700 million years ago
Crazy Mountains
this intrusion is about 49 million years old.
Fort Union fm.
0 5 10 mi
0 5 10 15 km
N
89
90
Livingston
Mill Creek Crystal Cave in the Meagher limestone has about 4000 feet of passages. The cave contains flowstone and a few stalactites and stalagmites.
Yellowstone Valley basalt flow in quarry is 8.4 million years old. northern limit of the Yellowstone Valley glacier
Canyon Mtn. 8038
M
Cretaceous sandstone and shale
Yellowstone R.
Andesite flows are about 48 million years old.
Mt. Delano 10138
basement rock
Emigrant Peak andesite complex is about 53 million years old.
petrified wood areas
Point of Rocks, volcanic plugs
Mt. Cowan 11205
Staurolite crystals can be found on Crystal Cross Mountain..
Emigrant
Jardine mining district, a major producer of gold, arsenic, and tungsten between 1894 and 1945. A large new gold mine opened in 1986.
Gallatin Range
Tertiary volcanics
Emigrant Peak 10691
Crown Butte petrified forest contains wood preserved in mudflow deposits of the Absaroka volcanic field.
Absaroka Range
granite
petrified wood
Dome Mtn. 8596
Gardiner Travertine Quarry. Rock is about 29000 years old.
Sawtooth Mtn. 9449
Corwin Springs
Devil's Slide, an upturned red layer of Triassic shale with "walls" of Phosphoria and Ellis sandstone.
columnar basalt here is about 1 million years old
MONTANA
Gardiner
Yellowstone National Park
Archean metamorphic rocks of the Beartooth Plateau on the northeast were uplifted against younger rocks on the southwest.
WYOMING
Gray Cretaceous beds with old coal prospects
U.S. 89
LIVINGSTON—YELLOWSTONE PARK
Four square mile area of active landslides. Tertiary volcanic rocks are sliding on Cretaceous shale.
Crevasse Mountain granite is about 2660 million years old. Gardiner fault zone. Archean metamorphic rocks of the Beartooth Plateau on the northeast were uplifted against younger rocks on the southwest.

View south from near Livingston into the north end of the Paradise Valley.
—Montana Dept. of Commerce photo

US 89: Livingston—Gardiner
53 mi./85 km.

Montana 89 follows the Yellowstone River between Livingston and Gardiner. The route passes through the entire magnificent length of the broad Paradise Valley between the Gallatin Range in the west and the Absaroka Range in the east. Narrow canyons guard the northern and southern ends of the Paradise Valley.

Just south of Livingston, the valley of the Yellowstone River abruptly narrows to a gorge cut through tilted Paleozoic sedimentary rock formations that include several thick sections of resistant limestone. Bold cliffs of pale gray Madison limestone rise boldly above the valley floor.

Most of the outcrops in the Gallatin Range west of the highway expose Absaroka volcanic rocks erupted about 50 million years ago from volcanic centers scattered from here south to the eastern part of Yellowstone National Park. The basement rock and folded Paleozoic sedimentary formations that lie beneath the volcanic cover in the Gallatin Range locally appear at the surface.

The Absaroka Range, east of the highway, is actually the western end of the Beartooth Plateau. It consists mostly of Precambrian

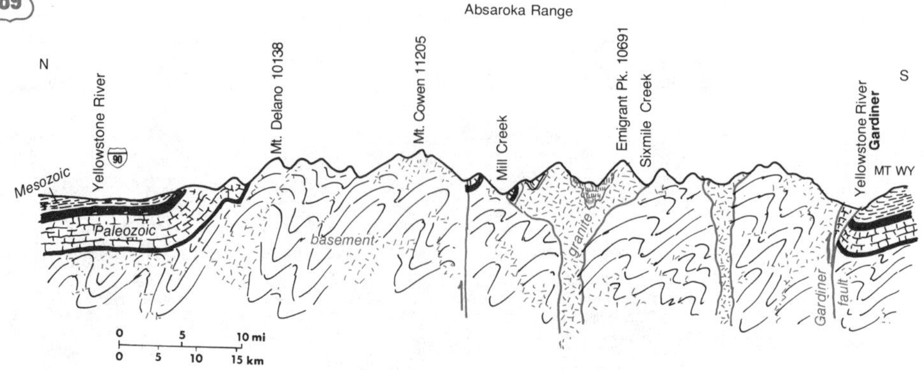

Section through the Absaroka Range along a line parallel to Montana 89

basement rocks with a local cover of volcanic rocks similar to those in the Gallatin Range.

Like other structural valleys in the northern Rockies, the Paradise Valley acquired a deep fill of basin sediment while it was a closed basin of internal drainage during most of the last 40 million years. Now, the Yellowstone River has cut an erosional valley into the upper part of those basin fill sediments. They appear in places as white bluffs of Renova formation rising above the river. Watch for the "Chalk Cliffs" a few miles north of Emigrant, especially to see the black basalt flow that caps them..

Volcanic Rocks

In parts of the Absaroka Range, the Absaroka volcanic rocks contain enormous quantities of petrified logs, the famous petrified forests of the Yellowstone region. Great mudflows similar to the one

A petrified tree, left, and an outcrop of volcanic mudflow conglomerate, right, stand next to each other in the southern part of the Absaroka Range.
—U.S. Forest Service photo by K. D. Swan

that poured down the Toutle River of Washington when Mount St. Helens erupted in 1980 swept rafts of logs along, and then buried them. Very few of the petrified logs appear to be the remains of trees buried where they grew. In fact, many of the log deposits contain pines and spruces mixed with tropical hardwoods. Evidently, the high volcanoes that erupted in this region 50 million years ago, like some modern volcanoes in central America, rose from a tropical lowland to elevations cool enough for spruce trees to thrive.

At Point of Rocks, about 6 miles north of Yankee Jim Canyon, the remains of an old volcanic vent make rounded hills west of the road. Andesite erupted here, probably while the Absaroka volcanoes were active about 50 million years ago. Other volcanic rocks in the Paradise Valley are much younger.

Several basalt lava flows lie on the Renova formation, which was still accumulating as recently as about 25 million years ago. Watch for black rim rock capping low mesas near the road—west of the highway just south of Emigrant, east of the highway along much of the route between Emigrant and Point of Rocks. A large quarry west of the road eight miles north of Point of Rocks provides a beautiful view of the basalt. An age date on that flow shows that it erupted a little more that eight million years ago, early Pliocene time.

It seems likely that the basalt flows erupted from a volcano in Yellowstone National Park, and poured down the Yellowstone Valley. It is possible to follow them south to the Gardiner area, but they are not easily visible from the road there. The river has cut its channel much deeper since the lava flows erupted, leaving the basalt-capped remnants of the valley floor of eight million years ago standing as low mesas.

Yankee Jim Canyon

About 42 miles south of Livingston, the road winds for three miles through Yankee Jim Canyon, a narrow gorge the Yellowstone River cut through hard Precambrian basement rocks. Roadcuts and the

Banded gneiss like that exposed in Yankee Jim Canyon.

canyon walls provide excellent exposures of complexly folded gneisses, mostly dark rocks full of swirling bands of lighter rock.

The Devil's Slide

About five miles north of Gardiner, a series of well exposed sedimentary layers stands almost vertically in the cliff on the opposite side of the river, at the base of Cinnabar Mountain. The bright red layer, the Devil's Slide, is the Chugwater mudstone, a Triassic formation about 200 million years old. Nearly everywhere in the world, Triassic sedimentary formations include far more than their share of bright red rocks; no one seems to know why. Some geologists have suggested that the atmosphere contained more oxygen then than at other times and reacted with iron to put more red iron oxide pigment in the rocks.

Cinnabar Mountain got its name after some early prospectors decided that it owed its red color to large concentrations of cinnabar, the mineral source of mercury. Their tunnel into the mountain produced no mercury. Later investigation showed that the red color is a clinker horizon that marks where a coal seam burned out of the sedimentary rocks.

Travertine

For many years, quarries near Gardiner have produced travertine for use as ornamental facing on buildings. The rock is a type of limestone that generally forms in caves and around large hot springs. Gardiner's travertine formed in hot springs similar to those now active at Mammoth, across the river in Yellowstone National Park. Broken or sawed surfaces reveal an intricate pattern of angular holes that impart an interesting textural effect to walls.

The Devil's Slide, an exposure of sedimentary rocks so steeply tilted that the layers stand almost on edge.
—U.S. Forest Service photo by E. M. Walton

The canyon of Blodgett Creek where it cuts through the Bitterroot mylonite zone northwest of Hamilton.

US 93: Missoula—Lost Trail Pass
95 mi./152 km.

Between Missoula and Lolo, the road follows the Bitterroot River past spectacular roadcuts in red and green mudstone. Those are Precambrian sedimentary formations, Belt rock. Watch for colorful bedding surfaces covered with mud cracks and ripple marks.

The route between Lolo and the area south of Darby passes through the lush Bitterroot Valley past the parade of gnarled gray peaks of the heavily glaciated and usually snowcapped Bitterroot Range that rise in the west. The northern end of the Bitterroot Range consists of metamorphic rocks, the peaks from Stevensville south, of granite—the eastern edge of the Idaho batholith.

The much lower Sapphire Range, never glaciated, rises in gently rounded hills east of the Bitterroot Valley. Those mountains are mostly Precambrian sedimentary rocks, along with a scattering of large intrusions of granite similar to that in the Bitterroot Range.

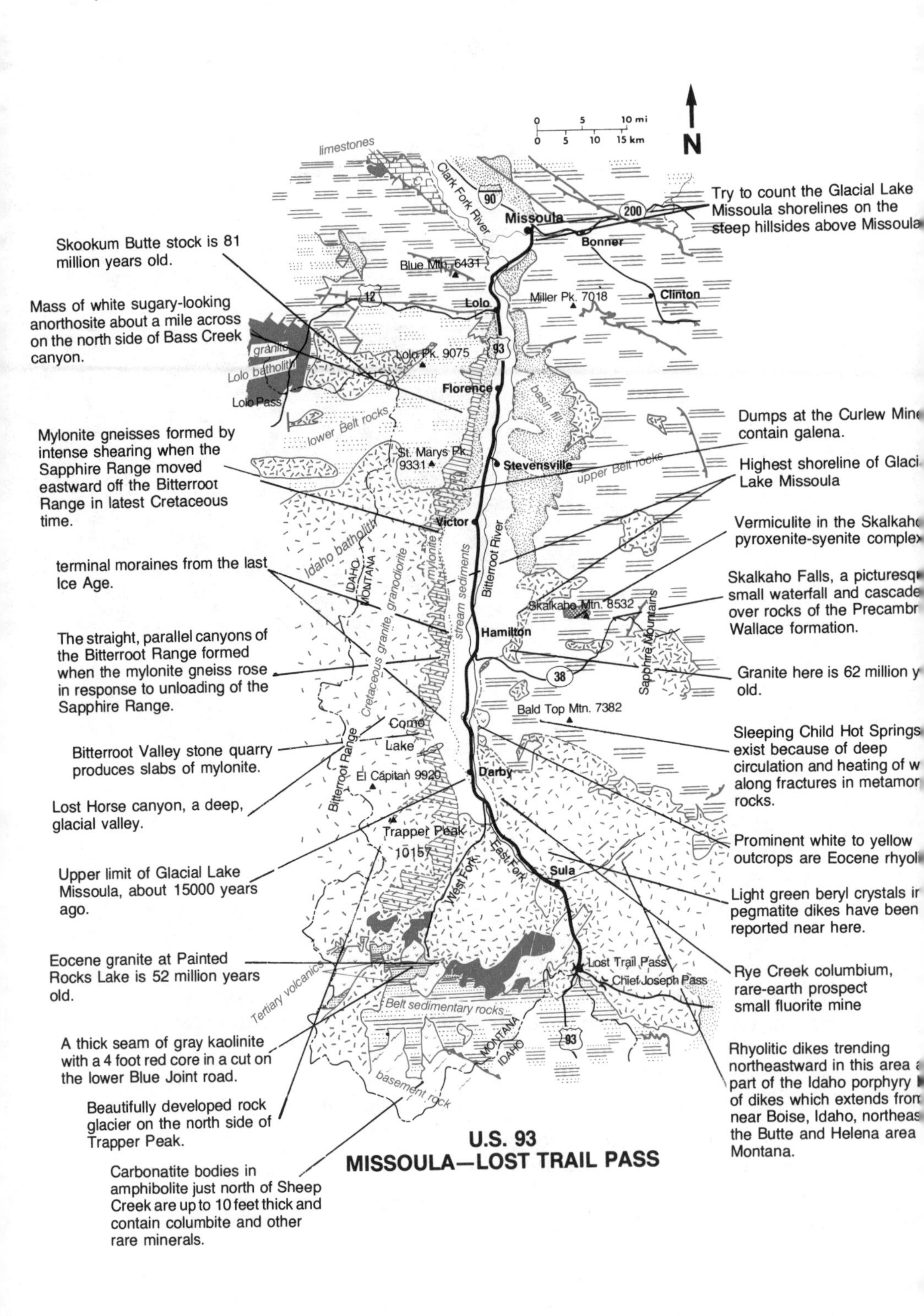

0 5 10 mi
0 5 10 15 km
N
limestones
Clark Fork River
90
Missoula
200
Bonner
Clinton
Blue Mtn. 6431
Miller Pk. 7018
12
Lolo
93
Lolo Pk. 9075
granite
Lolo batholith
Lolo Pass
Florence
basin fill
lower Belt rocks
St. Marys Pk. 9331
Stevensville
upper Belt rocks
Victor
Idaho batholith
IDAHO
MONTANA
Cretaceous granite, granodiorite
mylonite
stream sediments
Bitterroot River
Skalkaho Mtn. 8532
Sapphire Mountains
Hamilton
38
Bald Top Mtn. 7382
Como Lake
Bitterroot Range
El Capitan 9920
Darby
Trapper Peak 10157
West Fork
East Fork
Sula
Lost Trail Pass
Chief Joseph Pass
Tertiary volcanics
Belt sedimentary rocks
MONTANA
IDAHO
93
basement rock
Try to count the Glacial Lake Missoula shorelines on the steep hillsides above Missoula
Skookum Butte stock is 81 million years old.
Mass of white sugary-looking anorthosite about a mile across on the north side of Bass Creek canyon.
Mylonite gneisses formed by intense shearing when the Sapphire Range moved eastward off the Bitterroot Range in latest Cretaceous time.
Dumps at the Curlew Mine contain galena.
Highest shoreline of Glacial Lake Missoula
Vermiculite in the Skalkaho pyroxenite-syenite complex
terminal moraines from the last Ice Age.
Skalkaho Falls, a picturesque small waterfall and cascade over rocks of the Precambrian Wallace formation.
The straight, parallel canyons of the Bitterroot Range formed when the mylonite gneiss rose in response to unloading of the Sapphire Range.
Granite here is 62 million years old.
Sleeping Child Hot Springs exist because of deep circulation and heating of water along fractures in metamorphic rocks.
Bitterroot Valley stone quarry produces slabs of mylonite.
Lost Horse canyon, a deep, glacial valley.
Prominent white to yellow outcrops are Eocene rhyolite
Upper limit of Glacial Lake Missoula, about 15000 years ago.
Light green beryl crystals in pegmatite dikes have been reported near here.
Eocene granite at Painted Rocks Lake is 52 million years old.
Rye Creek columbium, rare-earth prospect small fluorite mine
A thick seam of gray kaolinite with a 4 foot red core in a cut on the lower Blue Joint road.
Rhyolitic dikes trending northeastward in this area are part of the Idaho porphyry belt of dikes which extends from near Boise, Idaho, northeast to the Butte and Helena area in Montana.
Beautifully developed rock glacier on the north side of Trapper Peak.
Carbonatite bodies in amphibolite just north of Sheep Creek are up to 10 feet thick and contain columbite and other rare minerals.

U.S. 93
MISSOULA—LOST TRAIL PASS

A fill of volcanic rocks, mostly rhyolite erupted about 50 million years ago, narrows the valley from the area north of Darby south to the pass. That rhyolite, along with some granite that probably dates from the same episode of igneous activity, forms most of the bedrock between Darby and Lost Trail Pass.

The Idaho Batholith

The granite of the Bitterroot Range extends west through more than 10,000 square miles of central Idaho. Geologists call it the Idaho batholith. The side road west into the spectacular glaciated valley of Lost Horse Canyon south of Darby provides good views of the granite of the Idaho batholith, which is mostly inaccessible despite its great expanse.

Most of the granite, including that in Montana, rose into the crust as molten magma during late Cretaceous time, between about 70 and 90 million years ago. In the northern end of the Bitterroot Range, that magma intruded Precambrian basement rocks to form complex assemblage of granite, gneiss, schist, and other rocks. The age of those basement rocks is uncertain. Some age dates suggest it may be 1.7 billion years old. Most other basement rock in western Montana is about a billion years older.

The main masses of granite magma crystallized about ten miles below the surface. So far as anyone knows, no volcanic rocks of the same age exist anywhere near the Idaho batholith, so it seems that none of the magma erupted. That deep granite might still be well

Cobbles of granite eroded from the Idaho batholith in the bed of Lost Horse Creek, south of Darby.

below the surface had it not been exposed almost as soon as it formed when the overlying slab of the Sapphire block moved east into western Montana.

About 50 million years ago, a second generation of granite magma invaded the older rocks. None of this later magma crystallized until it had risen almost to the surface, and much erupted to form enormous volumes of volcanic rock, including that in the south end of the Bitterroot Valley. Watch the bluffs east of the highway a few miles north of Darby for colorful exposures of crudely layered yellowish volcanic rock, rhyolite.

The Bitterroot Mylonite Zone

Stream canyons cut through the Bitterroot mylonite on the smooth eastern face of the Bitterroot Range.

Almost every vantage point that gives a good view of the Bitterroot Range also reveals that most of its eastern front is a smooth surface that tilts down to the east at an angle close to 25 degrees. A zone of distinctly platy and streaky looking rock more than a thousand feet thick called the Bitterroot mylonite lies beneath the smooth range front. Mylonite is a distinctive type of rock that forms where rocks deep below the surface are strongly sheared, typically in a deep fault zone. The easiest way to see the Bitterroot mylonite at close range is to drive to the mouth of any of the canyons south of Stevensville. Look for platy rock with a strong lineation on the surfaces of the slabs.

As the cross section shows, the Bitterroot mylonite zone is most strongly developed at the front of the Bitterroot Range, and fades at a depth of several thousand feet. It also wraps over the top of the range, and deep well cores show that it extends east beneath the Bitterroot Valley. The smooth surface on the lower part of the range front is so well preserved because it was buried for many millions of years beneath valley fill sediments of the Renova formation, now eroded off.

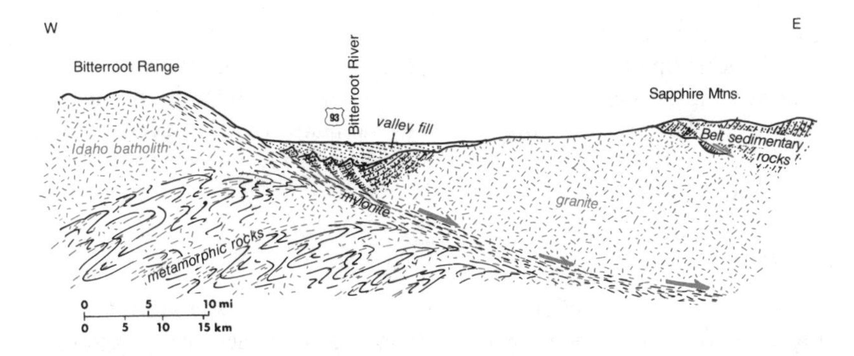

Section drawn across the line of the highway about midway between Victor and Hamilton. The metamorphic rocks are gneiss and schist that may be basement rock, or metamorphosed Belt rock, or both.

Close study of the unusual platy structure and complex texture in the Bitterroot mylonite shows that it developed through shearing as the rock above moved east over that below. Here, apparently, is the fault zone that carried the Sapphire Range east over the rocks beneath it. The distinct lineation on the slabby surfaces invariably points just south of due east—the direction the Sapphire block moved.

The west front of the Sapphire Range that faces the Bitterroot Valley is the trailing edge of the Sapphire block. Granite intrusions in those mountains probably formed as some of the magma that the block was sliding on rose into the rocks above. Like the older granite of this part of the Idaho batholith, those intrusions are about 70 to 75 million years old. That similarity in age probably means that the Sapphire block moved while some of the granite was still partially molten, still weak enough to shear easily, and to form the mylonite zone along the range front.

The Skalkaho Stock

The top of Skalkaho Peak, in the Sapphire Range about 10 miles directly east of Hamilton, contains an extraordinary igneous intrusion almost identical to the Rainy Creek complex near Libby, and to the much more famous igneous complex at Magnet Cove, in Arkansas. The lower part of the Skalkaho intrusion consists primarily of a wild assortment of strange pyroxenites, black rocks composed mostly of pyroxene. The white upper part of the intrusion is syenite, a rock composed almost entirely of feldspar. The original magma must have melted far below the earth's crust.

Besides the pyroxenites, the lower part of the Skalkaho intrusion also contains great volumes of rock composed mostly of giant crystals of black biotite mica. Much of the biotite is altered to vermiculite, the mica that puffs up like popcorn when strongly heated. Unfortunately, several attempts to mine vermiculite in the Skalkaho intrusion have done poorly. That is surprising because the rocks are almost identical to those in the Rainy Creek intrusion near Libby, which produces large amounts of vermiculite.

Glaciation

The serrated skyline of the Bitterroot Range speaks eloquently of heavy glaciation. Every jagged peak is a glacial sculpture, every valley an ice-gouged trough. Most of the glaciers descended to an elevation near 4000 feet before they finally got into a climate warm enough to melt the ice as fast as it advanced, and so stabilize the ice front. South of Hamilton, the glaciers emerged from the mountains onto the floor of the Bitterroot Valley before they reached that elevation. They left large moraines, the hummocky expanses of landscape between the road and the mountain front. North of Hamilton, the floor of the Bitterroot Valley is lower, and the glaciers melted before they could emerge from the mountains. In that area, a large alluvial fan of glacial outwash spreads down toward the road from the mouth of every canyon in the Bitterroot Range.

An Unruly Stream

Between Hamilton and Stevensville, the highway passes many sloughs, winding ponds that fill abandoned courses of the Bitterroot River. In that middle part of the Bitterroot Valley, the river deposits enormous quantities of sediment that choke the channel, forcing the stream to form a new channel. Then the river continues to deposit gravel that soon fills the new channel, forcing itself into yet another new channel. Canoeists and rafters find that they have to learn a new route down the river almost every year.

Such extreme channel instability is typical of streams that carry enormous loads of sediment. That doesn't seem to fit in this case because most of the Bitterroot watershed is forested or well tended farmland that is not eroding rapidly. There is no evidence that the generally clear river is receiving an impossible burden of sediment. This is quite a different problem.

Evidently, a large section of the Bitterroot Valley is dropping between two actively moving faults. Both trend northeast. One fault angles across the valley at Stevensville, the other just south of Hamilton. The northern fault moved during a strong earthquake in 1892 that displaced the ground surface near the Curlew Mine just west of Victor. The southern fault offsets a glacial moraine near Como

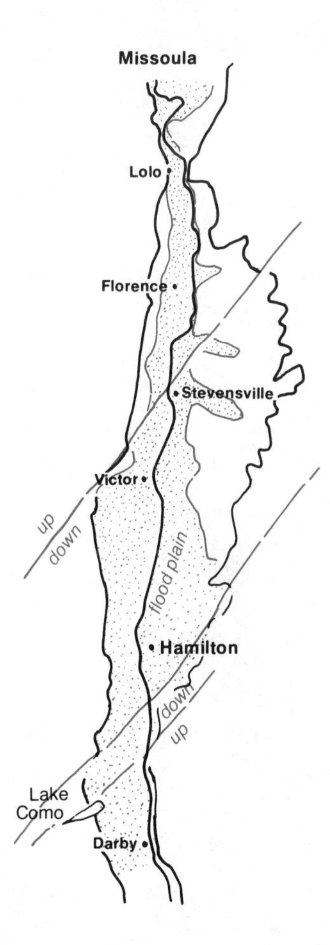

As the block between the active faults drops, the Bitterroot River fills that section of the valley floor with sediments. Notice the wide flood plain in the dropping section.

Lake northwest of Darby, so it must have moved considerably since the end of the last ice age. It seems that the Bitterroot River is depositing large quantities of sediment on the floor of the dropping fault block, filling the depression as fast as it sinks. If the river did not do that, the part of the Bitterroot Valley between Stevensville and Hamilton would be a lake.

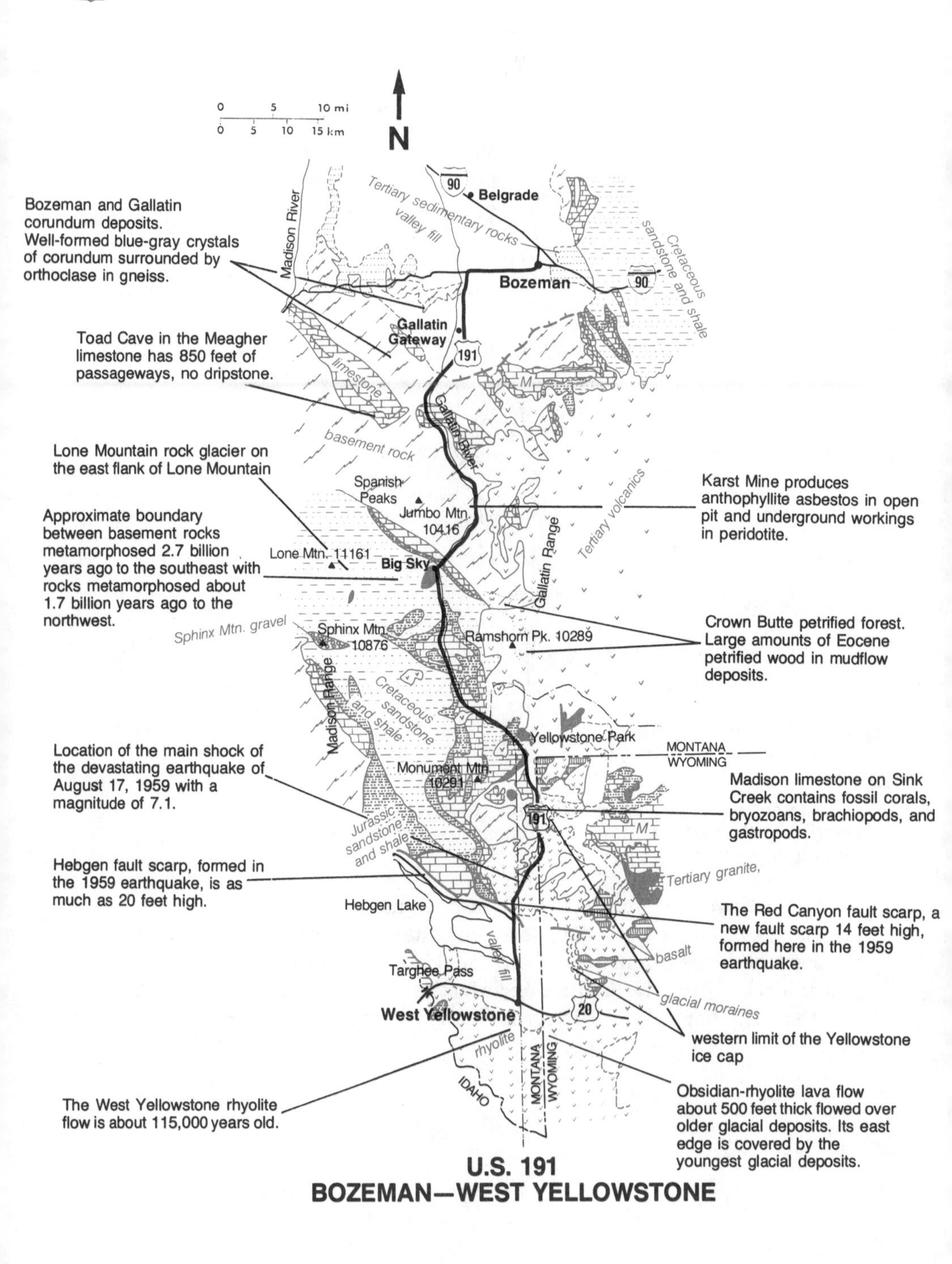

0 5 10 mi
0 5 10 15 km
N
Bozeman and Gallatin corundum deposits. Well-formed blue-gray crystals of corundum surrounded by orthoclase in gneiss.
Toad Cave in the Meagher limestone has 850 feet of passageways, no dripstone.
Lone Mountain rock glacier on the east flank of Lone Mountain
Approximate boundary between basement rocks metamorphosed 2.7 billion years ago to the southeast with rocks metamorphosed about 1.7 billion years ago to the northwest.
Location of the main shock of the devastating earthquake of August 17, 1959 with a magnitude of 7.1.
Hebgen fault scarp, formed in the 1959 earthquake, is as much as 20 feet high.
The West Yellowstone rhyolite flow is about 115,000 years old.
Karst Mine produces anthophyllite asbestos in open pit and underground workings in peridotite.
Crown Butte petrified forest. Large amounts of Eocene petrified wood in mudflow deposits.
Madison limestone on Sink Creek contains fossil corals, bryozoans, brachiopods, and gastropods.
The Red Canyon fault scarp, a new fault scarp 14 feet high, formed here in the 1959 earthquake.
western limit of the Yellowstone ice cap
Obsidian-rhyolite lava flow about 500 feet thick flowed over older glacial deposits. Its east edge is covered by the youngest glacial deposits.
Madison River
Tertiary sedimentary rocks
valley fill
90
Belgrade
Bozeman
Cretaceous sandstone and shale
Gallatin Gateway
191
limestone
M
Gallatin River
basement rock
Spanish Peaks
Jumbo Mtn. 10416
Lone Mtn. 11161
Big Sky
Tertiary volcanics
Gallatin Range
Sphinx Mtn. gravel
Sphinx Mtn. 10876
Ramshorn Pk. 10289
Madison Range
Cretaceous sandstone and shale
Yellowstone Park
MONTANA
WYOMING
Monument Mtn. 10291
Jurassic sandstone and shale
Tertiary granite,
Hebgen Lake
valley fill
basalt
Targhee Pass
glacial moraines
West Yellowstone
20
rhyolite
IDAHO
MONTANA
WYOMING

U.S. 191
BOZEMAN—WEST YELLOWSTONE

US 191: Bozeman—West Yellowstone
93 mi./149 km.

The highway crosses the southern end of the broad Gallatin Valley on deep deposits of valley fill sediments between Bozeman and the area south of Gallatin Gateway, a distance of about 15 or 20 miles.

A few miles south of Gallatin Gateway, the road enters the narrow canyon of the Gallatin River, which it follows almost to its headwaters in Yellowstone National Park. In the southernmost part of the route, the road follows Grayling Creek across the western edge of the Yellowstone Plateau. The wide variety of rocks exposed along the road tell of a complex series of geologic events.

Upper Gallatin Canyon

Mountains west of the river are the Madison Range, a raised block in which the rocks consist of basement rock still extensively covered by folded Paleozoic and Mesozoic sedimentary formations. Active faults along the western and southern edges of the block continue to raise the Madison Range higher.

Mountains east of the river are the Gallatin Range, another raised and actively rising block of the earth's crust. Basement rock is much

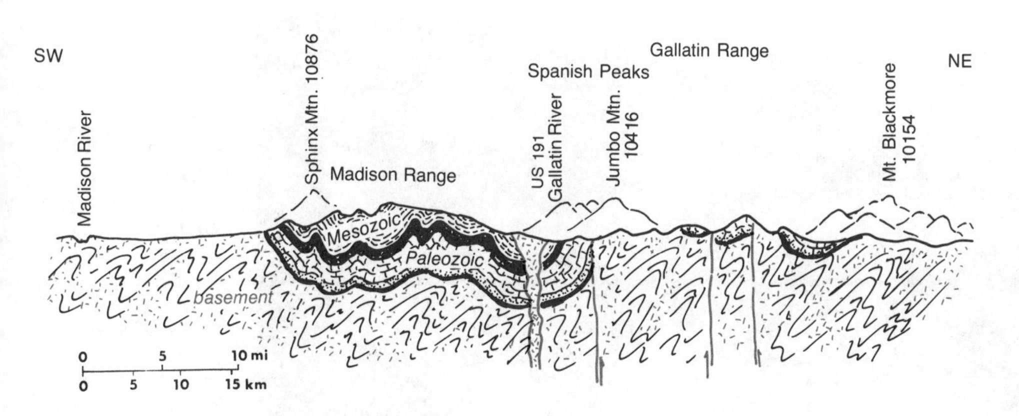

Section across the highway at the junction to the Big Sky resort area

more widely exposed in the Gallatin than in the Madison Range, and Paleozoic and Mesozoic formations are correspondingly less evident. Volcanic rocks erupted from the Absaroka volcano during Eocene time, about 50 million years ago, cover large areas of the Gallatin Range. Most of those rocks are rubbly andesite.

Spanish Peaks

The Spanish Peaks, one of the familiar landmarks of the region, form a tight cluster of craggy mountains that rises distinctly clear of the general profile of the Gallatin Range. They are easily visible from Bozeman, as from many other distant vantage points. The massif is a large block of Precambrian basement rock lifted high along the Spanish Peaks fault.

US 191 traces the eastern margin of the Spanish Peaks along the 20 miles between the mouth of the Gallatin Canyon and the vicinity of the Big Sky resort area, where the road crosses the Spanish Peaks fault. South of the fault, the road crosses sedimentary formations deposited during Paleozoic time. Rocks exposed from the Spanish Peaks fault north to the mouth of Gallatin Canyon are Precambrian basement rocks. Fault movement raised the basement rock block approximately 13,500 feet.

The highway passes especially nice exposures of the basement rock in the north end of the Gallatin Canyon, between its mouth and Spanish Creek Road. Watch for layered outcrops of streaky gneiss and schist. A closer look at the rocks shows that they consist largely of sparkling crystals of pink and white feldspar, easy to recognize by its milky look and tendency to break flat along cleavage surfaces. Black flakes of biotite mica abound, as do dark grains of quartz, and needles of glossy black hornblende.

Mirror Lake in the Spanish Peaks. —U.S. Forest Service photo by W. E. Steuerwald

Petrified tree stumps stand above weathered andesite mudflow deposits in the Gallatin Range.
—U.S. Forest Service photo by K. D. Swan

Absaroka Volcanic Rocks

The Absaroka volcanic center spread enormous volumes of andesite lava across the southern Gallatin Range during Eocene time, about 50 million years ago. The Absaroka andesites closely resemble the Elkhorn Mountains andesites that erupted in the area of the Boulder batholith some 20 million years earlier. So it seems likely that a large granite batholith may exist beneath the Absaroka volcanic pile, even though erosion has not yet bitten deeply enough to expose it.

Volcanic mudflow deposits in the Gallatin Range weathered into a natural arch.
—U.S. Forest Service photo by K. D. Swan

Andesite volcanoes typically produce numerous mudflows, so it is no surprise to find great thicknesses of mudflow deposits full of petrified logs in the southern Gallatin Range. Much of the wood is so well preserved that botanists can identify the kinds of trees; they turn out to be a mixture of tropical and temperate varieties. Evidently the mudflows swept from cool forests at high elevations near the tops of the volcanoes into tropical lowlands near their bases, ripping up the trees along the way.

Yellowstone Plateau

The southern third of the route follows the western edge of the Yellowstone Plateau, the edge of one of the biggest active volcanoes in the world. The center of Yellowstone National Park is an especially violent type of giant volcano called a resurgent caldera. Each major eruption of a resurgent caldera typically produces at least dozens of cubic miles of ash flows and opens an enormous collapse crater called a caldera, which then fills with lava.

The Yellowstone resurgent caldera has gone through at least three major eruptions at intervals of approximately 600,000 years, the last about that long ago. Each eruption produced extremely viscous rhyolite lava heavily charged with red hot steam. The eruptions were massive steam explosions that blasted clouds of finely shredded rhyolite lava into the air. Some of the lava drifted downwind as volcanic ash; most spread across the ground surface in a fast moving cloud of red hot steam and molten rock called an ash flow. When ash flows finally settle to the ground, large parts of them are still hot enough to weld the shreds of ash into solid rock. The pale rhyolites of the Yellowstone Plateau consist largely of such welded ash deposits.

There is every reason to consider the Yellowstone volcano active. Earthquakes that pass beneath the park come out the other side lacking the kind of seismic waves that can not pass through liquids—evidently they encountered some molten magma. Many of those earthquakes indicate that the center of the park is rising, probably evidence that the mass of magma just a few thousand feet below the surface is rising. Furthermore, some of the thermal areas are growing larger and hotter. The thing might erupt.

Whether the Yellowstone resurgent volcano does erupt depends upon what is happening in the hidden depths where the large mass of magma is almost certainly absorbing water. If the burden of rocks above is heavy enough to keep the lid on all that red hot steam, the magma will crystallize quietly into granite batholith. If not, it will explode more violently than anything in recorded human experience. The energy stored in that volume of red hot steam at shallow depth beneath the peak is comparable with that of a major nuclear arsenal. The last eruption of the Yellowstone volcano, for example, blew off a cloud of rhyolite ash that covered a large part of Wyoming to a depth of at least several feet, and blanketed everything all the way to southern Texas.

Glaciation

During the ice ages, the high Yellowstone and Beartooth plateaus collected so much snow that a continuous ice cap covered them both, probably to a depth of several thousand feet. Ice flowed off the plateau through the large valleys that drain it. The Gallatin River valley was full of ice everywhere south of the area just north of Emigrant. Watch for erratic boulders.

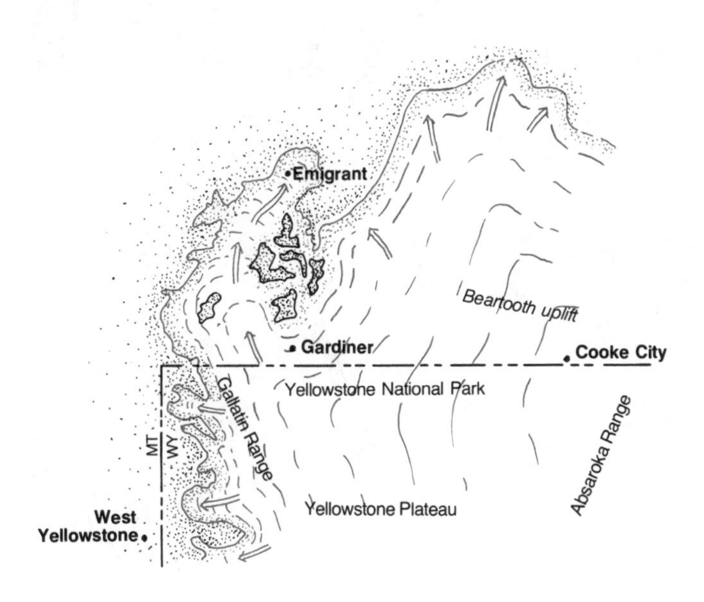

The approximate outline of the ice cap that covered the Yellowstone and Beartooth plateaus

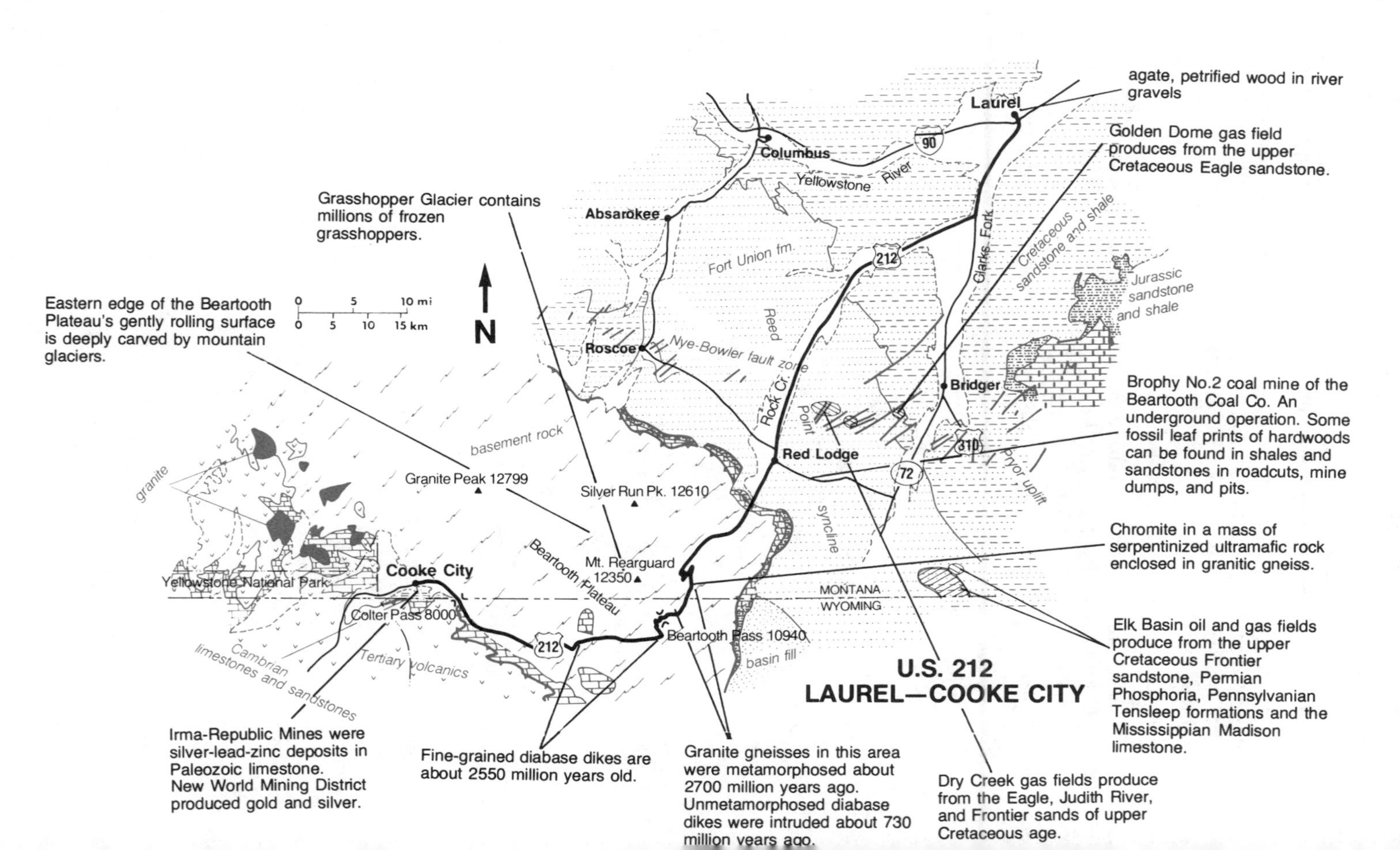
U.S. 212
LAUREL—COOKE CITY
agate, petrified wood in river gravels
Golden Dome gas field produces from the upper Cretaceous Eagle sandstone.
Brophy No.2 coal mine of the Beartooth Coal Co. An underground operation. Some fossil leaf prints of hardwoods can be found in shales and sandstones in roadcuts, mine dumps, and pits.
Chromite in a mass of serpentinized ultramafic rock enclosed in granitic gneiss.
Elk Basin oil and gas fields produce from the upper Cretaceous Frontier sandstone, Permian Phosphoria, Pennsylvanian Tensleep formations and the Mississippian Madison limestone.
Dry Creek gas fields produce from the Eagle, Judith River, and Frontier sands of upper Cretaceous age.
Granite gneisses in this area were metamorphosed about 2700 million years ago. Unmetamorphosed diabase dikes were intruded about 730 million years ago.
Fine-grained diabase dikes are about 2550 million years old.
Irma-Republic Mines were silver-lead-zinc deposits in Paleozoic limestone. New World Mining District produced gold and silver.
Eastern edge of the Beartooth Plateau's gently rolling surface is deeply carved by mountain glaciers.
Grasshopper Glacier contains millions of frozen grasshoppers.
0 5 10 mi
0 5 10 15 km
N
Laurel
Columbus
Absarokee
Roscoe
Bridger
Red Lodge
Cooke City
90
212
310
72
Yellowstone River
Clarks Fork
Cretaceous sandstone and shale
Jurassic sandstone and shale
Fort Union fm.
Reed
Point
Nye-Bowler fault zone
Rock Cr.
Pryor uplift
syncline
MONTANA
WYOMING
basin fill
Beartooth Pass 10940
Mt. Rearguard 12350
Silver Run Pk. 12610
Granite Peak 12799
basement rock
Beartooth Plateau
Colter Pass 8000
Yellowstone National Park
Tertiary volcanics
Cambrian limestones and sandstones
granite

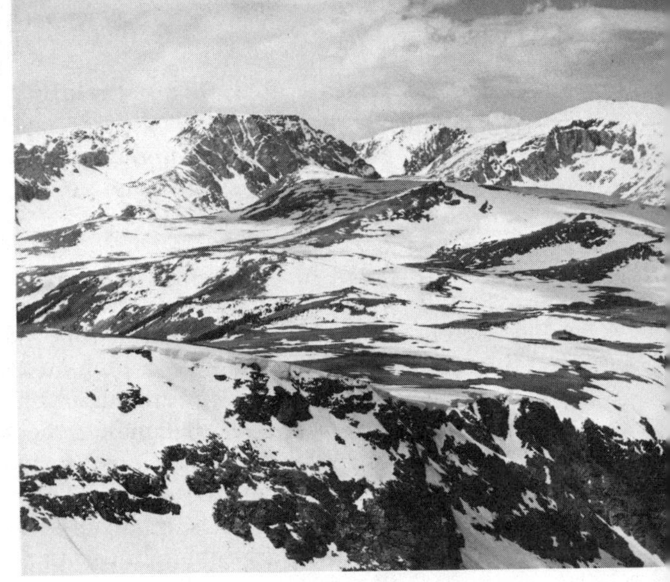

Late June on the Beartooth Plateau.

US 212: Laurel—Cooke City
109 mi./174 km.

The spectacular drive between Red Lodge and Cooke City, at the northeast entrance to Yellowstone National Park, is truly an adventure. The route crosses the high Beartooth Plateau at elevations near and above tree line, a land of alpine tundra, bare rocks, and little lakes that sparkle beneath the deep blue summer sky. Between Red Lodge and Laurel, the highway crosses broadly rolling plains eroded in the Fort Union formation near Red Lodge, in Cretaceous sedimentary rocks farther north.

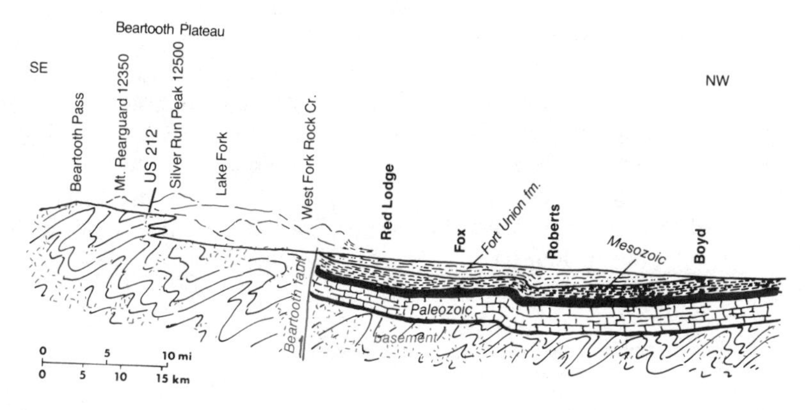

Section along US 212 through Beartooth Pass and Red Lodge

The Beartooth Plateau is a block of Precambrian basement rock lifted high by movement along deep faults. For about three miles southwest of Red Lodge, the highway passes large exposures of pale gray Madison limestone. The beds stand on edge because they drape down the side of the fault block.

Basement Rock

Except for local remnants of Paleozoic sedimentary formations, all the rocks on the high Beartooth Plateau belong to the Precambrian basement. These rocks were metamorphosed sometime around 3.2 billion years ago, a date that makes them the oldest in Montana, among the oldest in the world. Most are gneisses, coarsely granular gray or pink rocks that might look like granite if they were not so streaky. Here and there, the road passes exposures of black gabbro, and there are small areas of many other rock types, including a few valuable enough to mine.

An area on the Hellroaring Plateau north of the highway near Red Lodge contains very poor chromium ores that were mined during the second world war. Chromite, the only mineral source of chromium, occurs in dark rocks of the kind that form the earth's mantle and rarely appear near the surface. The chromite on the Hellroaring Plateau is in a dark greenish rock called peridotite.

Silver and Lead

Mines in the New World District in the hills just southeast of Cooke City have occasionally produced modest amounts of silver and lead from veins near a granitic intrusion that invaded Cambrian sedimentary formations about 50 million years ago.

Four trappers happened across the outcrop of a rich vein while fleeing some Indians in 1869. A few small mines started soon thereafter, but suffered a setback in 1877 when another party of Indians under Chief Joseph helped themselves to lead bullion for bullets. Despite complex ores, which were difficult to smelt, several mines went into more or less steady production in 1882. The silver crash of 1894 closed the mines, but some revived in 1905 and operated off and on until the 1920s. One estimate places the total production of copper, silver, gold, and lead at less than 250,000 dollars. A sidewalk lemonade stand in Cooke City might have done about as well, given as many years to operate.

The Migratory Sedimentary Cover

There is no doubt that sedimentary formations once covered the

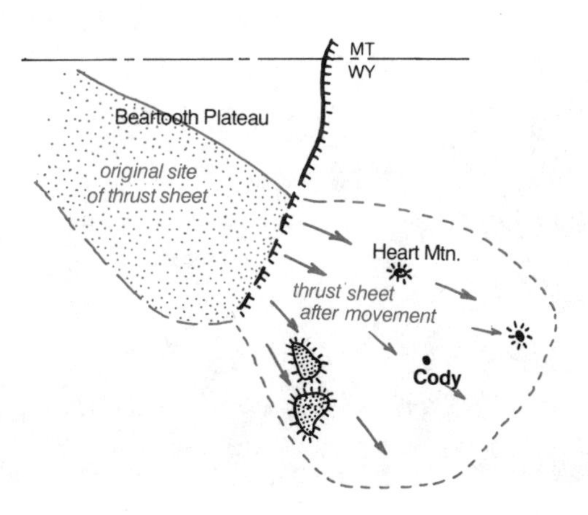

The Heart Mountain overthrust slid south off the Beartooth Plateau into Wyoming.

basement rock in the Beartooth Plateau, just as they still cover the basement in most of the surrounding region. In fact, some of the sedimentary rocks remain in isolated patches that rise above the general plateau surface to form the highest mountains, such as Beartooth Butte. The rest of the sedimentary cover slid off.

The oldest sedimentary rock in this region, the bottom of the stack, is the Flathead sandstone, which accumulated during Cambrian time, perhaps 550 million years ago. The Wolsey shale, another Cambrian formation immediately above the Flathead sandstone, consists largely of clay, and is very weak. Everything above the slippery Wolsey shale slid east down the gentle southward slope of the plateau surface into Wyoming, leaving a few patches of Flathead sandstone on top of the plateau. The displaced rocks are now in the Heart Mountain overthrust slab north of Cody, Wyoming. Eocene volcanic rocks that fill gaps in the overthrust slab show that it was moving while the nearby Absaroka volcanic center was active about 50 million years ago.

The Red Lodge Coal Field

Coal was discovered in the Fort Union formation near Red Lodge about 1882, and mining began by 1885. Production rose sharply after the railroads began using the coal in 1896, again in 1903 when coal fired generators began producing electricity for Red Lodge. More than 2500 miners were working when production was at its peak between 1918 and 1925. The mines went into steep decline in 1924 when big strip mines at Colstrip began supplying the railroads, nearly died during the depression, revived briefly during the second world war, then finally expired.

The Smith Mine in the Bear Creek area near Red Lodge, probably during the 1940s. —Montana Historical Society photo

Uplift of the Beartooth Plateau tilted the Fort Union formation, making its coal seams dip too steeply into the ground to permit strip mining. Underground coal mines can not compete against the large strip mines of eastern Montana, so many decades will probably pass before the Red Lodge coal field revives.

Alpine Tundra

Winter is never more than a few weeks away on the high top of the Beartooth Plateau, never more than a foot or two below the surface. The short summer there melts only the upper levels of the soil, leaving the deeper levels frozen. That kind of climate sponsors its own special kind of erosion.

Permanently frozen subsoil is watertight. And it automatically seals itself because any water that starts to seep through it freezes, plugging the channel it was following. Frozen subsoil traps surface water to make the thawed upper levels of the soil sloppy enough to flow down the slightest slope. Over the years, soil flows off the high places and fills the low places to level the landscape. Large areas of amazingly level alpine tundra on the high parts of the Beartooth Plateau testify to the effectiveness of that process.

A pattern of soil polygons, each about the size of a large room, covers the alpine tundras. They are most easily visible from above, but show up at ground level as lines of rocks that define on a much larger scale a pattern very much like that of shrinkage fractures in suncracked mud. Evidently, the rocks tumble into cracks that open as the ground shrinks in the extreme cold of winter. Year after year, the shrinkage cracks open in the same places as the frozen ground contracts in the deepest cold of winter, and each year more rocks tumble into the open cracks.

Glaciation

During the ice ages, an ice cap that covered most of the Beartooth Plateau drained down the valleys around its margins in enormous glaciers. West of Red Lodge, the highway follows the floor of the deeply glaciated valley of Rock Creek for almost 10 miles, then climbs the valley wall to the top of the plateau in a series of spectacular switchbacks. Numerous small lakes on top of the Beartooth Plateau fill basins the glacier gouged into hard basement rock. Many bedrock surfaces bear sets of parallel scratches, the marks of rocks embedded in the sole of the moving glacier that rasped the hard rock beneath as though they were particles of abrasive in a giant sheet of sandpaper.

View of the glacially gouged and straightened valley of Rock Creek near Red Lodge

287

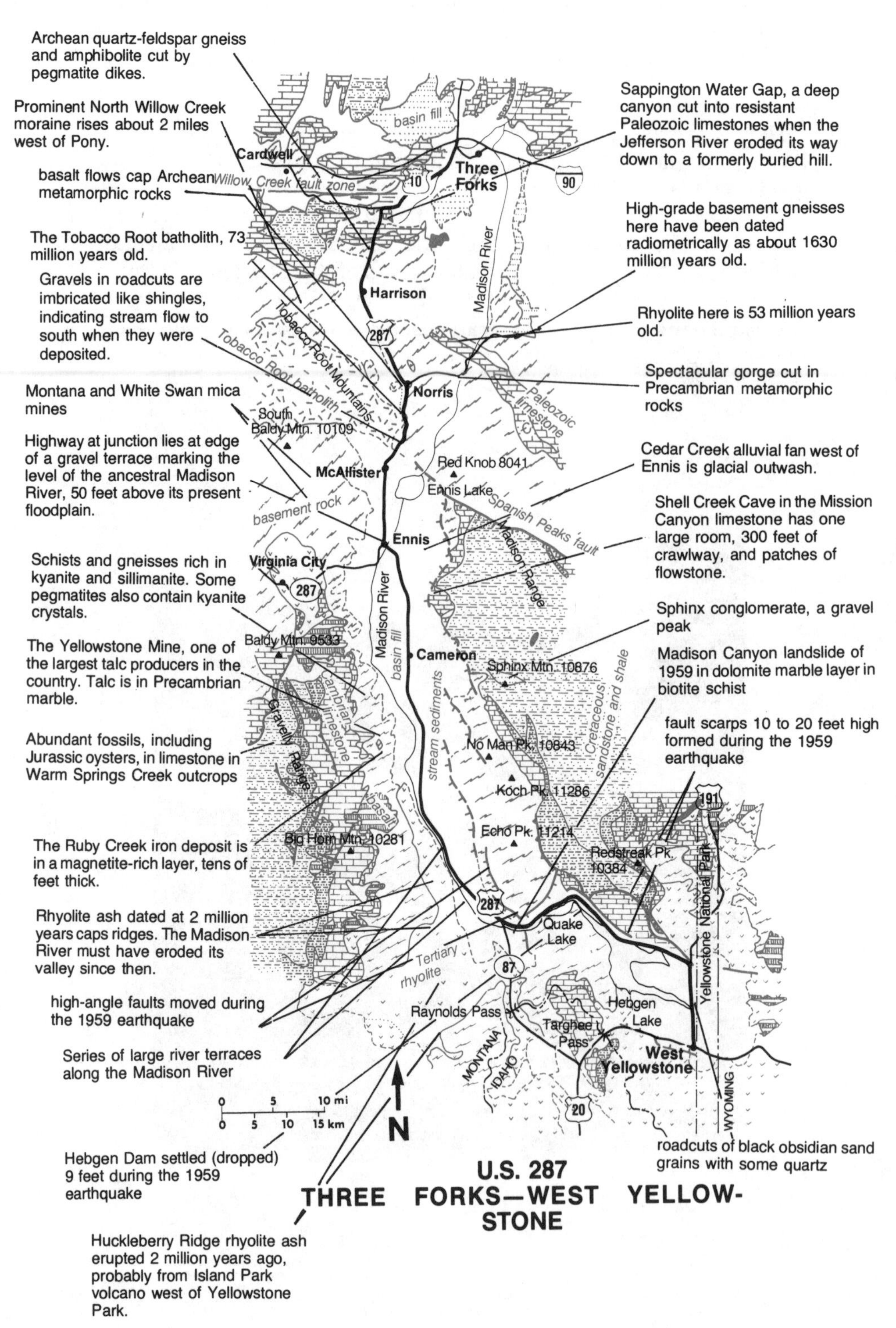
Archean quartz-feldspar gneiss and amphibolite cut by pegmatite dikes.
Prominent North Willow Creek moraine rises about 2 miles west of Pony.
basalt flows cap Archean metamorphic rocks
The Tobacco Root batholith, 73 million years old.
Gravels in roadcuts are imbricated like shingles, indicating stream flow to south when they were deposited.
Montana and White Swan mica mines
Highway at junction lies at edge of a gravel terrace marking the level of the ancestral Madison River, 50 feet above its present floodplain.
Schists and gneisses rich in kyanite and sillimanite. Some pegmatites also contain kyanite crystals.
The Yellowstone Mine, one of the largest talc producers in the country. Talc is in Precambrian marble.
Abundant fossils, including Jurassic oysters, in limestone in Warm Springs Creek outcrops
The Ruby Creek iron deposit is in a magnetite-rich layer, tens of feet thick.
Rhyolite ash dated at 2 million years caps ridges. The Madison River must have eroded its valley since then.
high-angle faults moved during the 1959 earthquake
Series of large river terraces along the Madison River
Hebgen Dam settled (dropped) 9 feet during the 1959 earthquake
Huckleberry Ridge rhyolite ash erupted 2 million years ago, probably from Island Park volcano west of Yellowstone Park.
Sappington Water Gap, a deep canyon cut into resistant Paleozoic limestones when the Jefferson River eroded its way down to a formerly buried hill.
High-grade basement gneisses here have been dated radiometrically as about 1630 million years old.
Rhyolite here is 53 million years old.
Spectacular gorge cut in Precambrian metamorphic rocks
Cedar Creek alluvial fan west of Ennis is glacial outwash.
Shell Creek Cave in the Mission Canyon limestone has one large room, 300 feet of crawlway, and patches of flowstone.
Sphinx conglomerate, a gravel peak
Madison Canyon landslide of 1959 in dolomite marble layer in biotite schist
fault scarps 10 to 20 feet high formed during the 1959 earthquake
roadcuts of black obsidian sand grains with some quartz
basin fill
Cardwell
Willow Creek fault zone
10
Three Forks
90
Harrison
Madison River
287
Tobacco Root Mountains
Tobacco Root batholith
Norris
Paleozoic limestone
South Baldy Mtn. 10109
McAllister
Red Knob 8041
Ennis Lake
Spanish Peaks fault
basement rock
Ennis
Madison Range
Virginia City
Baldy Mtn. 9533
Cameron
Sphinx Mtn. 10876
Cambrian limestone
Gravelly Range
basin fill
stream sediments
Cretaceous sandstone and shale
No Man Pk. 10843
Koch Pk. 11286
191
basalt
Big Horn Mtn. 10281
Echo Pk. 11214
Redstreak Pk. 10384
Yellowstone National Park
Quake Lake
Tertiary rhyolite
87
Raynolds Pass
Hebgen Lake
Targhee Pass
MONTANA
IDAHO
West Yellowstone
WYOMING
20
0 5 10 mi
0 5 10 15 km
N

U.S. 287
THREE FORKS—WEST YELLOWSTONE

US 287: Three Forks—West Yellowstone
117 mi./187 km.

The high and deeply glaciated peaks of the Tobacco Root Mountains parade along a jagged western skyline between US 10 in the Jefferson River Canyon and Ennis. South of Ennis, the much lower Gravelly Range lies west of the road. The Madison Range, another scenic spectacular, rises on the eastern skyline from the area north of Ennis to the Madison River Canyon at the southern end of the route. In the Madison Valley, between the Madison and Gravelly ranges, the road crosses deep deposits of basin fill sediment.

In and just south of the Jefferson River Canyon, the road passes prominent outcrops of Madison limestone. Watch for the great tilted slabs of pale gray rock gleaming among the dark green of the trees. That limestone formed in shallow sea water around 300 million years ago. It contains abundant fossils, most of them small fragments that appear to be the remains of animals that were eaten. Large fossils still in one piece include large corals that suggest honey combs, and smaller corals about the size and shape of a cow's horn.

The Sappington water gap. The Jefferson River began to flow on valley fill sediments, then eroded its way down onto a buried limestone hill and cut a small canyon through it.

Basement Rock

Although the road crosses large areas of basement rock on the east flank of the Tobacco Root Range and again along the Madison River Canyon, good exposures are few. Basement rocks weather too easily under the alkaline soils of areas as dry as southwestern Montana to form many conspicuous outcrops. Watch the roadcuts for pink, gray, and black gneiss—slabby and streaky rocks composed of large crystals that glitter in the sun. Along the east flank of the Tobacco Root Range, the road crosses a considerable expanse of granite, part of the Tobacco Root batholith that invaded the Precambrian basement rock about 70 million years ago.

The Pony Mining District, six miles west of Harrison, produced gold from deposits around the margin of the Tobacco Root batholith. As usual, placer miners started the operation, in this case after a flume brought water to the gulch in 1870. The placer deposits played out in 1875, and underground mines began to produce gold from Precambrian basement rocks. Gold mining continued, with its usual ups and downs, until after the second world war. One mine produced extremely modest quantities of tungsten.

Like most of the early gold districts, Pony mined free milling gold in quartz veins, gold that exists as specks of the native metal disseminated through quartz. A stamp mill pounded the quartz to a powder fine enough to break the particles of gold free so they could be washed out of the pulverized rock. When we visited Pony a few years ago, the remains of the mill with its machinery intact were still moldering in the woods.

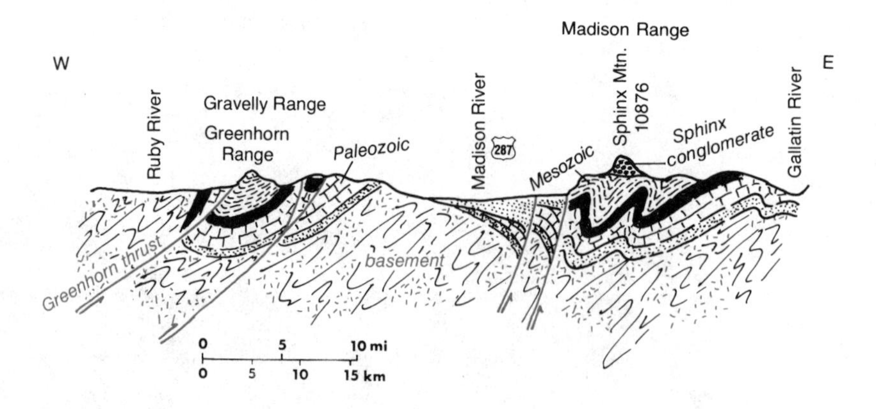

Section across the line of the highway south of Cameron. The cap of conglomerate on the Madison Range must have been laid down before the range rose along faults.

The Gravelly Range

The low Gravelly Range west of the Madison Valley is so named because deposits of coarse pebble conglomerate exist here and there along its crest. Neither the origin nor the age of the conglomerate are well understood. Many geologists think it is probably equivalent to the Sphinx conglomerate on top of the Madison Range and the Beaverhead conglomerate south of Dillon. The low hills along the east side of the range near the highway are Precambrian basement rock, as is the western margin of the range. In between, the Gravelly Range contains virtually all the Paleozoic and Mesozoic formations of southwestern Montana, all tightly folded and faulted.

Basement rock in the western part of Gravelly Range includes marble altered in some places to talc. Like other large talc deposits in southwestern Montana, those in the Gravelly Range contain no asbestos, pose no threat of cancer. The purity of their talc enables the mines to compete despite the high shipping costs to major markets. In fact, southwestern Montana is the world's largest talc producing district as measured in either tonnage or dollar value.

The Madison Range

The Madison Range consists largely of Precambrian basement rock, complex gneisses and schists that formed about 2.7 billion years ago through metamorphism of older rocks. Folded Paleozoic and Mesozoic sedimentary rocks still cover the basement rocks in large parts of the range. The squared jut of Sphinx Mountain, one of the most distinctive landmarks of southwestern Montana, rises on the crest of the Madison Range about ten miles east of Cameron. The reddish rock there is the Sphinx conglomerate, a thick deposit of coarse gravel that appears in no other part of the range.

North of Ennis, the Madison River abruptly leaves the broad Madison Valley to cut through the north end of the Madison Range in a narrow gorge, Beartrap Canyon. A dam there impounds Ennis Reservoir. The present stream course must have been the lowest available path when the river began to flow sometime about two million years ago. Erosion has since removed some of the deposits of basin fill sediment from the valleys, causing the more resistant rocks of the mountains to stand up in bold erosional relief. And the Madison Range has been rising across the path of the river as the fault along its eastern front continues to move, to a thundering accompaniment of occasional earthquakes.

Stream terraces rise from the Madison River like a broad flight of giant steps marching across the Madison Valley. They probably record changes in valley elevation caused by movements of the fault along the front of the Madison Range.

The Madison Canyon Earthquake

In the late night of August 17, 1959, sudden slippage on the Red Canyon fault in Madison Canyon released an earthquake of Richter magnitude 7.1, one of the most violent ever felt in the Rocky Mountains. People living as much as several hundred miles away woke in the night as the seismic waves shook their homes. Every town within 100 miles of Madison Canyon suffered damage to brick and stone structures. Had that earthquake struck a densely populated region, it would have caused a great natural disaster.

In the instant of the earthquake, the Madison Range rose and the canyon floor dropped to increase the vertical relief between mountain

Road broken by fault movement during Madison Canyon earthquake. —U.S. Forest Service photo by W. E. Steuerwald.

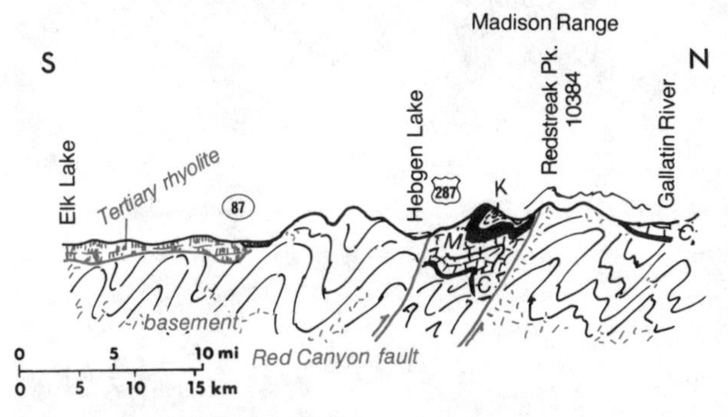

Section drawn across the line of the highway at Hebgen Lake. Sudden movement on the Red Canyon fault caused the 1959 earthquake.

top and stream about 15 feet. Movement on the fault broke the ground surface to create a low cliff, called a fault scarp. Although it has begun to mellow, the scarp is still easily visible. Look for it along the base of the mountains north of the road between Hebgen and Quake lakes. Older and less easily visible fault scarps tell of prehistoric earthquakes similar to the one in 1959.

The Huckleberry Ridge rhyolite ash flow that caps some of the ridges overlooking the Madison River Canyon erupted from the Island Park volcano several miles southwest of West Yellowstone about two million years ago. If the Centennial and Madison valleys had existed then, they would have trapped that ash before it could reach the crest of the Madison Range. So the Madison Range must have risen at least 5000 feet during the last two million years. One repetition of the 1959 earthquake every 6000 years could have done that.

Fault movement raised the south shore of Hebgen Reservoir, leaving former lakeside cabins there far from the shore, and dropped the north side, flooding cabins and parts of the old highway. Meanwhile, strong ground motion set the lake rocking back and forth in giant standing waves called seiches, the kind of waves small children love to start in the bathtub. Every time a wave crest rocked to the west end of the lake, a high tide of water sloshed over the dam, which cracked, but did not fail. Seismic ground waves and small landslides broke the old highway so badly that it required complete rebuilding, in places along a new route.

Rock Slide

Near the west end of the Madison Canyon, an enormous mass of fractured Precambrian basement rock broke off the south canyon wall, spilled down the slope, and across the river. Most of the mass

The Madison Canyon slide. —U.S. Forest Service photo by W. E. Steuerwald

came to rest lying against the north wall of the canyon where it buried a campground, along with an unknown number of people officially estimated at 28. The scar on the south canyon wall will be easily visible for centuries, as will the slide mass. A visitor center on top of the slide contains excellent exhibits.

The slide mass moved with enough momentum to carry its leading edge about 420 feet above river level on the north side of the valley, opposite where it started. Debris left on the canyon floor impounded the Madison River to create Quake Lake. As the new lake began to fill, many people feared that the slide mass might wash out, releasing an enormous flood. Those fears inspired hasty excavation of a spillway through the slide mass to limit and stabilize the level of the new lake. In hindsight, it seems most unlikely that the lake could have moved the massive pile of blocky slide debris in the canyon floor even it it had been allowed to fill to its maximum possible level.

Quake Lake begins to fill behind the Madison Canyon slide.
—U.S. Forest Service photo by W.E. Steuerwald

Slabs of Madison limestone tilted against the east edge of the Beartooth Plateau near Belfry.

US 310: Laurel—Wyoming
51 mi./82 km.

Between Laurel and the Wyoming line, US 310 passes between the Beartooth Plateau and the Pryor Mountains, through the north end of the Bighorn Basin. Except for some Fort Union formation near the Wyoming line, bedrock near the road consists entirely of late Cretaceous sedimentary formations. Thick sandstones in both the Fort Union and the Cretaceous formations make occasional picturesque outcrops.

Outcrops in the soaring Beartooth Plateau west of the highway expose Precambrian basement rock. The plateau is a great block of the earth's crust that rose at least 11,000 feet along a fault that follows its north edge, trending southeast. That fault disturbed the

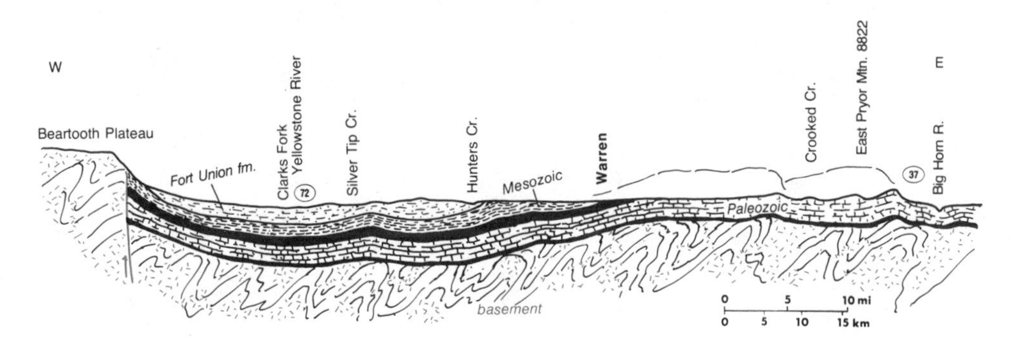

Section from the Beartooth Plateau across the highway near Warren

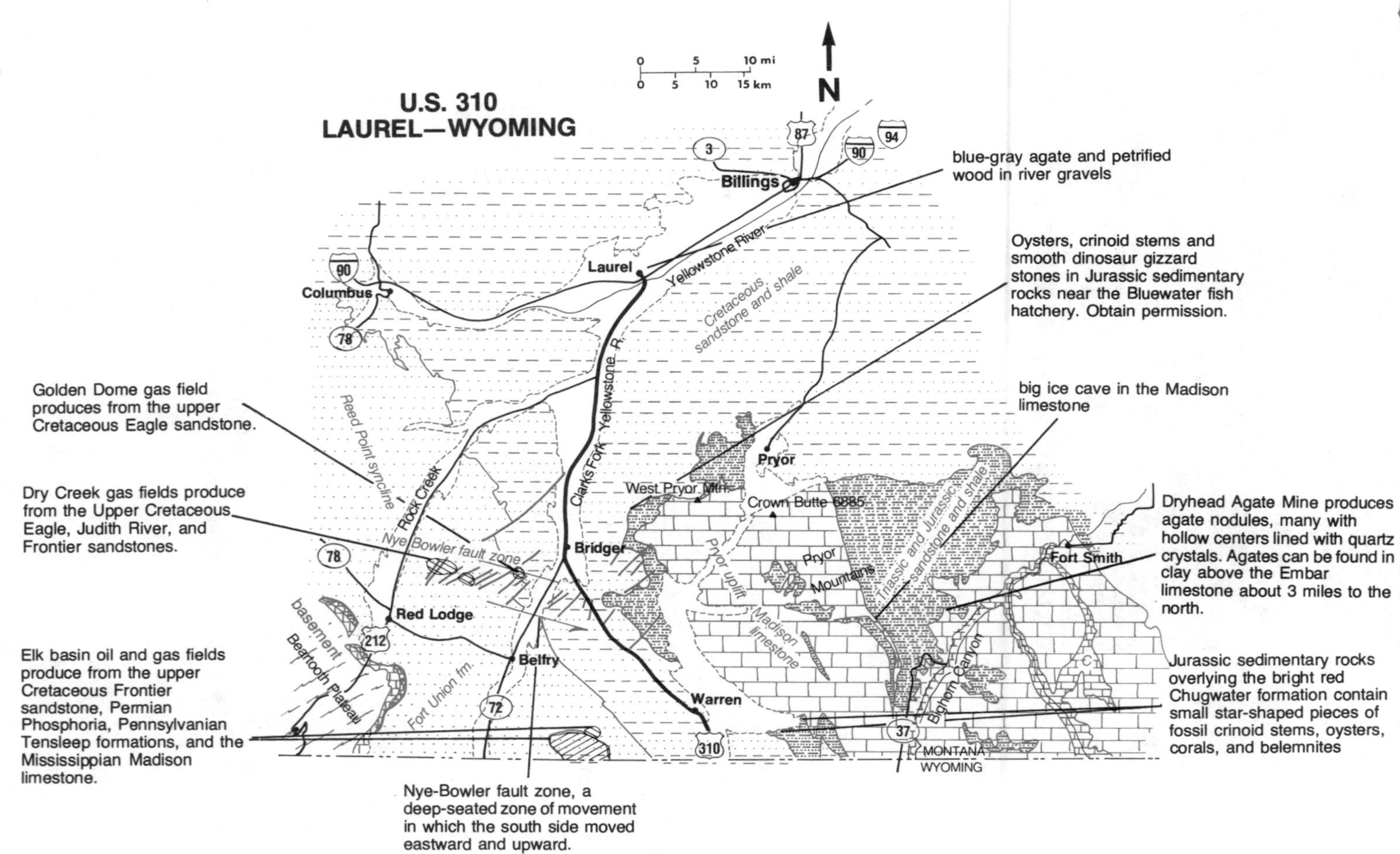
U.S. 310
LAUREL—WYOMING
0 5 10 mi
0 5 10 15 km
N
Billings
Laurel
Columbus
Yellowstone River
Cretaceous sandstone and shale
Clarks Fork Yellowstone R.
Reed Point syncline
Rock Creek
Nye-Bowler fault zone
Red Lodge
basement
Beartooth Plateau
Fort Union fm.
Belfry
Bridger
Warren
Pryor
West Pryor Mtn.
Crown Butte 6885
Pryor uplift
Madison limestone
Pryor Mountains
Triassic and Jurassic sandstone and shale
Fort Smith
Bighorn Canyon
MONTANA
WYOMING
3
87
90
94
78
212
72
310
37
blue-gray agate and petrified wood in river gravels
Oysters, crinoid stems and smooth dinosaur gizzard stones in Jurassic sedimentary rocks near the Bluewater fish hatchery. Obtain permission.
big ice cave in the Madison limestone
Dryhead Agate Mine produces agate nodules, many with hollow centers lined with quartz crystals. Agates can be found in clay above the Embar limestone about 3 miles to the north.
Jurassic sedimentary rocks overlying the bright red Chugwater formation contain small star-shaped pieces of fossil crinoid stems, oysters, corals, and belemnites
Golden Dome gas field produces from the upper Cretaceous Eagle sandstone.
Dry Creek gas fields produce from the Upper Cretaceous Eagle, Judith River, and Frontier sandstones.
Elk basin oil and gas fields produce from the upper Cretaceous Frontier sandstone, Permian Phosphoria, Pennsylvanian Tensleep formations, and the Mississippian Madison limestone.
Nye-Bowler fault zone, a deep-seated zone of movement in which the south side moved eastward and upward.

Fort Union formation, so it must have moved within the last 55 million years. Like so many other events in this part of Montana, that probably happened about 50 million years ago. The low Pryor Mountains in the distance east of the highway rise southward to become the high Bighorn Range in Wyoming, which also rose about 50 million years ago.

Oil in the Northern Bighorn Basin

About ten miles south of Bridger, a conspicuous arch of red rock appears in the low hills east of the highway. The red rock is the Chugwater formation, mudstone deposited during Triassic time, about 180 million years ago. Bright red mudstones and sandstones are so typical of sedimentary rocks formed then that many geologists utter the phrase "Triassic red beds," almost as though it were all one word. The arch is Red Dome, an anticline, exactly the kind of structure that should contain oil if any is around. But a well drilled there many years ago found only water. Wells drilled in other such folds in this northern end of the Bighorn Basin got better results.

The Golden Dome oil field, west of the road about seven miles south of Bridger, was discovered in 1953, after wildcat holes drilled in 1919 and 1932 failed to find oil. At the end of 1983, the field contained four gas wells and two oil wells, none in production. The much larger Elk Basin field straddles the state line a few miles west of the highway. It was discovered in 1915 after members of the U.S. Geological Survey found that the basin in the landscape was eroded into the top of a dome folded in the rocks—a fairly common thing. At the end of 1983, the Elk Basin field contained 67 wells producing oil and gas from several different formations. It had produced a total of almost 79 million barrels of oil.

Pryor Mountains

Rocks exposed in the Pryor Mountains are Paleozoic sedimentary formations, mostly Madison limestone except where streams have cut their canyons into older rocks beneath. Precambrian basement rock forms much of the Wyoming part of the range. The entire range appears to be a big slice of basement and younger sedimentary rocks that was shoved east, probably about 50 million years ago.

Several large ice caves in the Madison limestone northeast of Warren attract large numbers of visitors. The caves that fill with ice open vertically to the surface so air can enter only by sinking, which happens only when the air outside is colder, and therefore denser, than that inside. So air sinks into the cave on very cold days, and the

Devil Canyon, a tributary of the Bighorn Canyon. Rocks exposed in the upper canyon walls are Madison limestone.

inside stays cold all year because rock is an excellent insulator. If something, such as development of better visitor access, improves ventilation, the cave warms up, and the ice melts. The Forest Service regulates visiting hours to the caves in the Pryor Mountains to conserve the ice.

Dinosaurs

During the 1960s, a group of extremely interesting dinosaurs was found in early Cretaceous rocks a few miles southeast of Bridger near Red Dome. They were small animals, for dinosaurs, that were built somewhat along the general lines of oversized ostriches, and ran on two feet. They must have been meat eaters because the middle toe nail on each foot was elongated into a horrible long knife obviously designed to deliver slashing blows to their victims. These animals must have been extremely agile and quick to stand on one leg while slashing with the other. Many students of the subject interpret these fossils as evidence that at least some dinosaurs were warm-blooded animals, not at all like sluggish reptiles.

Montana 41 and 55: Dillon—Whitehall

54 mi./86 km.

The route threads a path through the valleys between mountain ranges, without actually crossing any mountains, or very many bedrock outcrops. Along almost the entire route, the road crosses deep deposits of basin fill sediments, mostly the silty Renova formation, laid down during the long dry periods of Tertiary time. Watch for them in roadcuts a few miles north of Dillon.

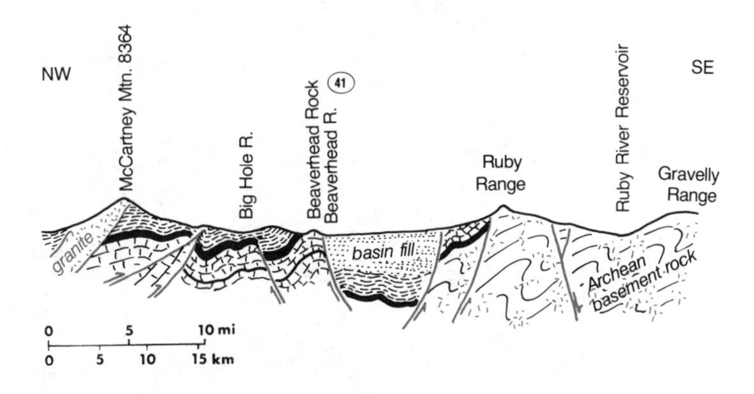

Section across Montana 41 about midway between Dillon and Twin Bridges

Ruby Range

Mountains east of the road between Twin Bridges and Dillon are the Ruby Range. Paleozoic sedimentary formations, largely Madison limestone, form the high, forested peaks in the northern end of the range. Precambrian basement rock is exposed in the lower and less forested southern part of the range.

A big talc mine produces from Precambrian rocks near the head of Stone Creek about ten miles north of Dillon, another from the same rocks just over the ridge. Heavy trucks haul talc from the mine on the west side of the ridge down the highway to the large mill beside Interstate 15 about ten miles south of Dillon. The other mine ships its talc by rail from Alder to mills in Nebraska and Belgium.

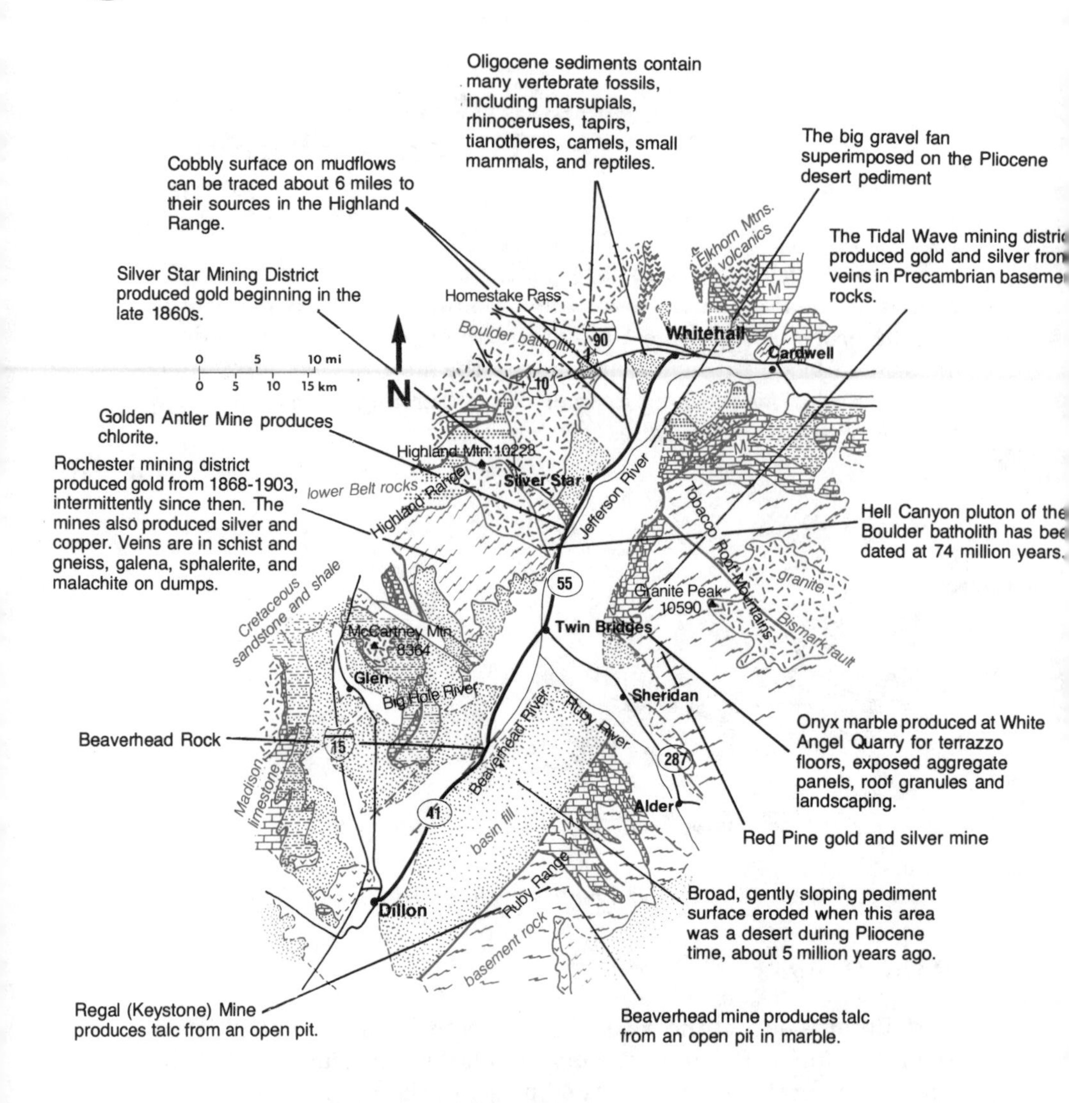
Oligocene sediments contain many vertebrate fossils, including marsupials, rhinoceruses, tapirs, tianotheres, camels, small mammals, and reptiles.
The big gravel fan superimposed on the Pliocene desert pediment
Cobbly surface on mudflows can be traced about 6 miles to their sources in the Highland Range.
The Tidal Wave mining distri produced gold and silver fro veins in Precambrian baseme rocks.
Silver Star Mining District produced gold beginning in the late 1860s.
Homestake Pass
Elkhorn Mtns. volcanics
Boulder batholith
Whitehall
Cardwell
90
10
0 5 10 mi
0 5 10 15 km
N
Golden Antler Mine produces chlorite.
Highland Mtn. 10228
Rochester mining district produced gold from 1868-1903, intermittently since then. The mines also produced silver and copper. Veins are in schist and gneiss, galena, sphalerite, and malachite on dumps.
lower Belt rocks
Highland Range
Silver Star
Jefferson River
Tobacco Root Mountains
Hell Canyon pluton of the Boulder batholith has bee dated at 74 million years.
Cretaceous sandstone and shale
55
granite
Granite Peak 10590
Bismark fault
Twin Bridges
McCartney Mtn. 8364
Glen
Big Hole River
Sheridan
Beaverhead Rock
15
Beaverhead River
Ruby River
Onyx marble produced at White Angel Quarry for terrazzo floors, exposed aggregate panels, roof granules and landscaping.
287
Madison limestone
41
basin fill
Alder
M
Red Pine gold and silver mine
Ruby Range
Dillon
basement rock
Broad, gently sloping pediment surface eroded when this area was a desert during Pliocene time, about 5 million years ago.
Regal (Keystone) Mine produces talc from an open pit.
Beaverhead mine produces talc from an open pit in marble.

Montana 41 and 55
DILLON—WHITEHALL

View from the air of the two big talc mines on the crest of the Ruby Range.
—Aero Tech Surveys photo

Tobacco Root Range

The craggy wall of the Tobacco Root Mountains rises east of the road between Whitehall and Twin Bridges. The range consists mostly of Precambrian basement rocks with a core of granite emplaced about 70 million years ago, at the same time the Boulder batholith was forming. Thick sections of Paleozoic and Mesozoic sedimentary rocks drape the flanks of the range. Those are the pale rocks exposed in great slabs that lie steeply tilted in the slopes facing the highway.

Prospectors of the 1860s found gold in the west flank of the Tobacco Root Mountains almost directly east of Twin Bridges, the Tidal Wave District. Mining began almost immediately, but with little success because the veins do not contain free gold. Later efforts to mine silver and lead were slightly more successful, and continued until about 1914. Over the years, the district probably produced something between one and two million dollars worth of metal, very little profit.

View east from about eight miles north of Twin Bridges across the Jefferson Valley to the Tobacco Root Mountains. The broad surface that slopes gently down from the mountains to the Jefferson River is the Parrott Bench.
—Montana Bureau of Mines and Geology photo by H. L. James

All along the west flank of the Tobacco Root Range, a spectacular surface called the Parrott Bench slopes gently westward toward the valley. Most of the Parrott Bench appears to be an old desert erosion surface developed during Pliocene time, when the region had an extremely arid climate. Much younger alluvial fans of glacial outwash spread across the bench from the mouths of the larger canyons that head high in the range where large ice-age glaciers gouged the valleys.

Highland Range

Mountains west of the road between Whitehall and Twin Bridges are the Highland Range. The hills visible from this highway consist mostly of Precambrian basement rock in the southern part of the range, 70 million-year-old granite of the Boulder batholith in the northern part. Some tightly folded Paleozoic and Mesozoic sedimentary formations appear in pale roadcuts near Silver Star.

Silver Star is near the southernmost reach of the Boulder batholith. Prospectors discovered gold west of town during the 1860s. By the 1870s, Silver Star had become one of the largest towns in the territory, a roaring mining camp producing gold, silver, and lead from ore bodies in the contact zones where the intruding granitic magma reacted with limestones. The district died about 1915, after having produced between two and three million dollars worth of ore.

Mines near Rochester, in the Highland Range about ten miles northwest of Twin Bridges, began to produce gold from bedrock veins about 1867. The town had a population of 800 by 1869, then foundered under a curse of too little and too much water. Lack of surface water foiled efforts to mine placer gold, and the underground mines went out of business as they flooded upon reaching the water table. When pumps were installed to drain the mines, all the wells in town went dry, and people had to haul their drinking water. Evidently the rocks in this area are full of fractures that permit ground water to flow freely.

Point of Rocks

A roadside historical marker about midway between Twin Bridges and Dillon identifies the large hill across the Beaverhead River west of the road as Beaverhead Rock. According to Meriwether Lewis, this was the first landmark the Indian woman Sacajawea recognized as the expedition entered the land of her people. According to William Clark, the real Beaverhead Rock is 27 miles farther south, beside Interstate 15 south of Dillon. Most of the local people agree with Clark, and call this hill "Point of Rocks."

Point of Rocks, or is it Beaverhead Rock?

Whatever its historical significance, there is no doubt that the cliff facing across the Beaverhead River toward the highway is Madison limestone, tilted gently down to the west. People who walk the crest of the hill westward find themselves crossing progressively younger sedimentary formations. It seems likely that the entire section was shoved east to where we see it along a large thrust fault, which must remain hypothetical because it is not exposed.

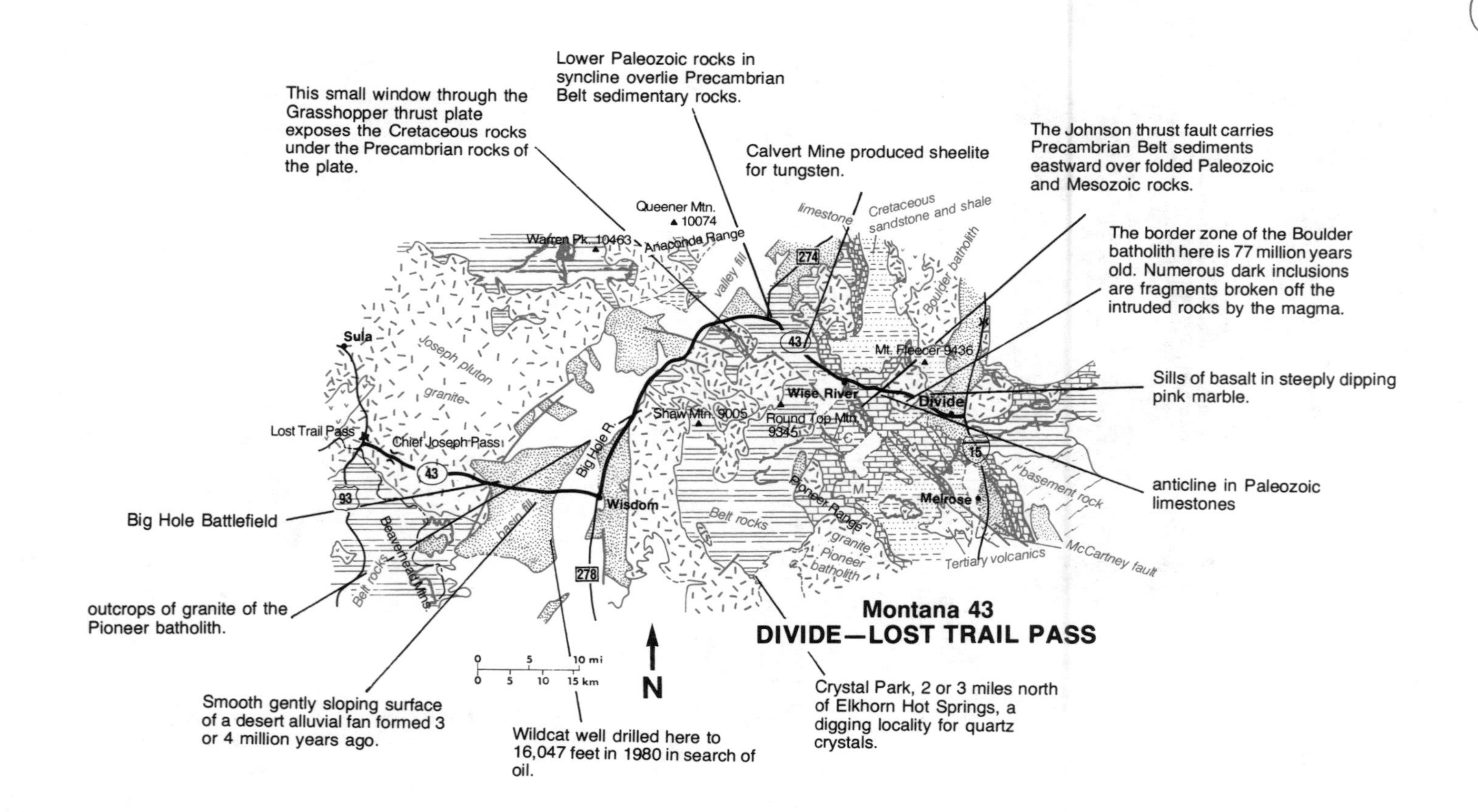
Montana 43
DIVIDE—LOST TRAIL PASS
This small window through the Grasshopper thrust plate exposes the Cretaceous rocks under the Precambrian rocks of the plate.
Lower Paleozoic rocks in syncline overlie Precambrian Belt sedimentary rocks.
Calvert Mine produced sheelite for tungsten.
The Johnson thrust fault carries Precambrian Belt sediments eastward over folded Paleozoic and Mesozoic rocks.
The border zone of the Boulder batholith here is 77 million years old. Numerous dark inclusions are fragments broken off the intruded rocks by the magma.
Sills of basalt in steeply dipping pink marble.
anticline in Paleozoic limestones
Big Hole Battlefield
outcrops of granite of the Pioneer batholith.
Smooth gently sloping surface of a desert alluvial fan formed 3 or 4 million years ago.
Wildcat well drilled here to 16,047 feet in 1980 in search of oil.
Crystal Park, 2 or 3 miles north of Elkhorn Hot Springs, a digging locality for quartz crystals.
Queener Mtn. 10074
Warren Pk. 10463
Anaconda Range
limestone
Cretaceous sandstone and shale
Boulder batholith
valley fill
Sula
Joseph pluton
granite
Lost Trail Pass
Chief Joseph Pass
Mt. Fleecer 9436
Wise River
Divide
Shaw Mtn. 9005
Round Top Mtn. 9345
Big Hole R.
Wisdom
Belt rocks
Pioneer Range
granite
Pioneer batholith
Melrose
basement rock
Tertiary volcanics
McCartney fault
basin fill
Beaverhead Mtns.
Belt rocks
274
43
15
93
278
0 5 10 mi
0 5 10 15 km
N

Montana 43: Divide—Lost Trail Pass
77 mi./123 km.

The road follows the Big Hole River as it traces its course around the north end of the Pioneer Range between the Beaverhead Valley and the Big Hole. Except for a small granite intrusion about five miles west of Divide, another about midway between Wise River and Wisdom, the road crosses folded Paleozoic sedimentary formations between Divide and Wisdom. Bedrock between Wisdom and Lost Trail Pass is valley floor sediment near Wisdom, granite in the hills east of the pass.

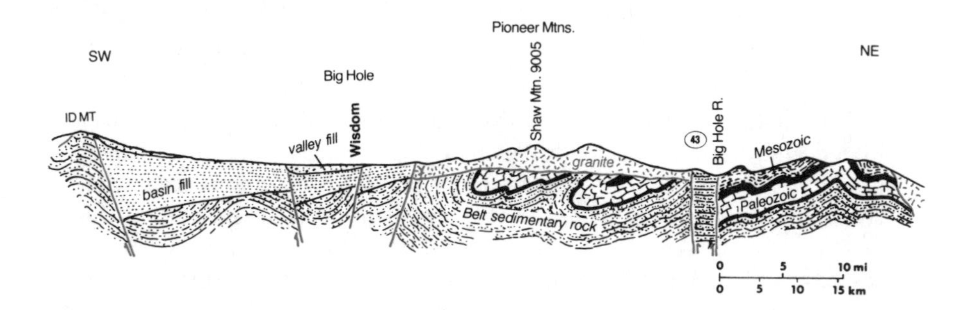

Section south of Montana 43 from Wisdom to Wise River, north of the road from Wise River east

The Big Hole

The Big Hole is the highest and widest of the broad mountain valleys of western Montana. Its spectacular sweep separates the Pioneer Mountains along its eastern margin from the southern Bitterroot Range on the west.

Like the Bitterroot Valley, the Big Hole is probably the gap opened behind a big block of the upper crust that detached from the top of the Idaho batholith and moved east about 70 million years ago. The Pioneer Range along the east side of the Big Hole is that block.

Like all the other structural valleys in the region, the Big Hole contains a deep fill of sediments accumulated during Tertiary time. Surface exposures of those sediments do not differ from those in the other valleys, the usual silty Renova formation capped here and there by younger gravels. But the story at depth is considerably different.

Two very deep wildcat exploratory wells drilled in the Big Hole during the early 1980s failed to find commercial quantities of oil or gas, but did reveal that the Big Hole is as extraordinary at depth as at the surface. The wells went some 14,000 feet through valley fill sediments before they penetrated older bedrock beneath the valley floor. That is far deeper than any other valley in the region. Volcanic rocks filled the bottom of the valley as it opened, and then the rest of it filled with sedimentary deposits, mostly with the Renova formation.

Pioneer Range

The north end of the Pioneer Range consists mostly of granite, the Pioneer batholith. Age dates show that it is about 70 million years old, the same age as both the Boulder and Idaho batholiths. Some geologists have argued that the Pioneer batholith is related to the Boulder batholith, which is only a few miles to the northeast. We prefer to associate it with the Idaho batholith which more closely resembles it in composition.

The eastern flank of the Pioneer Range consists of very tightly folded sedimentary rocks in several great slices that moved east along big thrust faults. The faults, and the folds associated with them, loosely resemble those in the ranges around the margin of the Sapphire block. And the Pioneer batholith, like the granites around the edge of the Sapphire block, appears to have been emplaced along the thrust faults. The resemblances are too numerous and too close to

Chunks of older rock suspended in granite exposed along the Big Hole River.

be a series of coincidences. So we think the Pioneer Range is another big piece of the earth's upper crust that detached from the top of the Idaho batholith and moved east on big thrust faults. If so, then the granite must logically have come from the Idaho batholith.

The western and eastern parts of the Pioneer Range really are distinctly separate. The Wise River flows between them. On a map, the two parts of the range look about like two elliptical slices of a boiled egg, one slightly overlapping the other. It looks as though the range moved in two big pieces, with a fault separating them.

The Southern Bitterroot Range

Between the area near the Big Hole Battlefield and Lost Trail Pass, the road crosses part of the southern Bitterroot Range. Much of the bedrock along this part of the route is granite, an incompletely known mass called the Joseph batholith. It was emplaced about 70 million years ago. This granite may well have crystallized from magma that was smeared eastward from the Idaho batholith in the sole of the Pioneer detachment block.

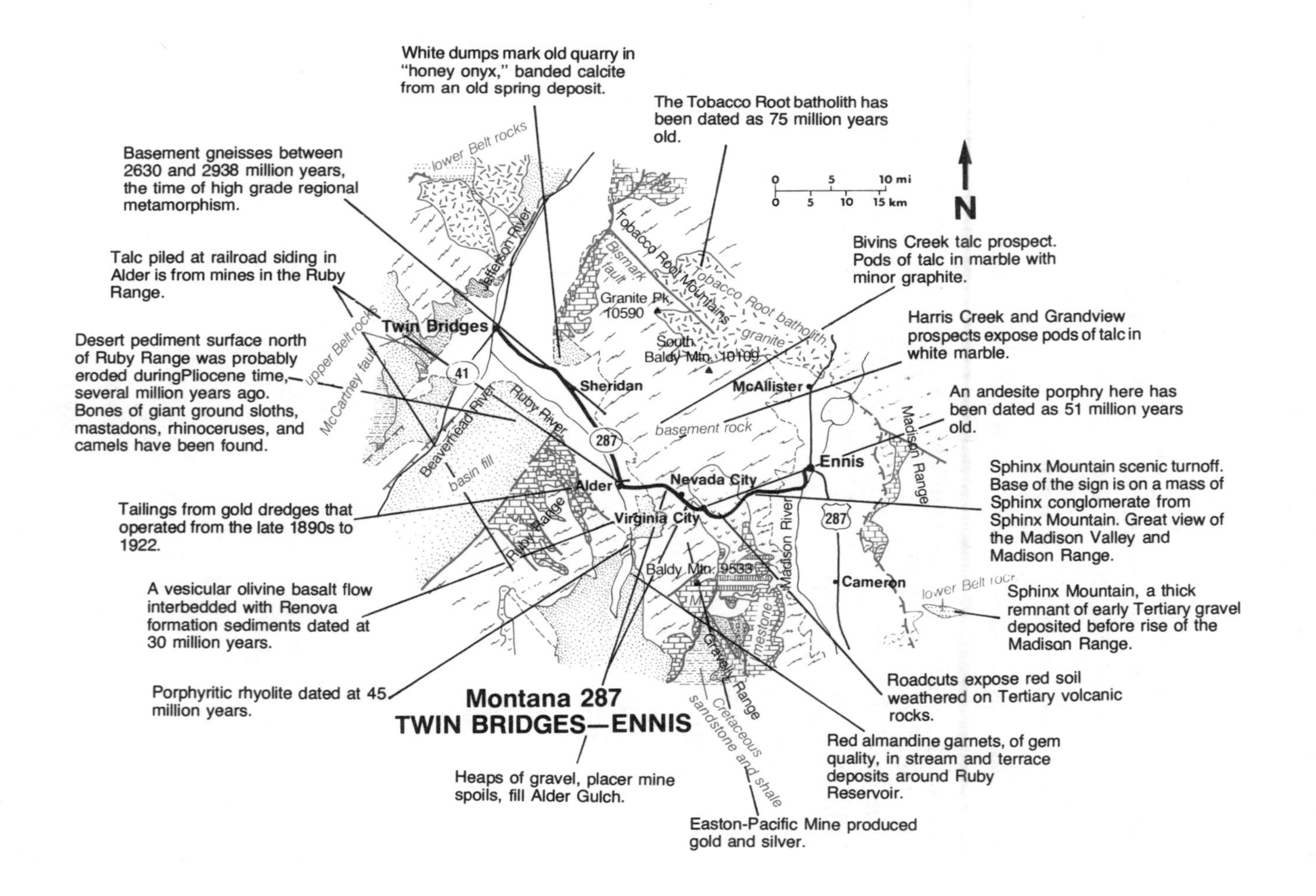
Montana 287
TWIN BRIDGES—ENNIS
White dumps mark old quarry in "honey onyx," banded calcite from an old spring deposit.
The Tobacco Root batholith has been dated as 75 million years old.
Basement gneisses between 2630 and 2938 million years, the time of high grade regional metamorphism.
Talc piled at railroad siding in Alder is from mines in the Ruby Range.
Desert pediment surface north of Ruby Range was probably eroded duringPliocene time, several million years ago. Bones of giant ground sloths, mastadons, rhinoceruses, and camels have been found.
Tailings from gold dredges that operated from the late 1890s to 1922.
A vesicular olivine basalt flow interbedded with Renova formation sediments dated at 30 million years.
Porphyritic rhyolite dated at 45 million years.
Heaps of gravel, placer mine spoils, fill Alder Gulch.
Easton-Pacific Mine produced gold and silver.
Red almandine garnets, of gem quality, in stream and terrace deposits around Ruby Reservoir.
Roadcuts expose red soil weathered on Tertiary volcanic rocks.
Sphinx Mountain, a thick remnant of early Tertiary gravel deposited before rise of the Madison Range.
Sphinx Mountain scenic turnoff. Base of the sign is on a mass of Sphinx conglomerate from Sphinx Mountain. Great view of the Madison Valley and Madison Range.
An andesite porphry here has been dated as 51 million years old.
Harris Creek and Grandview prospects expose pods of talc in white marble.
Bivins Creek talc prospect. Pods of talc in marble with minor graphite.
0 5 10 mi
0 5 10 15 km
N
lower Belt rocks
Jefferson River
Tobacco Root Mountains
Bismark fault
Tobacco Root batholith
granite
Granite Pk. 10590
South Baldy Mtn. 10109
Twin Bridges
41
upper Belt rocks
McCartney fault
Beaverhead River
Ruby River
Sheridan
McAllister
basement rock
287
basin fill
Alder
Nevada City
Virginia City
Ruby Range
Ennis
Madison Range
Madison River
287
Baldy Mtn. 9533
Cameron
Gravelly Range
Cretaceous sandstone and shale

Montana 287: Twin Bridges—Ennis
43 mi./69 km.

The highway between Twin Bridges and Alder follows the Ruby River past the high backdrop of the Tobacco Root Range on the eastern skyline. Mountains west of the road are the high northern third of the Ruby Range where the bedrock is tightly folded Paleozoic sedimentay formations, most conspicuously the Madison limestone.

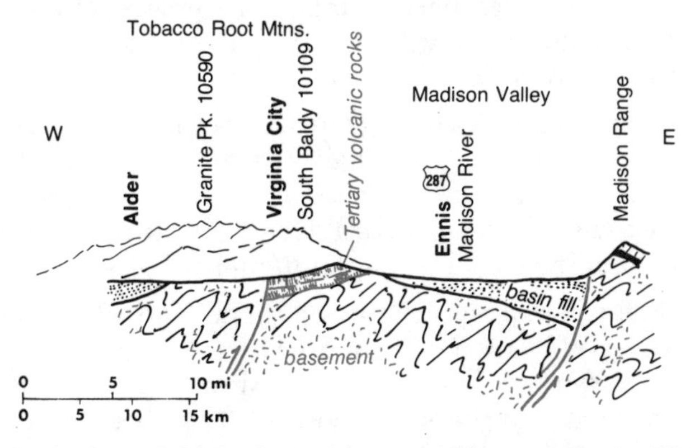

Section along the line of the highway between Alder and Ennis. Blocks of basement rock moved up and down along big faults to form the mountain ranges and valleys.

Alder is at the northern end of the Ruby Valley, which separates the Ruby Mountains on the west from the Gravelly Range on the east. Both ranges are important sources of talc. Large stockpiles of raw talc mined from a bit open pit in basement rock near the crest of the Ruby Range usually stand beside the railroad in Alder.

Between Alder and Virginia City, the road follows Alder Gulch past mile after mile of old gold dredge tailings. Virginia City is in a low saddle that separates the southern end of the Tobacco Root Range from its southern continuation in the Gravelly Range. The road between Virginia City and Ennis crosses the long eastern slope of the Tobacco Root Range with the Madison Range cutting its jagged profile high into the eastern skyline. Bedrock in this area is all Precambrian basement; coarsely crystalline schists, gneisses, and pegma-

287

tites glitter colorfully from the roadcuts.

Placer Mining

By 1864, within a year after prospectors found placer gold in Alder Gulch, miners were washing gravel all the way from Alder to Virginia City. People called the gulch "Fourteen Mile City." Of the several small communities embedded in that early example of urban sprawl, only Virginia City survives. Now, the road follows a deserted wasteland of placer mine tailings and stagnant dredge ponds, all that remains of the flood plain of Alder Creek.

Streams function as natural sluice boxes in which particles of gold settle through the gravel bed, and lodge behind irregularities in the bedrock surface beneath. That is why the richest mines in most gold districts work placer deposits in stream gravels, not the bedrock mother lode. Think of placer mining as a process of cleaning out a large natural sluice box with smaller man-made contraptions that work on the same principle.

Small scale placer miners generally washed gold out of stream gravels with a sluice box, a long wooden trough with cleats nailed crosswise to its floor. Miners shoveled gravel into one end of the sluice box, and washed it through with a stream of water. Gold, being heavy, lodged behind the cleats while the lighter pebbles washed through.

Although a few irregular piles of hand worked gravels remain, most of the spoil heaps are in windrows left by big steam and electric dredges that worked these gravels between the late 1890s and 1922. Most of the dredge workers and their families, about 500 people, lived in a nearly vanished town called Ruby, at the mouth of Alder Creek. The dredges recovered some 9 million dollars worth of gold at the cost of virtually the entire floodplain.

Gold dredges float on ponds of their own making as they navigate

Panning for shows of gold, the first step in locating a claim.
—Montana Historical Society photo

through the floodplain by digging the gravel ahead and dumping it in their wake, moving the pond with them as they go. A chain of buckets scoops the gravel into a big sluice box, and the washed pebbles land on a conveyor belt that dumps them behind the dredge. The dumping conveyor swings back and forth to make the little cross ridges on the long tail of gravel that the dredge leaves behind. Most of the fine grained sediment washes into the stream.

It is possible to design a placer mine to trap the fine grained sediment, instead of flushing it down the stream, and to level the spoils, instead of leaving them in ridges. If that is done, the mined stream and flood plain may return to some semblance of their natural productivity. But the old miners had no thought of reclamation. Many thousands of years will pass before natural processes can begin to restore Alder Creek.

Dredge spoils in Alder Gulch.

High Gravel

A roadside sign beside the long slope between Virginia City and Ennis points out the distinctive blocky profiles of Sphinx Mountain and The Helmet, two prominent peaks on the crest of the Madison Range east of Ennis. The upper 3000 feet of those mountains is a coarse pebble conglomerate as enigmatic as the Sphinx itself. It seems likely that the stream gravels on top of Sphinx Mountain may correlate with equally enigmatic gravel conglomerates on the Gravelly Range and the Beaverhead conglomerate in the mountains along Interstate 15 south of Dillon. If so, they were probably laid down in big alluvial fans during late Cretaceous time, perhaps about 70 million years ago. Obviously, the rocks that now lie along the crest of the Madison Range were then on the valley floor.

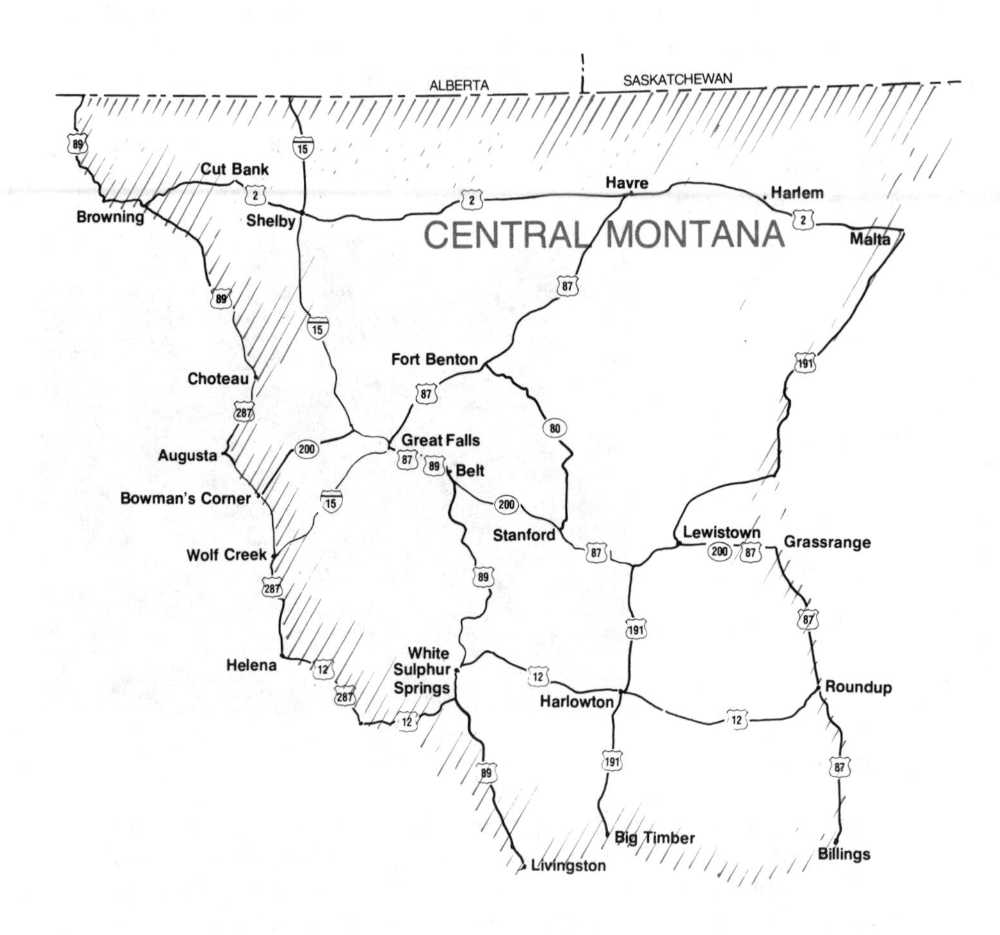

Roads covered in this section.

IV
Central Montana, A Land of Scattered Roads

The Rocky Mountains end at the eastern front of the Sawtooth Range, so all those mountains in central Montana are different. They are 20 million years younger than the Rocky Mountains, and they formed for reasons of their own that seem not to have anything to do with the Rocky Mountains. Most are broad arches bulged in the earth's crust, and most contain igneous rocks that formed sometime close to 50 million years ago. The association in space between crustal arches and igneous rocks, although far from perfect, is close enough to suggest that they are probably associated in time as well, and owe their existence to some common cause still not well understood.

The Sedimentary Veneer

Sedimentary rocks dominate the geologic landscape of central Montana. Geologists devote a great deal of attention to them because some contain oil and gas, others coal. Although neither of those commodities has the mystique of gold, their economic potential is infinitely greater. Production from the coal mines and oil and gas fields of central and eastern Montana dwarfs that of the gold mines in the western part of the state. In addition to their great mineral wealth, the sedimentary rocks of central Montana are an historical archive, the repository of the earth's records.

A Paleozoic fossil crinoid from the Little Belt Mountains. —Larry French photo

The oldest sedimentary rocks in central Montana are the Belt formations deposited a billion or more years ago, during Precambrian time. The Big and Little Belt Mountains contain large areas of Belt rock, as does the region around Helena. Elsewhere in the western part of central Montana, those rocks lie beneath a thick cover of younger sedimentary formations. They do not extend into the eastern part of the region.

During Paleozoic time, between the beginning of the Cambrian period about 670 million years ago and the end of the Permian period about 200 million years ago, central Montana lay near sea level. During long periods when the region was below sea level, layers of sediment accumulated on the floor of a vast inland sea. Of the many formations laid down during those 375 million years, the Madison limestone is the most conspicuous. It is actually a thick sequence of several formations laid down during Mississippian time, around 300 million years ago. Wherever it appears, the Madison limestone makes boldly conspicuous outcrops so pale gray that they are almost white. You don't have to be a geologist to recognize it. Besides its role as a major part of mountain landscapes, the Madison limestone is source of much central Montana oil and gas.

Shallow sea water again covered central Montana during long intervals of Mesozoic time, between the beginning of the Triassic period about 240 million years ago and the end of the Cretaceous period 65 million years ago. Dinosaurs dominated the animal world then, and the Rocky Mountains formed during the later half of Mesozoic time. The shorelines of the inland sea shifted back and forth across the region, weaving a complex pattern in the deposits of sand and mud. The sandstones and shales that formed then contain most of the region's oil and gas, all of its coal.

Sometime near the end of Cretaceous time, the crust beneath central Montana began to rise, and the sea to retreat for the last time into easternmost Montana and beyond. Enormous deposits of sand and mud and peat were laid down in the shallow margin of the retreating sea, along its coast, and on the coastal plain that grew steadily wider as the inland sea withdrew. Then an incredible catastrophe destroyed the dinosaurs, along with many other large and important groups of animals that lived both on land and in the sea.

The sudden passage of the dinosaurs marks the end of Mesozoic time, the beginning of the Tertiary period. Sedimentary rocks that were then forming in Montana contain little sign of the catastrophe except one very thin layer of dark silt and the abrupt disappearance of dinosaur fossils. Geologists call the rocks above the dinosaur fossils the Fort Union formation, and apply several other names to the similar Cretaceous rocks that do contain dinosaur bones.

After the last of the Fort Union formation had accumulated, perhaps about 55 million years ago, few sediments were deposited in central Montana for at least ten million years. Then the Renova formation accumulated in the broad valleys and in patches across the plains. Toward the end of Tertiary time, the Flaxville gravels spread across the entire region on a land surface that still survives in large areas. Meanwhile, central Montana had acquired its mountains, probably about 50 million years ago.

Folds and Faults

It is extremely difficult to find simple patterns in the confusing jumble of central Montana mountain ranges. We

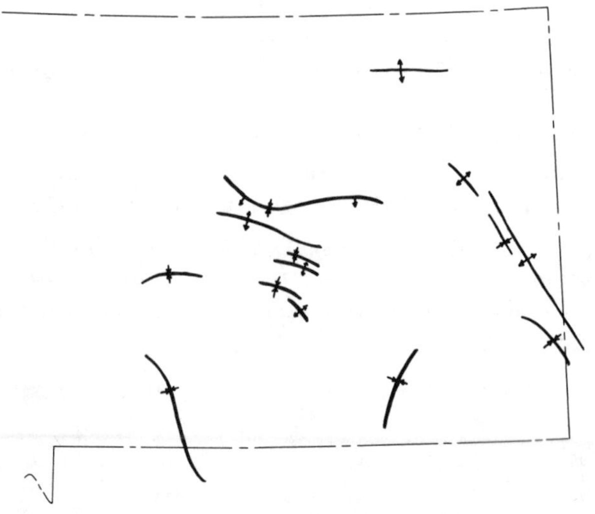

The major folds of central and eastern Montana.

suggest that much of the confusion can be resolved into two sets of structures, one that trends southeast, another that trends very slightly west of north. But that grouping may be a bit too simple because it doesn't embrace some folds in central Montana. The structure of the Big Snowy Mountains, for example, is shaped so nearly like a dome that it does not clearly belong to either set of folds. And the Bull Mountain basin equally lacks clear direction.

The major southeast trending fault zones of central Montana.

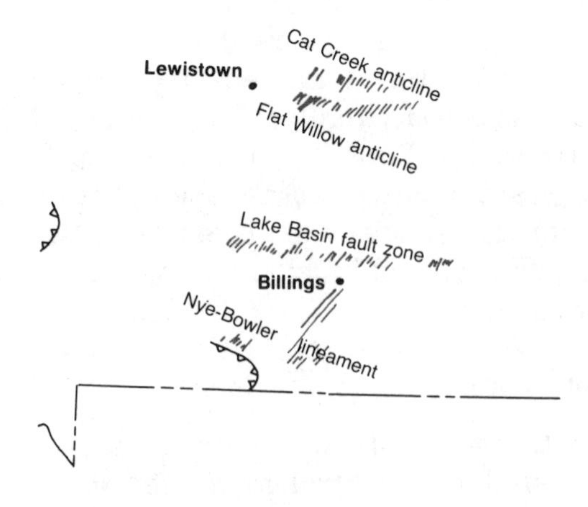

The southeast trending folds tend to associate with narrow zones of parallel faults that align with and therefore appear to be related to the Lewis and Clark fault zone. In general, each block south of the fault zone moved east, and also rose relative to the block north of the fault. The Crazy Mountain basin, a block of the earth's crust dropped between southeast trending faults, provides the major exception.

The set of folds that trends slightly west of north probably formed at the same time, therefore in response to the same stress. Consider the Bridger Range. The Livingston formation, a late Cretaceous deposit of volcanic debris, stands steeply tilted in the sharp northwest trending fold of the Bridger Range. But the Renova formation, an Oligocene and early Miocene deposit in the Gallatin Valley west of the Bridger Range, is not similarly tilted. Evidently, the Bridger Range was folded sometime after Cretaceous time had ended and before Oligocene time began, sometime between 65 and 40 million years ago. We suggest that about 50 million years ago is a good guess simply because so many other things were happening then in central Montana.

IGNEOUS ROCKS

Most of the mountain ranges of central Montana contain at least some igneous rocks, several contain little else. Although some may be as much as 65 million years old, most of those igneous rocks formed approximately 50 million years ago, at the same time the younger igneous rocks were forming in the Rocky Mountains.

The igneous activity of 50 million years ago extended northeastward through Idaho and Montana on an extremely broad trend that extends through central Montana. There, north of the Lewis and Clark fault zone, the igneous rocks become much more widely scattered and relatively modest in volume. And they include some extremely rare kinds of rocks unlike any that formed in the Rocky Mountains. Those rare rocks are peculiar in containing uncommonly large amounts, two or three times the normal amount, of the alkali elements sodium and potassium, especially potassium. The alkalic rocks are typically enriched in either sodium or potassium, but a few contain large amounts of both elements.

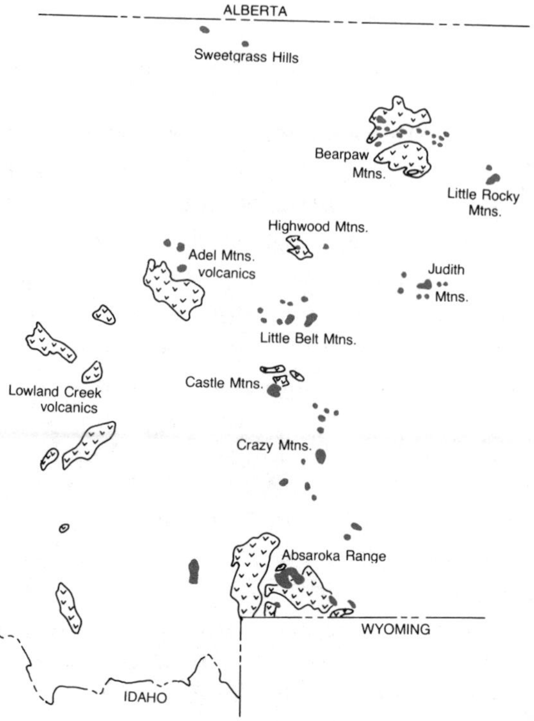

Major centers of 50 million-year-old igneous rocks in central Montana. Areas in color are intrusions.

So far, no one knows why the igneous rocks of central Montana exist. It seems unreasonable to try to associate them with the slab of oceanic crust that was sinking under what was then the western edge of North America, in central Washington. Five hundred miles is too far. Some geologists have suggested that the slab of oceanic crust was sinking at such a gentle angle that it went as far as central Montana before it got deep enough to start generating molten magmas. The main trouble with that idea is the existence of a perfectly good 50 million-year-old volcanic chain, the right age, in central Washington and Oregon, the right place. So the sinking slab could not have been descending at such a gentle angle.

Several major centers of more or less similar igneous activity, most notably those in east Africa, are clearly associated with continental rifting. But there is little evidence to suggest that something was pulling the North American continent apart in central Montana some 50 million years ago. The meaning of the sodium- and potassium-rich rocks of central Montana remains unclear.

Dark Igneous Rocks

The Adel, Highwood, and Bearpaw mountains lie along a nearly straight northeast trend, the farthest end of the long northeast trend of 50 million-year-old igneous activity. They are very similar volcanic centers in which activity began with eruption of large volumes of perfectly ordinary light-colored magma, then ended with eruption of shonkinite, a peculiar dark igneous rock that would be basalt if it were not so greatly enriched in potassium. In all three centers some of the shonkinite magma crystallized below the surface to form interesting intrusions.

Shonkinite contains numerous blocky crystals of black augite set in a fine-grained matrix composed of minute crystals of augite and potassium feldspar. Large volumes of shonkinite exist in the Adel, Highwood, and Bearpaw mountains, very little in the other mountains of central Montana. The rock was named for Shonkin, in the Highwood Mountains, and is something of a Montana specialty, even though it does exist in a few other places here and there around the world.

The Adel, Highwood, and Bearpaw mountains.

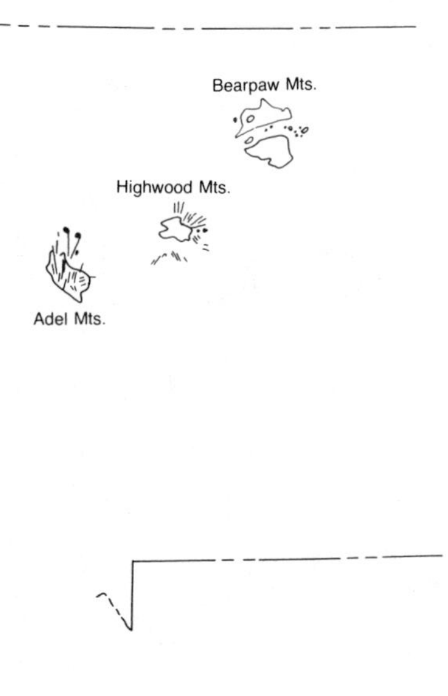

An outcrop of shonkinite.

The Adel, Highwood, and Bearpaw mountains contain great swarms of shonkinite dikes, vertical fractures that filled with magma. They radiate from centers like spokes radiating from the hub of a wheel. In all three ranges, many of the dikes end in

A schematic diagram of a radial dike swarm with laccoliths at the ends of some of the dikes.

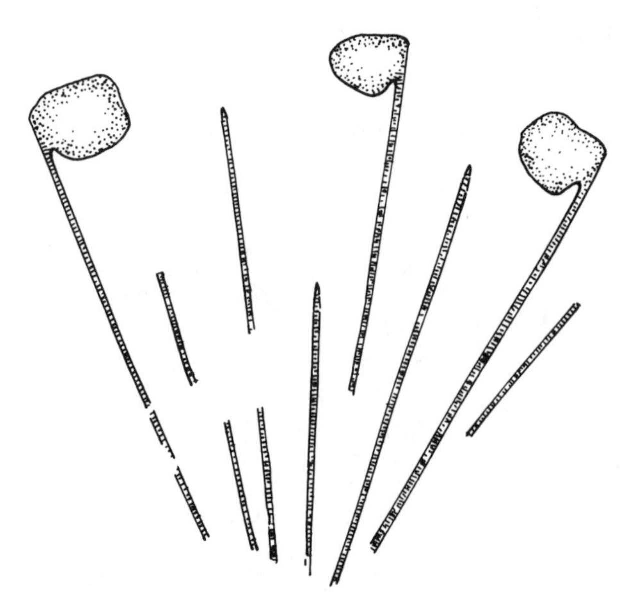

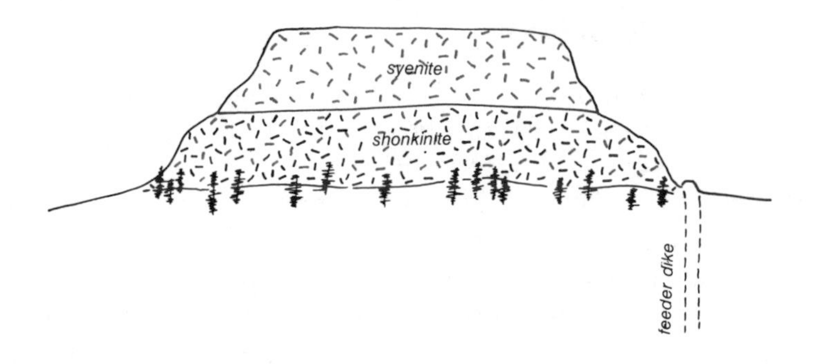

Section through a shonkinite laccolith with a syenite cap.

laccoliths, intrusions that formed as great blisters of magma injected between layers of sedimentary rock. The laccoliths are shaped about like cookies with flat bottoms and bulging tops, and the feeder dikes that injected the magma into them enter at one edge. It seems that the magma flowed through a vertical fracture to form the dike, then ponded beneath a layer of strong rock to form the laccolith.

Laccoliths in the Highwood and Bearpaw mountains typically consist of a large mass of dark shonkinite beneath a much smaller cap of white syenite composed almost entirely of potassium feldspar. Close examination of the shonkinite generally reveals that it contains widely scattered globs of syenite about the size of lemons. Evidently, the syenite and shonkinite magmas separated like oil and water. The lighter syenite floated to the top of the intrusion to form the syenite cap, which rests on the shonkinite like cream on milk. A few straggling globs of syenite were left trapped in the shonkinite as it crystallized.

The edge of the Lost Lake laccolith in the Highwood Mountains. Dark rock in the foreground is shonkinite. The magma injected between the layers of pale Cretaceous sandstone to form sills, the dark ribbons.

The Crazy Mountains

The Crazy Mountains are so different and so peculiar that a frustrated geologist might have named them. The alkalic igneous rocks are rich in sodium, as well as in potassium. Some of the rocks are not alkalic. All the igneous rocks of whatever kind crystallized below the surface to form intrusions. There are no volcanic rocks. None of the intrusions formed a cap of white syenite. Furthermore, the Crazy Mountains stand in the middle of the Crazy Mountain basin, a part of the crust that was folded down instead of up, as with most of the other igneous centers.

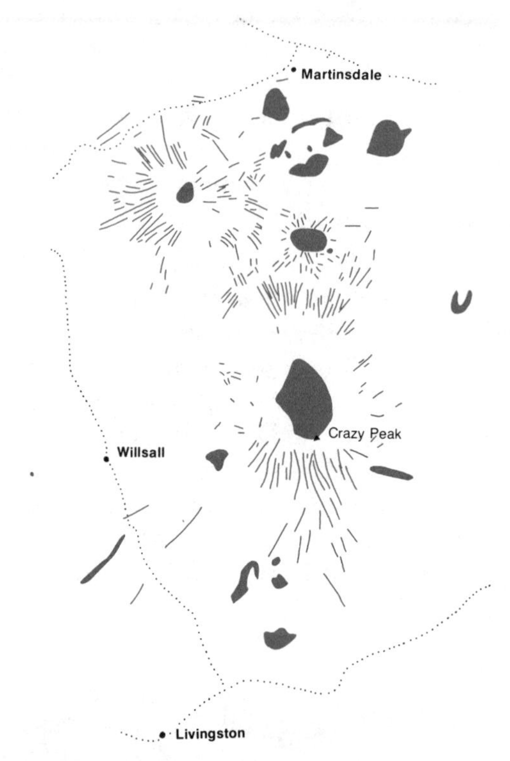

The Crazy Mountains contain several radial dike swarms.

The high southern end of the Crazy Mountains contains a single enormous igneous intrusion, the Big Timber stock, which is not especially rich in alkali metals. The lower northern end of the range is a swarm of much smaller intrusions most of which are rich in sodium and potassium. All were emplaced in the youngest sedimentary rocks in the region, the

Fort Union formation, so the magma must have crystallized at such extremely shallow depth that it is surprising to find no volcanic rocks. Maybe they are all lost to erosion.

The magmas that formed the Big Timber stock, as well as some of those in the smaller intrusions in the northern Crazy Mountains, baked the surrounding Fort Union formation to form wide haloes of metamorphosed rock. Those contact metamorphic rocks are as hard and resistant to erosion as the igneous rocks they enclose.

Pale Igneous Rocks

Igneous rocks abnormally rich in sodium, and to some extent in potassium, exist in the Judith and Moccasin mountains, the north end of the Little Belt Mountains, the Little Rocky Mountains, the Sweetgrass Hills, and several lesser centers. Most of those rocks are varieties of syenite, igneous rock that

The major centers of syenitic and granitic igneous activity of 50 million years ago.

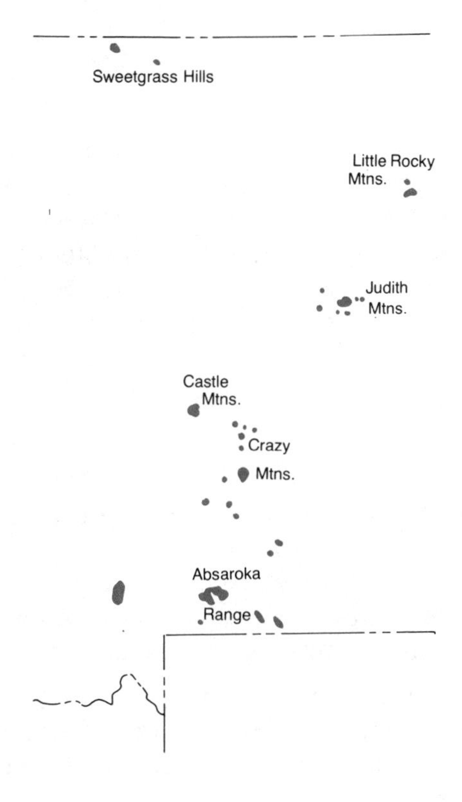

consists mostly of feldspar. Some of those centers also contain light colored igneous rocks, granites, that are not enriched in sodium or potassium. In fact, some igneous centers, such as the Castle Mountains, contain no alkalic rocks.

The syenitic igneous centers show a general tendency to start their activity with rocks of fairly normal composition, varieties of granite, and then finish by producing magmas strongly enriched in sodium. Most of the magmas in the syenitic centers crystallized at depth to form intrusions; very little erupted to make volcanic rocks. Emplacement of those intrusions probably caused a slight bulging of the land surface. Fifty million years of erosion have since stripped away the several thousand feet of relatively soft sedimentary rocks that originally enclosed the intrusions, leaving the more resistant igneous rocks standing as mountains.

In general, the syenitic igneous rocks tend to form intrusions called stocks that are more or less circular in map view, and appear to extend to great depth with no sign of a base. If we could see the entire stock, it might look like a giant vertical cylinder. But there are exceptions. The syenitic igneous rocks in the northern Little Belt Mountains form a radial dike swarm with laccoliths on the ends of some of the dikes.

Most of the syenitic intrusions brought gold, as well as silver and lead, into central Montana. In a few cases, the ore minerals are in the igneous rock, but they more commonly occur in veins around the margins of the intrusion. Nearly every large syenitic intrusion sponsored a gold rush during the early years of settlement. A few actually became the scenes of serious mining.

Diatremes

The alkali-rich igneous rocks of central Montana also include several dozen diatremes, very small igneous intrusions composed of extremely unusual rocks that could only have melted deep within the earth's mantle. Areas that contain alkalic igneous rocks enriched in sodium or potassium generally also contain swarms of diatremes.

Diatremes are typically dikes or vertical pipes of very dark and dense igneous rock that generally contain numerous fragments of the older rocks they intrude. In most cases, the

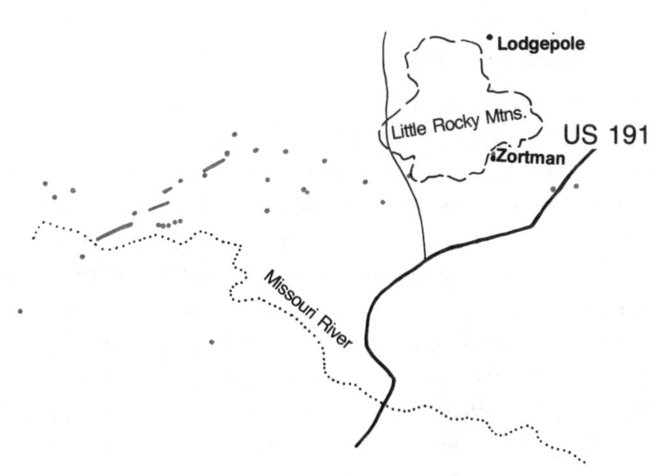

Major diatremes in central and eastern Montana. Diatremes are small and hard to find, so it seems likely that more await discovery.

inclusions come from rock formations both above and below the level of the outcrop—rocks from all levels mixed together within the diatreme. And the igneous rock is also much broken. All that internal stirring suggests that the diatreme was emplaced as broken chunks of rock suspended in a great blast of escaping gas, probably carbon dioxide. No doubt, the gas reached the surface, where it must have shot a great column of volcanic ash and broken rocks of all sorts high into the air. No such eruption has occurred in historic time, so we have no eyewitness accounts; but it seems safe to assume that emplacement of a diatreme must be a noisy affair.

Most diatremes are no more than a quarter of a mile across, and few of them resist weathering and erosion well enough to form hills, or even good outcrops. Despite their small size and generally inconspicuous appearance, diatremes are scientifically important because they provide a glimpse of the earth's deep interior. Several of those in Montana have also produced large quantities of gem sapphires. Rumor has it that diamonds have been found in one Montana diatreme. If so, that is no surprise; many diatremes in other parts of the world do contain diamonds.

Oil and Gas

Wildcat wells found oil in 1903 in what is now the eastern part of Glacier National Park. The park has recovered nicely, but Montana has not been the same since.

Oil and gas development began slowly in Montana with many discouraging failures. Then, during the 1920s, a series of new discoveries in several parts of central Montana inspired renewed interest in the region that soon led to rapid development of many oil fields, enormous increases in production. Crude oil production in central Montana reached a maximum of about 28,000 barrels of oil per day in 1960, and hovered close to that level through the next decade. In those days, new discoveries were adding known oil and gas reserves faster than annual production was depleting them. That happy state of affairs can't last indefinitely. Oil is, after all, a finite and non-renewable resource that exists in limited quantity.

The inherent limitations of oil development came to roost in central Montana about 1973 as the rate of new discoveries began to lag far behind depletion. Oil production began a steady decline that continued despite intense exploration for new reserves during the oil shortages of the 1970s. Output was down to about 15,000 barrels of oil per day by 1983, and the downward trend seems likely to continue.

Oil and gas fields of central Montana. Oil fields black, gas fields color.

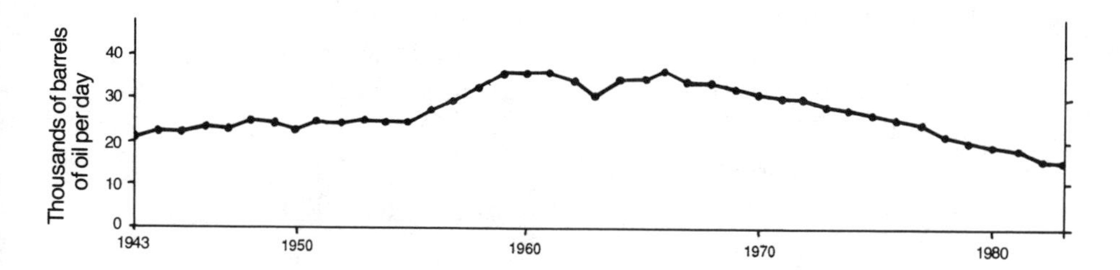

Central Montana oil production.

Continued exploration of any oil producing region must finally reach the point at which most of the fields have been found. Even though there is no way to count undiscovered oil fields, we can know they are getting scarce when it becomes increasingly difficult to find them, just as fishermen know a pond is about fished out when they start having increasing difficulty catching anything. The failure of increased exploration during the 1970s to reverse the downward trend of oil production figures for central Montana suggests that the region probably contains relatively few undiscovered oil fields—that it is about fished out. Nevertheless, many decades will pass before the last pump stops. And there is always hope that new exploration techniques may help locate a few more new oil fields.

Coal

All those shorelines that shifted back and forth across central Montana during Mesozoic and earliest Tertiary time left a legacy of coal. Swamps and marshes that thrive today along tropical shorelines, such as the west coast of Florida, lay down deposits of peat, a mucky sediment that consists mostly of partially decayed plant material. If later sediments bury the peat and accumulate to enough depth to put it under considerable pressure, it turns into coal. That happened during

Mesozoic time along the coasts and the floodplains of rivers that emptied into the coasts of central Montana.

Trees are scarce in much of central Montana, so the first white settlers were glad to find coal, and quick to begin mining it for domestic use. Even poor coal is better than buffalo chips. Demand for central Montana coal increased drastically with the coming of the railroads with their fuel-hungry steam engines during the 1880s, still further with the proliferation of smelters to serve the mining districts. By the end of the nineteenth century, several large coal fields were active, and production increased steadily until the 1920s. Central Montana was a major coal producing region. Then cheap oil intervened.

Diesel engines, oil furnaces, gas furnaces, and gas fired smelters all contributed to the downfall of central Montana coal. Production went into a steep and steady decline that closed one mine after another until the end came during the 1950s. It seemed unlikely then that central Montana would ever again produce coal. Now it seems that it probably will, someday.

Prospects for renewal of big coal mining in central Montana probably lie in the distant future. Almost all of the old mines worked underground because they produced from coal seams that were thin, deeply buried, steeply tilted, or some combination of those problems. Central Montana contains very little coal at shallow enough depth and in thick enough beds to make strip mining feasible, so there is no alternative to underground mining. It is difficult for underground mines to compete with large strip mines like those in eastern Montana.

Glaciation

The craggy peaks of the high southern Crazy Mountains are as spectacularly glaciated as any range in Montana. But those and the Bridger Range are the only mountains in central Montana that look that way. No others snatched enough snow out of the passing clouds to support valley glaciers of their own. The other mountains of central Montana look as though the ice ages had never happened. Compare the craggy peaks, broadly gouged valleys, and sharp ridges in the southern Crazy Mountains to the more softly rounded forms of the others.

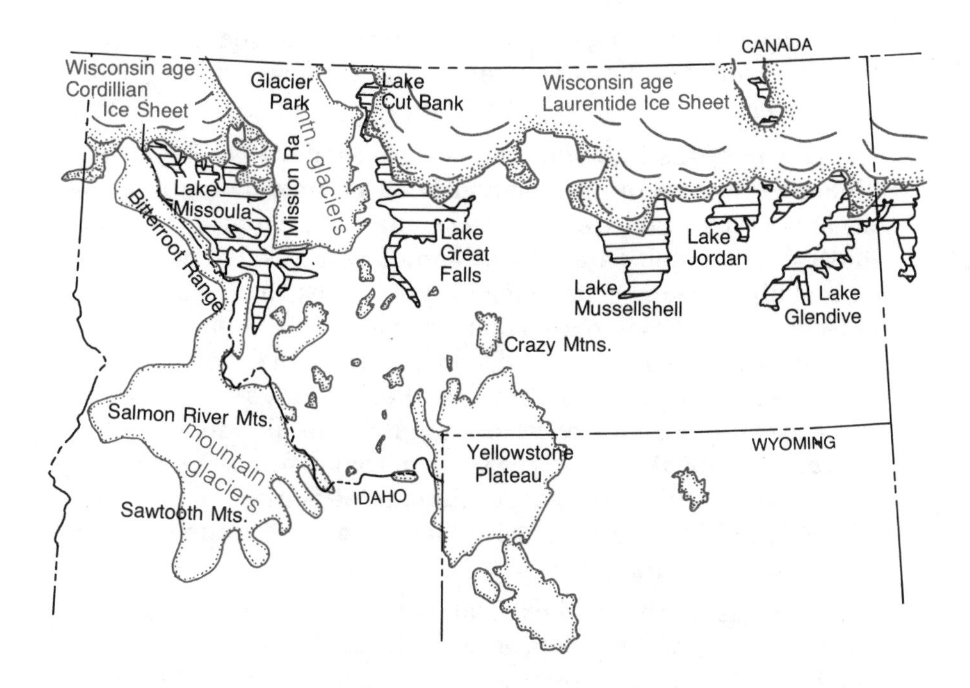

The approximate maximum extent of continental ice cover in central Montana during the Bull Lake and Pinedale ice ages.

The rest of the story of glaciation in central Montana involves the big continental ice sheets that formed in central Canada and slowly spread in all directions. As the Bull Lake and Pinedale glaciers reached their maximum extent, continental ice moving southwestward from the vicinity of Hudson Bay finally advanced into central and eastern Montana. Ice lapped onto the northern edges of the Little Rocky, Highwood, and Bearpaw mountains, as well as the Sweetgrass Hills, but did not cover any of those ranges. The southern margin of the glacier appears to have been extremely irregular, but followed close to the line of the Missouri River.

That was the thin edge of the glacier that reached into Montana, and the ice was moving very slowly. It probably reached Montana only as the ice age approached its climax, then began to melt a few thousand years later. So the effects of glaciation on the landscape of central Montana are relatively minor, surprisingly so. In most areas, the only obvious evidence of glaciation is widely scattered erratic boulders. Here and there, hummocky tracts of morainal topography form long ridges that precisely record the ice margin, and there are a few abandoned valleys that once carried torrents of glacial meltwater.

It just happens that the land surface tends to get generally higher southward in the glaciated parts of central Montana. So the big continental ice sheet faced slightly higher ground to the south. Glacial meltwater and stream flow trapped between the ice and high ground beyond its margins formed a series of large lakes connected by streams along the edge of the glacier. Had you lived in central Montana during the maximum of Bull Lake glaciation, you could have paddled a canoe along the edge of the ice almost all the way from Cut Bank to Glendive—sailed a large steamboat along a good part of that distance. Generally similar but smaller and less continuous lakes existed along the edge of the Pinedale ice sheet when it reached its maximum about 15,000 years ago.

The largest of those lakes, Glacial Lake Great Falls, covered a vast expanse of the plains between Great Falls and Cut Bank and flooded the site of Great Falls to a depth of some 600 feet. If that lake still existed today, we would think it an inland sea comparable to the Great Lakes.

Imagine that scene.

Glacially transported and scratched boulder.

In most places, those lakes were almost too big to see across. Icebergs that weighed thousands of tons broke off the glaciers that formed high cliffs of blue ice along the northern shores of the lakes, smashed into the water with a thundering roar, then drifted across the lake looking like floating islands of ice. Herds of mastodons and horses filtered through dense forests and across prairies covered with tall grass as they came down to the shore to drink. All that remains now is the old shoreline where mastodons once stood, scattered boulders dropped from melting icebergs, and level plains floored with deep deposits of fine sediment laid down on the bottom of the lake.

The Shonkin Sag, a valley eroded through the Highwood Mountains by drainage from Glacial Lake Great Falls. H. L. James photo

Even where they are not conspicuous, old lake shorelines are fairly easy to recognize as horizontal benches that look exactly like modern lake shorelines except that there is no lake. Imagine the water, and everything else is in its place. Glacial Lake Great Falls left two prominent shorelines: one at an elevation of 3900 feet, the other at 3600 feet. It also left the other signs of a vanished glacial lake: a landscape littered with boulders, and vast level expanses of former lake bed. We can easily reconstruct a map of Glacial Lake Great Falls in its glory simply by tracing the elevations of the two shorelines on a topographic map.

Glacial Lake Great Falls drained through a spillway that wound through the northern part of the Highwood Mountains, thence east through the present valley of Arrow Creek. That must have been quite a torrent of water; it eroded a valley more than large enough to carry a river the size of the modern Missouri. The meltwater river is long gone, but the valley survives as the Shonkin Sag, one of Montana's geologic spectaculars.

While the overflow from Glacial Lake Great Falls was thundering through the Shonkin Sag, the old course of the Missouri River from the Fort Benton area to that near Havre lay under ice and lake water, and was filling with sediment. When the ice melted, the Missouri River, deprived of its old channel, poured across the plains to establish a new course from near Fort Benton to near Fort Peck. That new channel has since developed into a spectacularly deep and narrow canyon obviously much younger than the rest of the Missouri River valley.

Meanwhile, the Milk River began to flow through the old Missouri River valley as soon as the ice melted back far enough to uncover it. So now the Milk River flows from near Havre to near Fort Peck in an oversized valley eroded by a much larger stream. The narrow canyon of the broad Missouri and the broad valley of the narrow Milk River complement each other. Big Sandy Creek follows another long segment of the former valley of the Missouri River between Great Falls and Havre.

Farther east, the ice of the Bull Lake glaciation impounded Glacial Lake Musselshell in the area around and south of the Little Rocky Mountains. Imagine the mountains as they were then, ice jammed onto their northern edge, a lake nearly surrounding the rest of the range. Where that lake was, boulders dropped from drifting icebergs litter the wheat fields. Still farther east, water backed up behind the same glacier formed another large lake north of Glendive, more in North Dakota.

Interstate 15: Helena—Great Falls

89 mi./142 km.

Helena is near the south end of the Helena Valley, between mountains eroded in granite of the Boulder batholith of western Montana and the Big Belt Mountains to the northeast. Between Helena and Wolf Creek, the highway angles across the southern end of the northern Montana portion of the overthrust belt. That part of the route crosses big slabs of rock, mostly colorful Precambrian sedimentary formations, that moved eastward into the overthrust belt. Between Wolf Creek and Cascade, the highway crosses part of the Adel Mountains, a volcanic pile that covers Cretaceous sedimentary rocks of the high plains. The route between Cascade and Great Falls crosses the same Cretaceous formations still lying nearly as flat as when they were laid down in the floor of a shallow inland sea more than 80 million years ago.

The Scratchgravel Hills

The Scratchgravel Hills rise west of the road about 4 miles north of Helena. Watch for an isolated group of hills scantily covered with trees a bit east of the main mass of mountains. They contain a granitic intrusion that crystallized about 85 million years ago, timing that probably relates it to the earliest stages of the nearby Boulder batholith. A thin sheet of gravel that covers part of the surface on the north side of the Scratchgravel Hills contains gold nuggets that early settlers collected by ploughing and raking the land, hence the name. Several mines worked meager bedrock deposits of gold in the Scratchgravel Hills from about 1870 until the First World War.

Gates of the Mountains

About 21 miles north of Helena, a side road leads to Gates of the Mountains, an impressive canyon cut through white Madison lime-

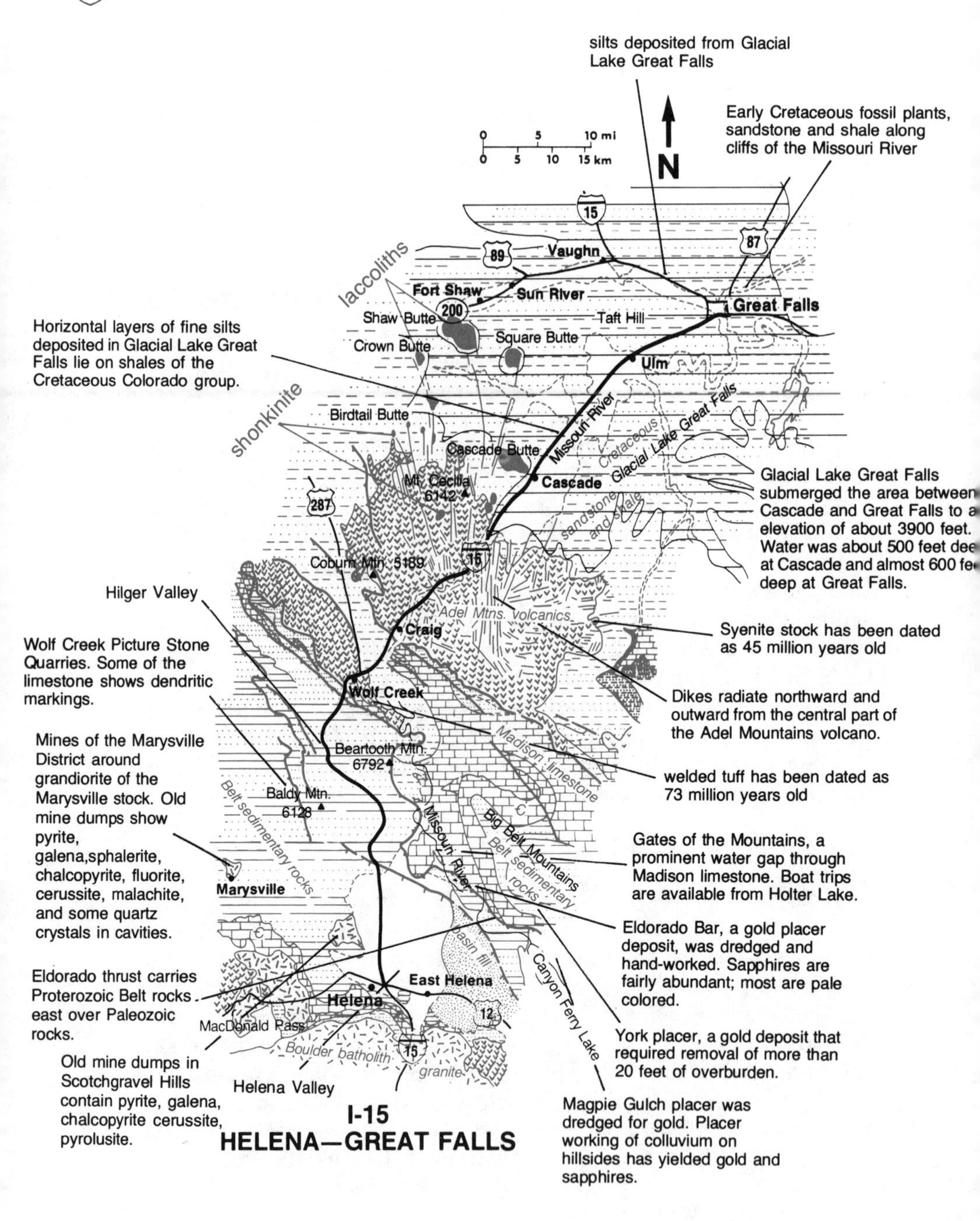
15
silts deposited from Glacial Lake Great Falls
Early Cretaceous fossil plants, sandstone and shale along cliffs of the Missouri River
N
0 5 10 mi
0 5 10 15 km
15
89
Vaughn
87
laccoliths
Fort Shaw
Sun River
Great Falls
200
Shaw Butte
Taft Hill
Crown Butte
Square Butte
Ulm
Horizontal layers of fine silts deposited in Glacial Lake Great Falls lie on shales of the Cretaceous Colorado group.
shonkinite
Birdtail Butte
Missouri River
Glacial Lake Great Falls
Cretaceous
Cascade Butte
Cascade
Mt. Cecilia 6142
sandstone and shale
287
Glacial Lake Great Falls submerged the area between Cascade and Great Falls to a elevation of about 3900 feet. Water was about 500 feet dee at Cascade and almost 600 fe deep at Great Falls.
Coburn Mtn. 5189
15
Hilger Valley
Adel Mtns. volcanics
Craig
Syenite stock has been dated as 45 million years old
Wolf Creek Picture Stone Quarries. Some of the limestone shows dendritic markings.
Wolf Creek
Dikes radiate northward and outward from the central part of the Adel Mountains volcano.
Mines of the Marysville District around grandiorite of the Marysville stock. Old mine dumps show pyrite, galena,sphalerite, chalcopyrite, fluorite, cerussite, malachite, and some quartz crystals in cavities.
Beartooth Mtn. 6792
Madison limestone
welded tuff has been dated as 73 million years old
Baldy Mtn. 6128
Belt sedimentary rocks
Missouri River
Big Belt Mountains
Belt sedimentary rocks
Gates of the Mountains, a prominent water gap through Madison limestone. Boat trips are available from Holter Lake.
Marysville
Eldorado Bar, a gold placer deposit, was dredged and hand-worked. Sapphires are fairly abundant; most are pale colored.
basin fill
Eldorado thrust carries Proterozoic Belt rocks east over Paleozoic rocks.
East Helena
Canyon Ferry Lake
Helena
12
MacDonald Pass
York placer, a gold deposit that required removal of more than 20 feet of overburden.
Boulder batholith
15
granite
Old mine dumps in Scotchgravel Hills contain pyrite, galena, chalcopyrite cerussite, pyrolusite.
Helena Valley
I-15
HELENA—GREAT FALLS
Magpie Gulch placer was dredged for gold. Placer working of colluvium on hillsides has yielded gold and sapphires.

Except that the water no longer rushes because it is impounded behind Holter Dam, Gates of the Mountains look very much as they did to Lewis and Clark.
—Montana Department of Commerce photo by Gary Wunderwald

stone. Historians identify this as the canyon that Lewis and Clark named "Gates of the Mountains" because they considered it their point of entry into the Rocky Mountains. We think that identification is wrong because Lewis and Clark described a canyon cut through black rocks. That would fit the canyon through the volcanic rocks of the Adel Mountains that the road crosses about midway between Cascade and Wolf Creek, certainly not this gorge eroded through white limestone.

Gates of the Mountains is one of those canyons that formed as a river that began flowing well above its present level entrenched its course into resistant rocks, in this case, the Madison limestone. During the summer months, excursion boats provide regular trips through the canyon past spectacular winding ledges of tightly folded Madison limestone, real overthrust belt structures. And there are exposures of soft sediment left when Glacial Lake Great Falls flooded the canyon.

A Valley That Lost Its River

Between about 20 and 30 miles north of Helena, the highway passes through the Hilger Valley. It looks exactly like a stream valley, complete with tributaries, and it is big enough to hold a stream larger than the Missouri River, but the Hilger Valley contains no stream worth mentioning. We think it was probably eroded during late Miocene time, when the region enjoyed a wet climate that must have maintained large rivers.

Outcrops of Precambrian Belt rock in Little Prickly Pear Canyon. —Montana Bureau of Mines and Geology photo by H. L. James

North of the Hilger Valley, about 35 miles north of Helena, the highway follows Little Prickly Pear Creek through a marvelously picturesque canyon cut in colorful red and green mudstones. Most of those rocks belong to the Precambrian Spokane shale, a Belt formation named for the Spokane Hills east of Helena.

Adel Mountains

Between Wolf Creek and Cascade, the highway passes through the northern fringe of the Adel Mountains, the eroded ruins of a volcano

View from the highway into the Missouri River Canyon. Rocks in the hills are black shonkinite of the Adel Mountains. We think this is the canyon Lewis and Clark called "Gates of the Mountains."

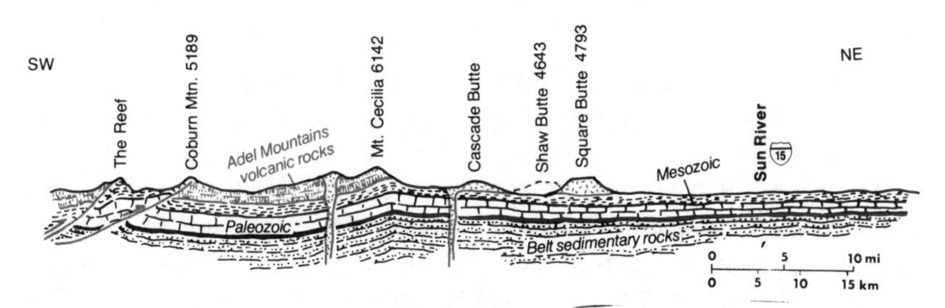

Section west of the highway between Wolf Creek and Great Falls. The thrust faults at the southwest end are the eastern edge of the Rocky Mountains.

that was active about 50 million years ago. Most of the rocks in the Adel Mountains are shonkinite. A close view of the rocks reveals large crystals of shiny black augite set in a dark gray matrix composed of very small crystals of augite and feldspar. Augite typically has a dull surface, so the glossy crystals in shonkinite are a bit of a surprise.

Many of the roadcuts reveal crude layering that formed as eruptions laid new layers of ash and lava on the sloping flanks of the volcano. Fifty million years of erosion have eliminated every vestige of the original volcanic landscape. Nevertheless, it should be possible to reconstruct the original locations and shapes of the Adel Mountains volcanoes by systematically measuring the attitudes of those layers throughout the Adel Mountains, and plotting them on a map.

Cascade Butte,west of the highway at Cascade, is a large laccolith, an intrusion of igneous rock that formed as a large blister of magma between layers of sedimentary rock. It is part of the same group of laccoliths that appears south of Montana 200 west of Great Falls. Roadcuts south of Cascade expose some of the big dikes that fed magma into Cascade Butte and its neighbors to the northwest. They appear as vertical masses of dark shonkinite that cut across the layering of the Cretaceous sedimentary rocks.

Shonkinite dikes look like ruined walls converging toward Three Sisters Mountain. View southeast from the highway a few miles south of Cascade.

Several of the big roadcuts between Cascade and Wolf Creek reveal dark shonkinites of the Adel Mountains lying on pale sandstone or soft shale. Those cuts expose the base of the old volcano where it rests on the older Cretaceous sedimentary rocks that lie beneath this part of the central Montana plains. This part of Montana was still a shallow inland sea when the sedimentary rocks accumulated between 100 and 80 million years ago, some millions of years before the volcanic rocks covered them.

The Rainbow Falls at Great Falls as they looked before the dam was built.
—Mansfield Library, University of Montana

Interstate 15: Great Falls—Sweetgrass
119 mi./190 km.

Bedrock between Great Falls and the Canadian border is almost entirely late Cretaceous sedimentary formations deposited between 80 and 65 million years ago. Glaciers and glacial lakes left the landscape so thoroughly plastered with their debris that bedrock outcrops are rare along most of the route.

Glaciation

Widely scattered blocks of pink granite and streaky gray or pink gneiss, basement rock carried in from northern Manitoba, dot the fields along much of the way. They speak of ice. Nothing but a glacier could have carried such big rocks so far. And the glaciers did a lot more than litter the countryside with boulders.

Between Great Falls and the area about three miles south of Dutton, the route crosses the old floor of Glacial Lake Great Falls. There are two shorelines, at elevations of about 3900 and 3600 feet. In a few exceptional places the old lake shorelines are easy to see as a distinct little scarp rising above a narrow bench. The higher shoreline probably formed during the Bull Lake ice age, sometime between about 70,000 and 130,000 years ago. At that time, the lake must have been about 600 feet deep in the Great Falls area. The lower shoreline probably marks the water level of about 15,000 years ago, during the Pinedale ice age. The tract of low hummocky hills just south of Dutton is the moraine that marks the farthest reach of that younger glacier.

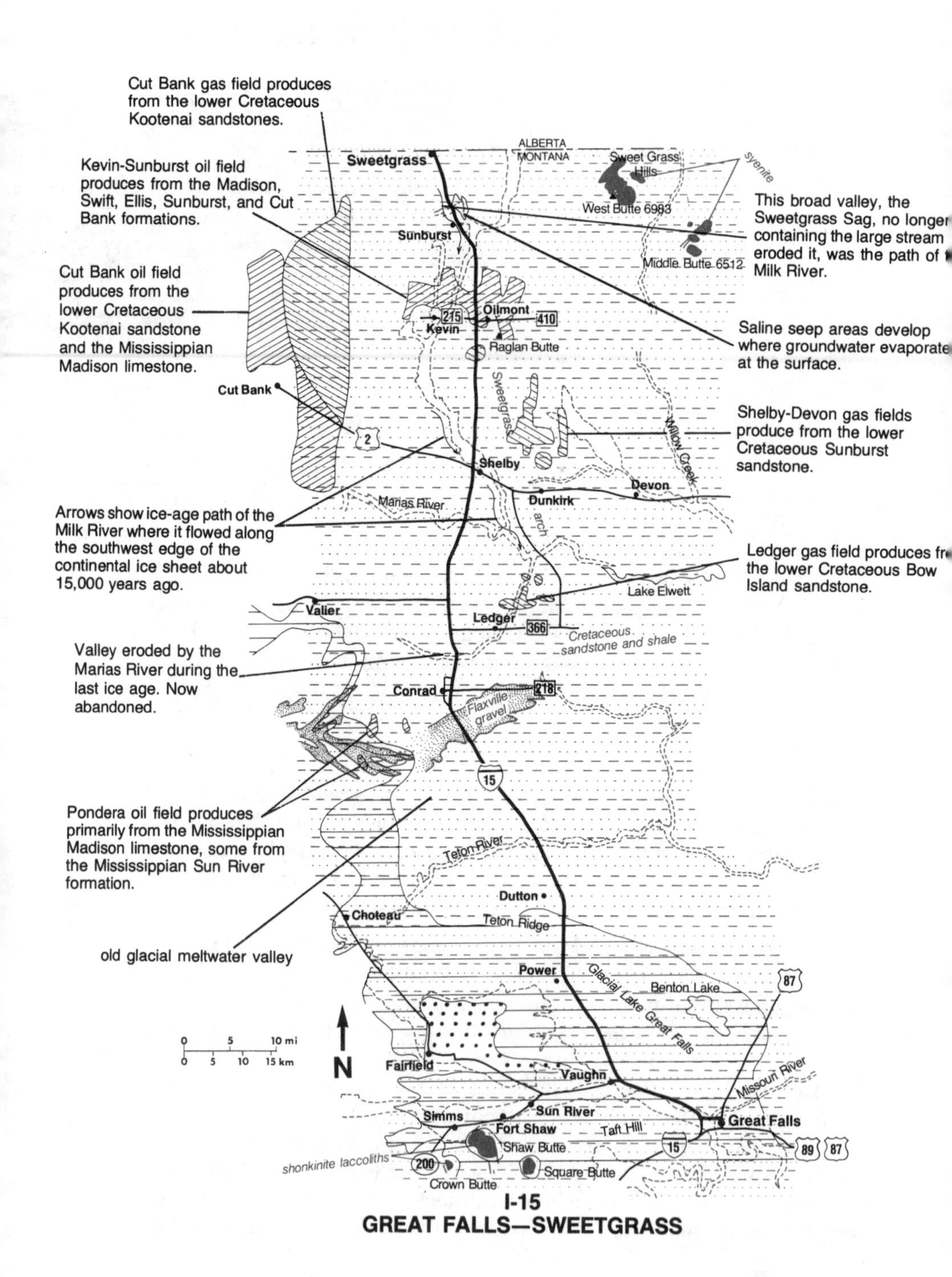

Cut Bank gas field produces from the lower Cretaceous Kootenai sandstones.
Kevin-Sunburst oil field produces from the Madison, Swift, Ellis, Sunburst, and Cut Bank formations.
Cut Bank oil field produces from the lower Cretaceous Kootenai sandstone and the Mississippian Madison limestone.
ALBERTA
MONTANA
Sweetgrass
Sweet Grass Hills
syenite
West Butte 6983
Middle Butte 6512
This broad valley, the Sweetgrass Sag, no longer containing the large stream eroded it, was the path of Milk River.
Sunburst
Kevin
215
Oilmont
410
Raglan Butte
Saline seep areas develop where groundwater evaporate at the surface.
Cut Bank
2
Sweetgrass arch
Willow Creek
Shelby-Devon gas fields produce from the lower Cretaceous Sunburst sandstone.
Shelby
Dunkirk
Devon
Marias River
Arrows show ice-age path of the Milk River where it flowed along the southwest edge of the continental ice sheet about 15,000 years ago.
Ledger gas field produces fr the lower Cretaceous Bow Island sandstone.
Lake Elwett
Valier
Ledger
366
Cretaceous sandstone and shale
Valley eroded by the Marias River during the last ice age. Now abandoned.
Conrad
218
Flaxville gravel
15
Pondera oil field produces primarily from the Mississippian Madison limestone, some from the Mississippian Sun River formation.
Teton River
Dutton
Choteau
Teton Ridge
old glacial meltwater valley
Power
Glacial Lake Great Falls
Benton Lake
87
0 5 10 mi
0 5 10 15 km
N
Fairfield
Vaughn
Missouri River
Sun River
Simms
Fort Shaw
Great Falls
Taft Hill
Shaw Butte
15
89
87
shonkinite laccoliths
200
Square Butte
Crown Butte

I-15
GREAT FALLS—SWEETGRASS

Glacial Lake Great Falls certainly backed deeply against the glaciers of both ice ages. Swarms of icebergs broke off the floating edges of the glaciers, and sailed out into Glacial Lake Great Falls carrying boulders embedded in them. As the drifting ice melted, it dropped its burden of rocks where we now see them scattered. Small groups of boulders here and there between Great Falls and Dutton probably mark places where grounded icebegs melted, dumping their whole load of rocks in one place. Don't confuse such natural groupings with the piles of rocks farmers have picked out of their fields.

The old course of the Milk River?

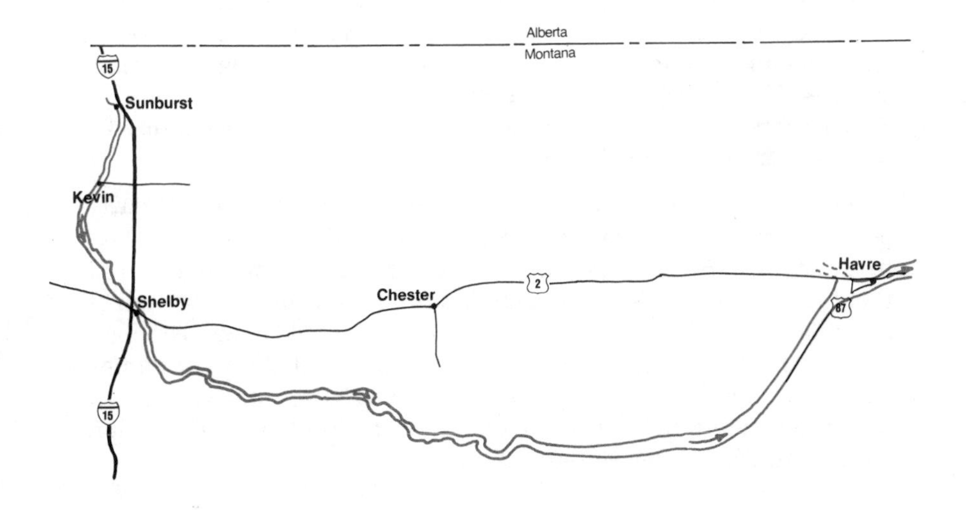

Shelby shelters from the wind in the floor of a clearly defined river valley that contains no worthy stream. The dry valley is exactly the right size to hold the Milk River. It continues north through Kevin and Sunburst to Sweetgrass, and then to Milk River, Alberta, where it connects with the present valley of the Milk River. Some geologists argue that the Milk River came this way before the big ice age glaciers drowned its valley in Glacial Lake Great Falls. Other geologists contend that the dry valley is an old glacial meltwater channel.

The Sweet Smell of Sour Crude Oil

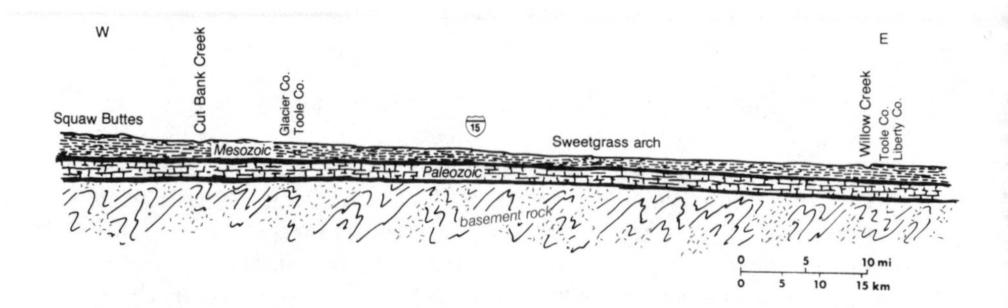

Section across the line of the highway south of Shelby. The crustal movements that created the Rocky Mountains left these rocks untouched, so they lie almost as flat as they were laid down. Even the big petroleum trap of the Sweetgrass arch hardly shows in this section drawn to true vertical scale.

From Conrad to the Canadian border, the highway travels along the length of the Sweetgrass arch. Petroleum migrating upward in the tilted rocks accumulated at and near the crest of the arch to form many oil and gas fields, including several with inspiring production records. Most of the wells in those fields have been producing for many years now, and have been down to a barrel or two a day for years—"stripper wells."

Several hundred stripper wells steadily pump their small daily portion of oil, an average of slightly more than three barrels per day, from the Pondera field a few miles west of Conrad. Oil was discovered there in 1927 in the Mississippian Madison limestone, at a depth of approximately 2000 feet. That is shallow, by oil field standards. You can see the shallowness of the wells by looking at the light pumps with their small counterweights. The value of the total production from the Pondera field until the end of 1983 was almost 112 million dollars, all from a few square miles.

Wildcat wells discovered oil near both Kevin and Sunburst in 1922. Further drilling soon revealed that both discoveries were producing from the same reservoir, that they had found an extremely large oil field. Almost 1000 wells, most of them about 1500 feet deep, produce quite a lot of oil and small amounts of gas from the field. The wells produce an average of about two barrels of oil per day. Most of that comes from the Mississippian Madison limestone, lesser amounts from Jurassic and Cretaceous formations at much shallower depths. By the end of 1983, the total value of the oil amounted to some 215 million dollars. The highway passes through part of the Kevin-Sunburst field between Shelby and the Canadian border.

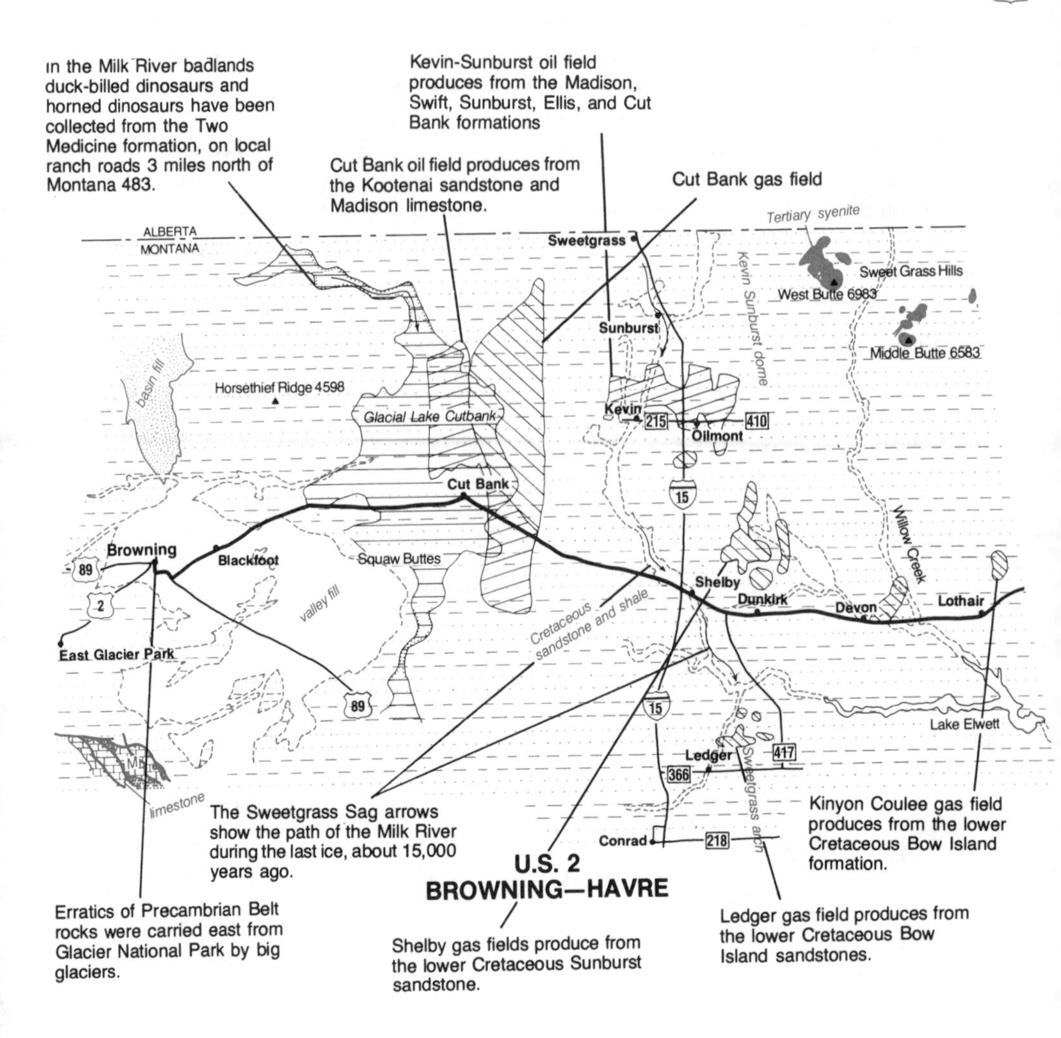

US 2: Browning—Havre

161 mi./258 km.

The long stretch of highway between Browning and Malta crosses a vast expanse of nearly level plains with far more geologic interest than a first view of the landscape might suggest. This is one of the most active and productive oil provinces in Montana: the bedrock

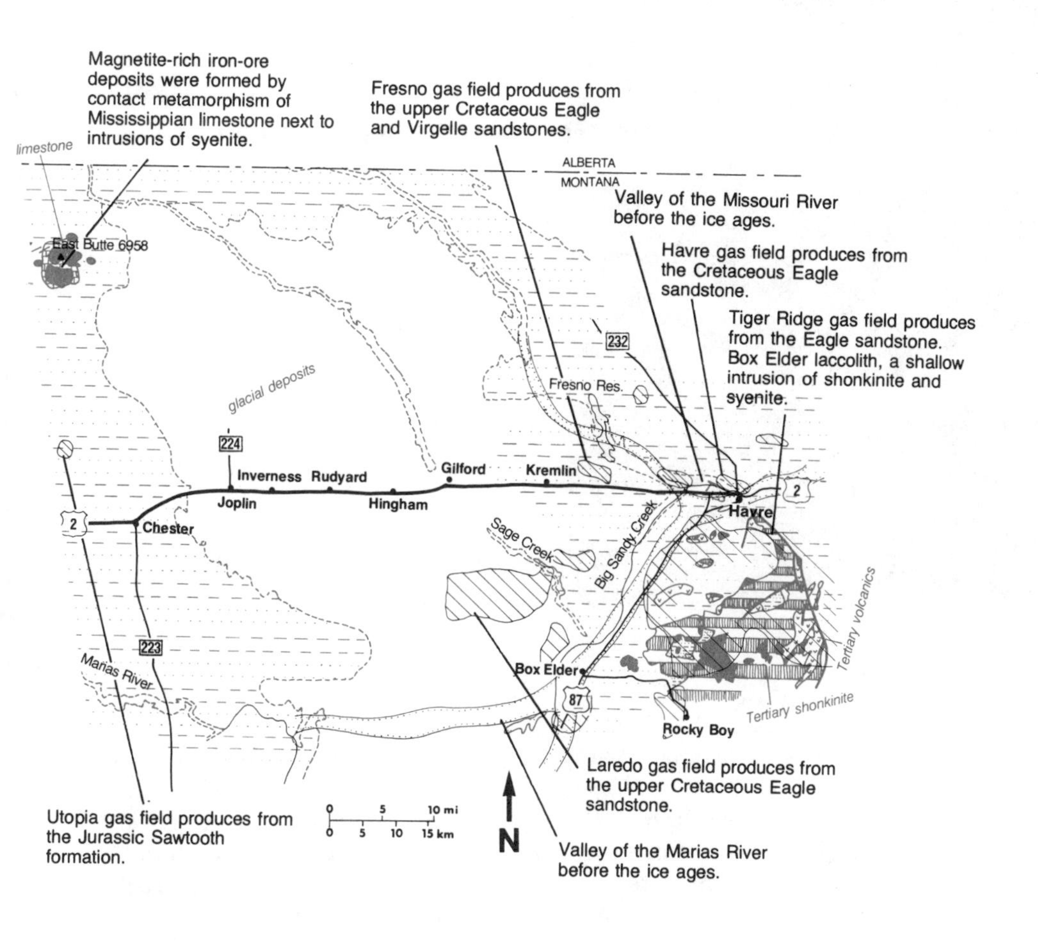
Magnetite-rich iron-ore deposits were formed by contact metamorphism of Mississippian limestone next to intrusions of syenite.
limestone
East Butte 6958
Fresno gas field produces from the upper Cretaceous Eagle and Virgelle sandstones.
ALBERTA
MONTANA
Valley of the Missouri River before the ice ages.
Havre gas field produces from the Cretaceous Eagle sandstone.
Tiger Ridge gas field produces from the Eagle sandstone. Box Elder laccolith, a shallow intrusion of shonkinite and syenite.
232
Fresno Res.
glacial deposits
224
Inverness
Rudyard
Gilford
Kremlin
Joplin
Hingham
Havre
2
Chester
Sage Creek
Big Sandy Creek
Tertiary volcanics
223
Marias River
Box Elder
87
Rocky Boy
Tertiary shonkinite
Laredo gas field produces from the upper Cretaceous Eagle sandstone.
Utopia gas field produces from the Jurassic Sawtooth formation.
0 5 10 mi
0 5 10 15 km
N
Valley of the Marias River before the ice ages.

contains interesting fossils, unusual igneous rocks exist in the Sweetgrass Hills, and interesting glacial features abound.

All the bedrock exposed along the road is sedimentary formations deposited in shallow sea water during late Cretaceous time, during the period between 100 and 65 million years ago. But good bedrock exposures are rare in this part of Montana because a deep upholstery of glacial debris covers most of the landscape.

The section vividly illustrates how flat the layers of sedimentary rock remain where they lay beyond the reach of the crustal movements that raised the Rocky Mountains. Even the Kevin-Sunburst Dome, which trapped so much oil and gas, is hardly visible.

A leg bone of a duck-billed dinosaur in the badlands north of Rudyard.
—Larry French photo

Oil and Gas

In the area around Shelby and Cut Bank, the highway crosses a large structural warp in the bedrock called the Sweetgrass arch. Oil and gas are both lighter than water, so they tend to move upward through the rocks until they come to the crest of an anticlinal arch, where they can migrate no farther. In some cases, they come to rocks too impermeable to let the oil move before they reach the crest of the arch. Then the oil and gas accumulate in what is called a stratigraphic trap on the flanks of the structure. The Sweetgrass arch contains a number of oil and gas fields of both types.

Around Cut Bank, the highway passes many wells, separators, and storage tanks, the Cut Bank field. The discovery well found natural gas in 1926, but was abandoned for lack of a market. Production finally began in 1931 after another hole struck both oil and gas. Most of the wells produce from Cretaceous sandstones in the Kootenai formation at a depth of approximately 3000 feet. Several dozen deeper wells produce from Mississippian limestones in the Madison group. In 1983, the field produced more than a million barrels of oil from about 900 wells, and had produced a total of more than 600 million dollars worth of oil since the first wells were drilled.

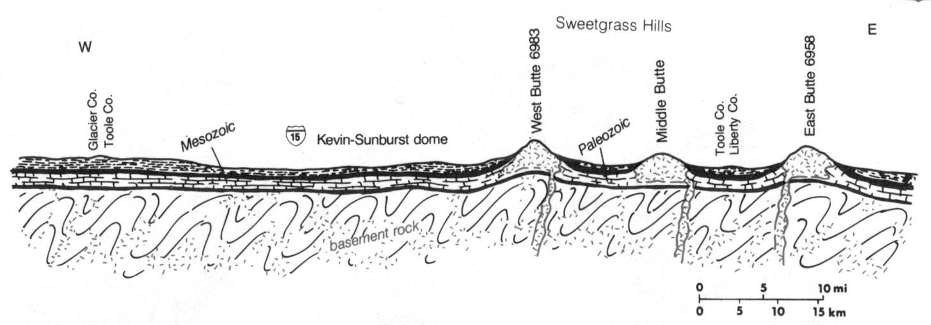

Section north of US 2 across the Kevin-Sunburst dome and the Sweetgrass Hills.

The Sweetgrass Hills

West, Middle, and East buttes punctuate the distant skyline north of the road between Shelby and Chester. Each is a miniature mountain range composed of a complex of igneous intrusions and the older sedimentary rocks that surround them. Grassy and Haystack buttes, much smaller hills hardly visible from the highway, are single igneous intrusions. During the Bull Lake ice age, the continental glacier flowed around the three big buttes of the Sweetgrass Hills, leaving them standing like islands above a sea of ice.

Igneous rocks in the Sweetgrass Hills are mostly pale syenites composed mainly of feldspar and rich in sodium and potassium. The rocks closely resemble those in the Judith and Moccasin mountains north of Lewistown and those in the Little Rocky Mountains south of Chinook. All those igneous rocks formed during the great spasm of activity of approximately 50 million years ago.

Igneous rocks in the Sweetgrass Hills. Thin lines are sills and dikes.

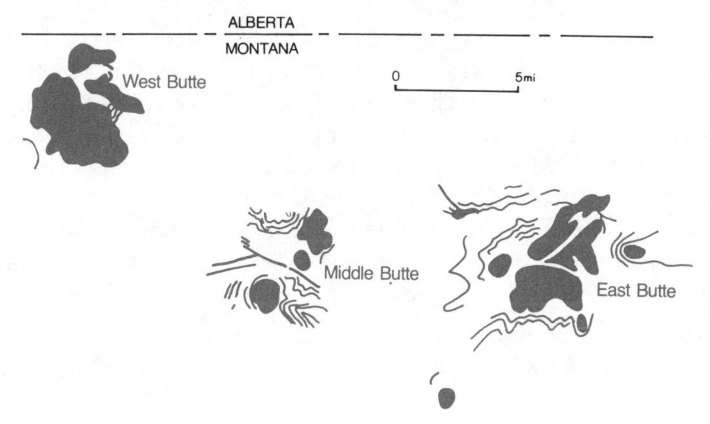

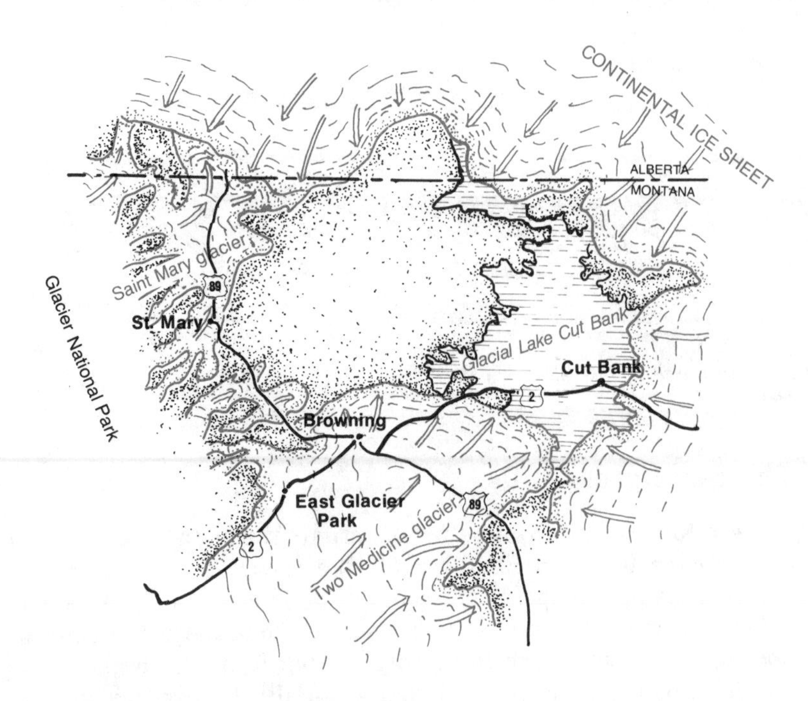

Glaciers and Glacial Lake Cut Bank when the Bull Lake ice age was near its maximum.

Glaciation

During the Bull Lake glaciation of 70,000 to 130,000 years ago, ice pouring east from the Rocky Mountains ponded on the plains along the mountain front to create a series of big piedmont glaciers. One of those, the Two Medicine Glacier, stretched almost ten miles east of Browning and left, as it melted, a vast expanse of moraine, chaotically hummocky landscape with little ponds snuggled among lumpy little hills.

Meanwhile, the great continental glacier that started in central Canada reached almost to Cut Bank, and covered almost all of northern Montana from there east. The road crosses more hummocky morainal landscape between Cut Bank and Shelby. That ice almost met the Two Medicine Glacier moving east from the Rocky Mountains south of Cut Bank, and did meet the St. Mary Glacier, another piedmont ice sheet, near the Canadian border. Glacial Lake Cut Bank flooded the area between the two glaciers to an elevation as high as 3900 feet. It overflowed into Glacial Lake Great Falls.

Although the moraines left by the continental and piedmont glaciers look like moraines everywhere, the erratic boulders that litter their surfaces differ. The Two Medicine and St. Mary piedmont glaciers left Precambrian sedimentary rocks eroded from formations exposed in Glacier National Park: gray limestone, and red and green mudstones. The continental glacier left boulders of Precambrian basement rock, pink granite, and pink and gray gneiss that must have come from northern Manitoba.

Except for the conspicuous moraine between Cut Bank and Shelby, the continental glacier left remarkably little mark on the landscape. This area was near the farthest extent of glaciation, so the ice did not arrive until the glaciers grew to their maximum size. At its farthest reach, the continental glacier was thin, and moving slowly, so it lay gently on the land. Along most of the route east of Shelby, occasional erratic boulders of granite or gneiss, rounded rocks lying in the fields, are the only obvious souvenirs of the ice age.

Badland exposures of late Cretaceous shales north of Rudyard. The dark layer is a coal bed. —Larry French photo

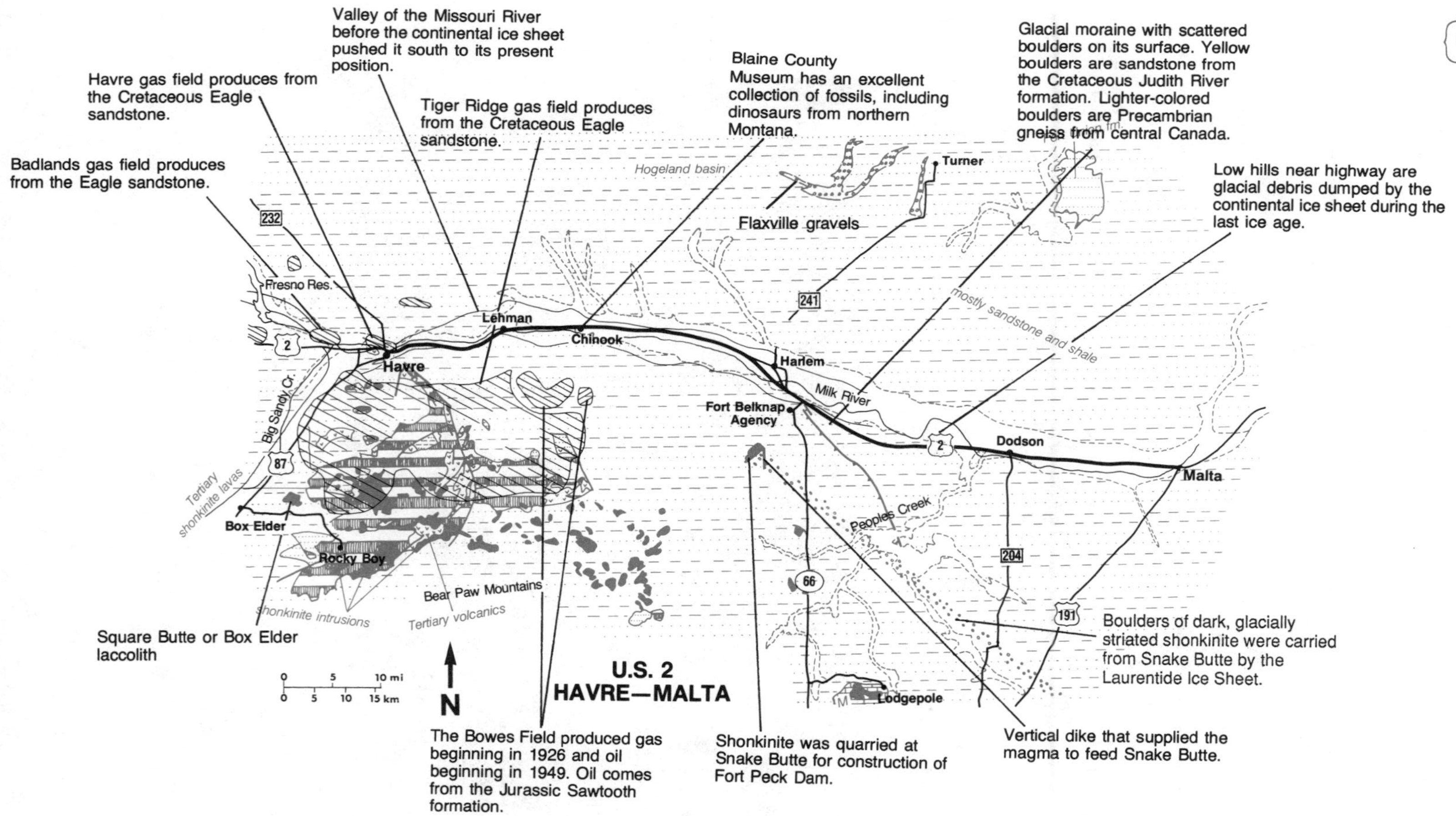

2
Valley of the Missouri River before the continental ice sheet pushed it south to its present position.
Havre gas field produces from the Cretaceous Eagle sandstone.
Tiger Ridge gas field produces from the Cretaceous Eagle sandstone.
Badlands gas field produces from the Eagle sandstone.
Blaine County Museum has an excellent collection of fossils, including dinosaurs from northern Montana.
Glacial moraine with scattered boulders on its surface. Yellow boulders are sandstone from the Cretaceous Judith River formation. Lighter-colored boulders are Precambrian gneiss from central Canada.
Low hills near highway are glacial debris dumped by the continental ice sheet during the last ice age.
Hogeland basin
Flaxville gravels
Turner
232
241
Fresno Res.
2
Lehman
Chinook
Havre
Harlem
mostly sandstone and shale
Milk River
Fort Belknap Agency
Big Sandy Cr.
87
Tertiary shonkinite lavas
Box Elder
Rocky Boy
Bear Paw Mountains
shonkinite intrusions
Tertiary volcanics
Dodson
Malta
Peoples Creek
204
66
191
Lodgepole
Square Butte or Box Elder laccolith
0 5 10 mi
0 5 10 15 km
N
U.S. 2
HAVRE—MALTA
Boulders of dark, glacially striated shonkinite were carried from Snake Butte by the Laurentide Ice Sheet.
The Bowes Field produced gas beginning in 1926 and oil beginning in 1949. Oil comes from the Jurassic Sawtooth formation.
Shonkinite was quarried at Snake Butte for construction of Fort Peck Dam.
Vertical dike that supplied the magma to feed Snake Butte.

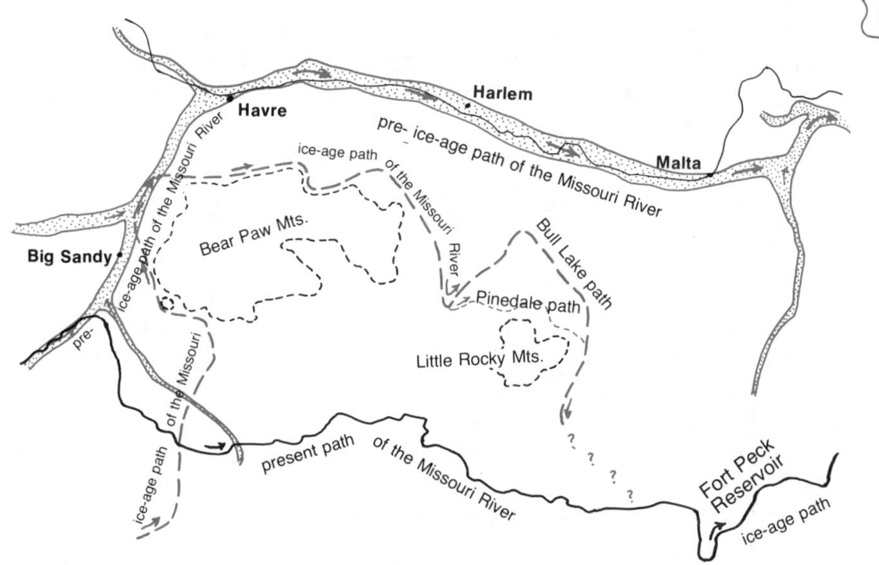

The old and modern courses of the Missouri and Milk rivers.

US 2: Havre—Malta
88 mi./141 km.

US 2 follows the Milk River all the way between Havre and Malta. Bedrock in this part of Montana is sedimentary formations deposited during Cretaceous time, some in shallow sea water, others on land near the coast. Exposures are rare because the road follows the Milk River Valley, and the rocks do not resist erosion.

The Misplaced Milk River

A few miles east of Havre, the valley of the Milk River abruptly becomes much too broad for the stream. From there to its mouth, the little river wanders aimlessly in the spacious floor of a broad valley that it could not have eroded. There is no corresponding change in resistance of the bedrock, so the explanation must involve some event in the history of the river. The first thing to consider is ice. All the countryside north of the Missouri River lay beneath glacial ice at the end of the Bull Lake ice age.

Shortly after the turn of the century, an early geologist pointed out that the broad valley of the lower Milk River is about the size of the

Missouri River Valley below Fort Peck, and that the Missouri River flows through a narrow canyon for a long distance between Fort Benton and Fort Peck. He suggested that the Missouri River may have occupied the broad valley of the lower Milk River until the ice sheet pushed it south. When the ice melted, the Milk River started flowing through the old valley of the Missouri River, which continued to flow in a channel it had established along the edge of the glacier.

If that explanation is valid, we should be able to find the abandoned valley of the Missouri River that connects the modern river in the area between Great Falls and Fort Benton with the broad stretch of the Milk River Valley that starts near Havre. The grossly oversized valley of Big Sandy Creek south of Havre is probably part of the abandoned segment of the Missouri River, now deeply filled with glacial debris. Another segment of the abandoned valley lies just north of Havre, and is also partly filled with glacial sediments.

As the most recent ice age reached its maximum about 15,000 years ago, continental ice creeping southwest from central Canada reached almost as far south as the Milk River between Harlem and Malta, and did reach the Milk River not far east of Malta. The moraines form long tracts of lumpy hills a few miles north of the road.

The Bearpaw Uplift

The Bearpaw Mountains, a volcanic pile erupted about 50 million years ago, sprawl along the distant horizon directly south of Havre. Rocks there are mostly shonkinite. The mountains stand on the crest of the Bearpaw uplift, a broad arch in the plains that also trapped enormous volumes of natural gas, along with some oil.

Section across US 2 through the Bearpaw Mountains. Barber Butte is an igneous intrusion that stands high because it resists erosion.

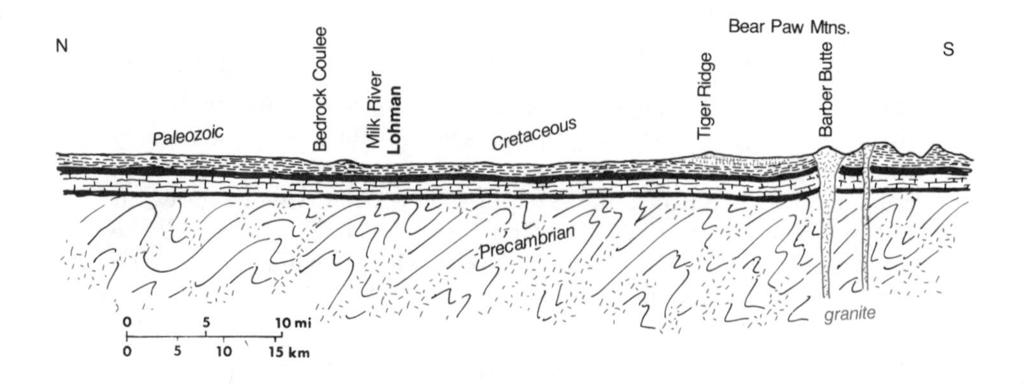

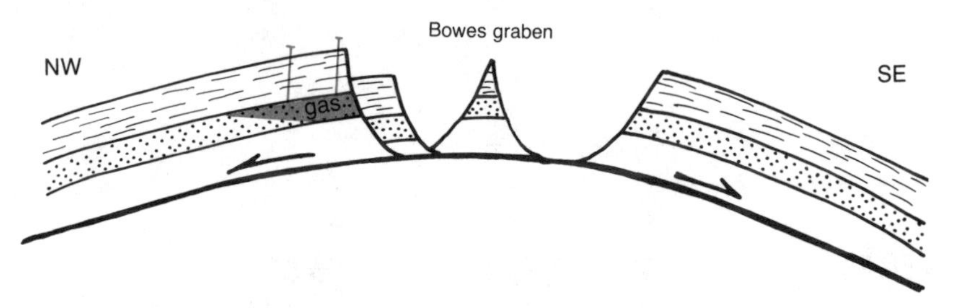

How blocks of Eagle sandstone sliding on slippery bentonite trap natural gas in the Bowes field.

The Bowes gas field produces from a bizarre reservoir. The Eagle sandstone there rests on beds of bentonite, an extremely slippery clay that forms through weathering of volcanic ash. The Elkhorn Mountains volcanoes were erupting between Helena and Butte while the volcanic ash was accumulating on the Bearpaw uplift, so it seems likely that they were the source. The Eagle sandstone broke into blocks that slid on the slippery bentonite, as the Clagget shale above it slid across the broken edge of the sandstone. Then the impermeable shale trapped gas moving up through the sandstone.

The Tiger Ridge gas field, the largest in Montana, also produces from the Eagle sandstone in the crest of the Bearpaw uplift. The unbelievably productive Eagle sandstone was deposited in beaches and barrier islands along the coast during late Cretaceous time. It is a rich source of gas, oil, and water because the open pore spaces between sand grains hold fluids.

The Little Rocky Mountains

The isolated cluster of low peaks on the distant skyline directly south of Harlem is the Little Rocky Mountains, a dome punched in the rocks of the plains. The core of the range contains the only exposures of basement rock in this part of Montana, as well as large masses of igneous rock that intruded them about 50 million years ago. The nearly vertical Madison limestone stands up in erosional relief around the edge of the range to form a wall that surrounds the range.

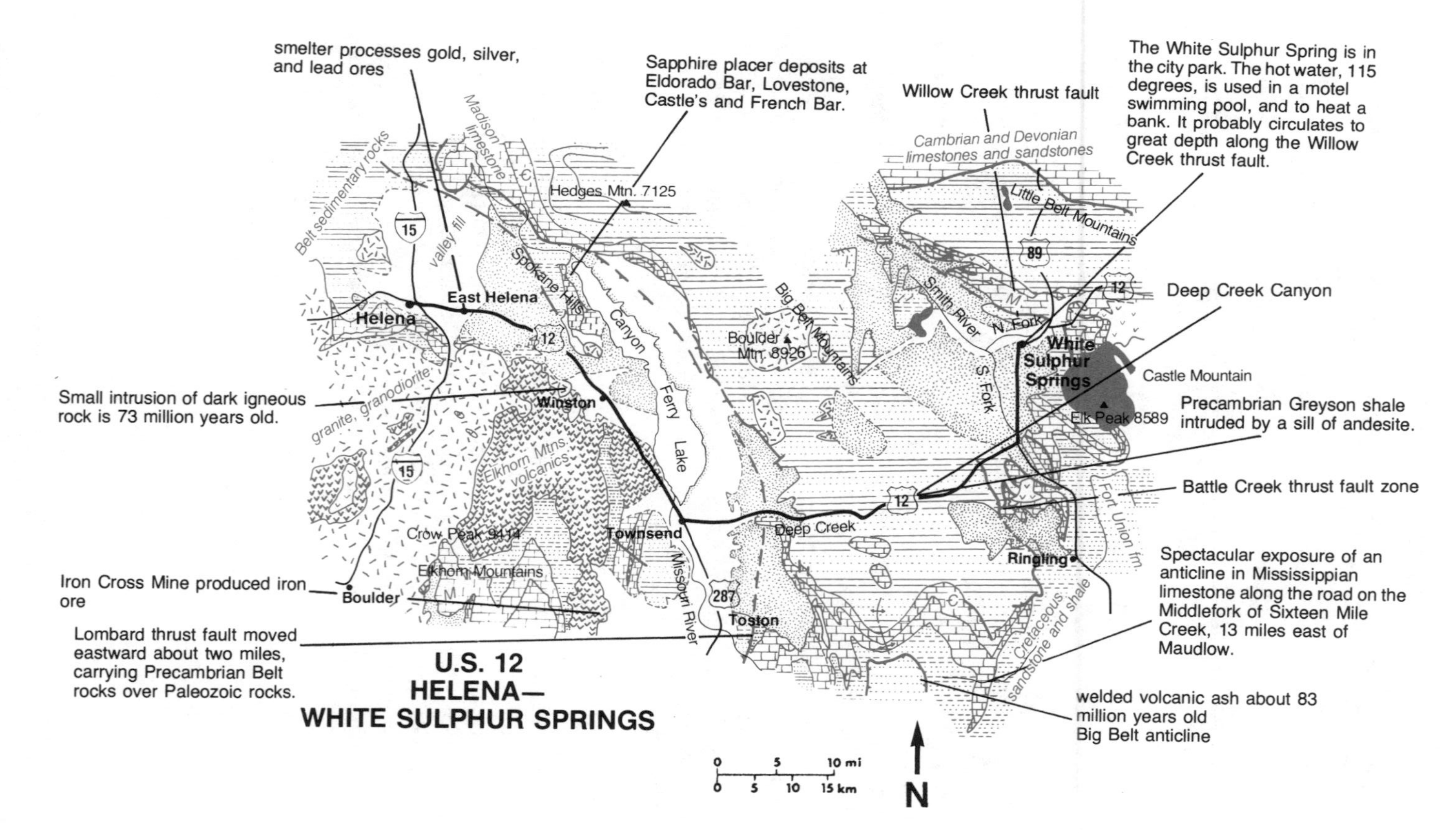

smelter processes gold, silver, and lead ores
Sapphire placer deposits at Eldorado Bar, Lovestone, Castle's and French Bar.
Willow Creek thrust fault
The White Sulphur Spring is in the city park. The hot water, 115 degrees, is used in a motel swimming pool, and to heat a bank. It probably circulates to great depth along the Willow Creek thrust fault.
Cambrian and Devonian limestones and sandstones
Madison limestone
Belt sedimentary rocks
Hedges Mtn. 7125
Little Belt Mountains
15
89
valley fill
East Helena
Helena
Spokane Hills
12
Canyon
Ferry
Lake
Big Belt Mountains
Boulder Mtn. 8926
Smith River
N. Fork
S. Fork
Deep Creek Canyon
White Sulphur Springs
Castle Mountain
Elk Peak 8589
Precambrian Greyson shale intruded by a sill of andesite.
Small intrusion of dark igneous rock is 73 million years old.
granite, granodiorite
Winston
Elkhorn Mtns. volcanics
Battle Creek thrust fault zone
Fort Union fm.
Deep Creek
Crow Peak 9414
Townsend
Ringling
Spectacular exposure of an anticline in Mississippian limestone along the road on the Middlefork of Sixteen Mile Creek, 13 miles east of Maudlow.
Iron Cross Mine produced iron ore
Boulder
Elkhorn Mountains
Missouri River
287
Toston
Cretaceous sandstone and shale
Lombard thrust fault moved eastward about two miles, carrying Precambrian Belt rocks over Paleozoic rocks.
U.S. 12
HELENA—
WHITE SULPHUR SPRINGS
welded volcanic ash about 83 million years old
Big Belt anticline
0 5 10 mi
0 5 10 15 km
N

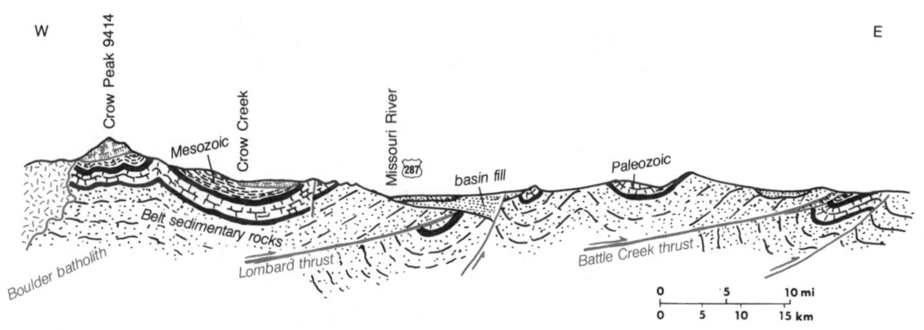

Section drawn along a line south of the highway from the area west of Townsend to that north of Ringling. The big thrust faults moved rocks east several miles from the area of the Boulder batholith.

US 12: Helena—White Sulphur Springs

74 mi./118 km.

About ten miles east of Helena, the road crosses the Spokane Hills, which separate the Helena and Townsend valleys. The reddish undertone in the landscape of the Spokane Hills expresses the red mudstones of the Precambrian Spokane formation, Belt rock. The route between the Spokane Hills and Townsend follows the broad valley that separates the Big Belt Mountains to the east from the Elkhorn Mountains on the west. The Missouri River is impounded in the valley floor to form Canyon Ferry Reservoir.

The Elkhorn Mountains lie south and west of the highway most of the way between Helena and Townsend. They are a complex, range consisting of a foundation of folded and faulted sedimentary rock formations that supports an extensive cover of volcanic rocks erupted from the Boulder batholith.

About seven miles east of Townsend, the road crosses the line of the Lombard thrust fault. Some geologists argue that this should be considered the eastern boundary of the northern Rocky Mountains, simply because it is the easternmost noteworthy thrust fault in a region renowned for that kind of structure. The Lombard fault is part of a horseshoe pattern of such faults with a displacement that is probably no more than a few miles, quite modest by the standards of northern Rocky Mountain thrust faults.

Between Townsend and the junction with Montana 89, the route crosses the Big Belt Mountains through picturesque Deep Creek

Canyon cut through Precambrian Belt rocks. The road between the junction with Montana 89 and White Sulphur Springs crosses deep valley fill sediments with the west flank of the Castle Mountains rising in the east.

Uranium and Thorium

In 1955, prospectors discovered uranium along the shore of Canyon Ferry Reservoir about five miles northeast of Winston. It is in the Renova formation, desert basin fill sediment deposited in the Townsend Valley during Oligocene time, about 35 million years ago. Here, as in many places, the volcanic ash deposits in the Renova formation contain abundant plant remains. Uranium dissolved in groundwater tends to deposit around organic matter, mostly as the yellow uranium oxide mineral carnotite. The deposits near Winston are too small and lean to mine.

Small but extremely high grade deposits of thorium exist on the east side of Canyon Ferry Reservoir almost directly across from Winston, and in several other places in the Big Belt Mountains. It occurs in and around small intrusions of granitic rock into Precambrian sedimentary formations, Belt rocks. If someone ever finds any use for thorium, these deposits may someday support small mines.

The Big Belt Mountains

Between Townsend and the junction with Montana 89, the road passes through the southern end of the Big Belt Mountains in the narrow canyon of Deep Creek. The Big Belt Mountains consist essentially of a broad crustal arch that exposes Precambrian sedimentary rocks, Belt formations, in its core. The road through Deep Creek Canyon passes marvelous exposures of Belt rock. Patches of Tertiary valley fill, Renova formation, exist on top of the range, so we must conclude that these rocks were under a valley floor as recently as 25 million years ago, and have been buckled into mountains sometime since then.

Steeply tilted layers of Precambrian sedimentary rock exposed in a roadcut in Deep Creek Canyon.
—Montana Bureau of Mines and Geology photo by H. L. James

Belt rock exposed in Deep Creek Canyon consists mostly of a great thickness of rather pale gray mudstone, the Greyson formation, and white limestone of the Newland formation. Both show a well developed platy structure, slaty cleavage, which is typical of strongly deformed and slightly recrystallized rocks. Thin slabs of slate spall off nearly every exposure, and in some places the sedimentary layers they cut across make ribbons of light and dark gray on their flat surfaces.

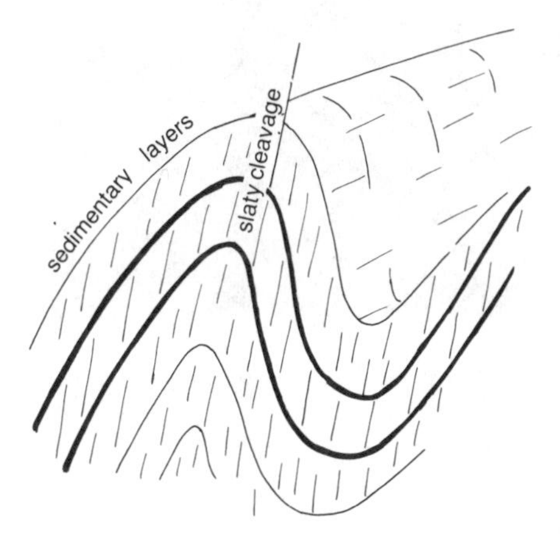

Slaty cleavage splits the rock across the original sedimentary layers along surfaces parallel to one that would slice the fold in half.

The Incredible Gravels in Confederate Gulch

Two former Confederate soldiers, paroled prisoners of war, found placer gold in Confederate Gulch near the center of the Big Belt Mountains in 1864. If we can believe the reports of the people who were there, at least some of whom cherished a reputation for

The placers were worked out and Diamond City was already a ghost town when this picture was taken about 1870.
—Montana Historical Society photo

Hydraulic miners at work washing gravel with giant streams of water. This picture was taken near Gardiner instead of in Confederate Gulch, but it illustrates the same technique. —Montana Historical Society photo

occasional sobriety, the gravels in Confederate Gulch were among the richest ever washed anywhere. They were freakishly rich. Single pans are said to have yielded as much as a thousand dollars worth of gold—at a time when an ounce of gold was worth something less than $20. Several accounts are that no less than two and one half tons of gold were recovered in a glorious final clean up of the sluice boxes. Total production from Confederate Gulch may have been worth about 16 million dollars—late 1860s money.

The gold-bearing gravels of Confederate Gulch existed only within the two acres of Montana Bar, and everyone left as soon as that ground had been worked. By 1870, the population of Confederate Gulch had declined from more than 10,000 to 255. Hardly a trace now remains of the roaring town of Diamond City, and the gulch yields grudging little to occasional prospectors who come seeking an overlooked scrap of the bonanza. The old timers got it all. Despite the incredible placer bonanzas, bedrock gold deposits near Confederate Gulch proved uninspiring. Evidently, the natural sluice of the stream concentrated gold weathered out of a very large volume of lean bedrock ore.

Confederate Gulch also saw hydraulic mining on a large scale. The technique involves bringing water to the site in a high flume, and then down through several hundred feet of pipe to huge nozzles that look about like old muzzle-loading cannon. The massive jets of high pressure water from hydraulic nozzles were said to have enough force to smash a brick building with a single pass at a range of several hundred feet—we don't know that anyone actually did that. The hydraulic nozzles certainly had enough force to wash down high gravel banks and flush their contents through sluice boxes. Hydraulic mining was especially effective in places like Confederate Gulch, where the gold-bearing gravels were in terraces well above stream level.

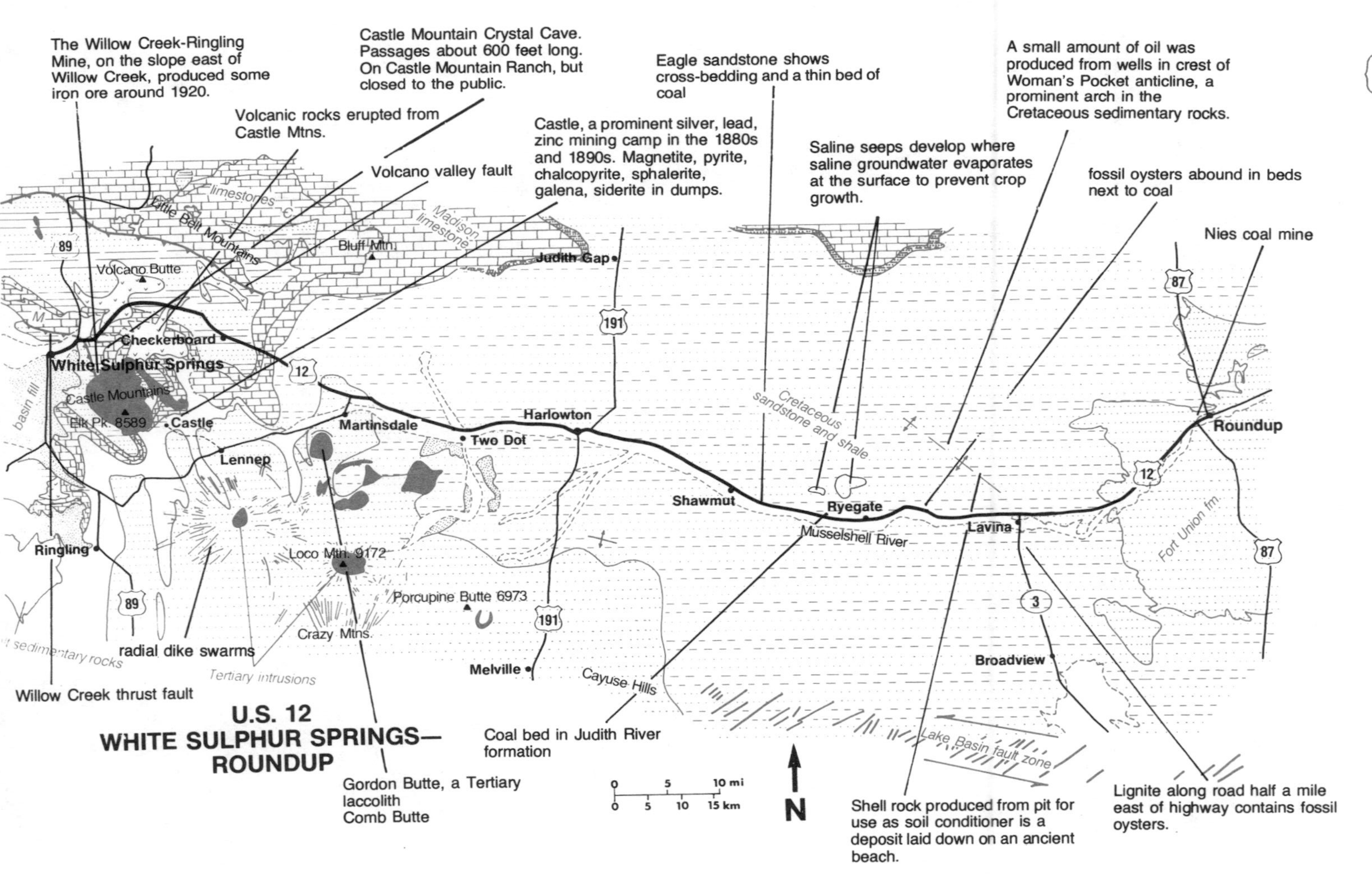

The Willow Creek-Ringling Mine, on the slope east of Willow Creek, produced some iron ore around 1920.
Castle Mountain Crystal Cave. Passages about 600 feet long. On Castle Mountain Ranch, but closed to the public.
Eagle sandstone shows cross-bedding and a thin bed of coal
A small amount of oil was produced from wells in crest of Woman's Pocket anticline, a prominent arch in the Cretaceous sedimentary rocks.
Volcanic rocks erupted from Castle Mtns.
Castle, a prominent silver, lead, zinc mining camp in the 1880s and 1890s. Magnetite, pyrite, chalcopyrite, sphalerite, galena, siderite in dumps.
Volcano valley fault
Saline seeps develop where saline groundwater evaporates at the surface to prevent crop growth.
fossil oysters abound in beds next to coal
Nies coal mine
limestones
Little Belt Mountains
Madison limestone
Bluff Mtn.
Judith Gap
89
Volcano Butte
191
87
Checkerboard
White Sulphur Springs
12
Castle Mountains
Elk Pk. 8589
Castle
basin fill
Martinsdale
Harlowton
Two Dot
Cretaceous sandstone and shale
Roundup
Lennep
Shawmut
Ryegate
Lavina
Musselshell River
Fort Union fm.
Ringling
Loco Mtn. 9172
Porcupine Butte 6973
3
89
191
Crazy Mtns.
radial dike swarms
sedimentary rocks
Melville
Broadview
Tertiary intrusions
Cayuse Hills
Willow Creek thrust fault
Lake Basin fault zone
U.S. 12
WHITE SULPHUR SPRINGS—ROUNDUP
Coal bed in Judith River formation
Gordon Butte, a Tertiary laccolith
Comb Butte
0 5 10 mi
0 5 10 15 km
N
Shell rock produced from pit for use as soil conditioner is a deposit laid down on an ancient beach.
Lignite along road half a mile east of highway contains fossil oysters.

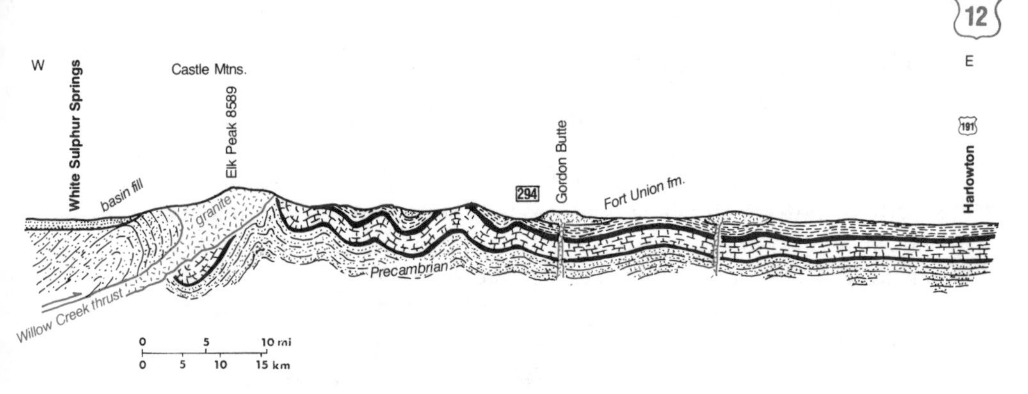

Section slightly south of the line of the highway between White Sulphur Springs and Harlowton.

US 12: White Sulphur Springs—Roundup

126 mi./202 km.

The road between White Sulphur Springs and Checkerboard follows the headwaters of the Musselshell River through a picturesque canyon between the Castle Mountains south of the highway and the Little Belt Mountains to the north. Large white outcrops in the canyon are Madison limestone, a sedimentary formation deposited in shallow sea water about 300 million years ago.

This is mountainous country. Along most of the route between Checkerboard and Harlowton, the jaggedly glaciated crest of the southern Crazy Mountains, snowcapped most of the year, dominates the southern skyline. A few good vantage points provide glimpses of the sprawling mass of the Gallatin Range and Beartooth Plateau far to the south. And the Little Belt Mountains cut their low and irregular profile out of the skyline north of the road. Between Harlowton and Roundup, the low bulge of the Big Snowy Mountains, a broad arch in the plains, rises in the north.

Except for a 12-mile stretch east of Lavina, the highway follows close to the Musselshell River all the way from Harlowton to Roundup. Older sedimentary formations along most of the route consist largely of shale deposited in shallow sea water during Cretaceous time, between about 135 and 65 million years ago. Those soft rocks erode into a relatively flat landscape. Cliffs along the river near Roundup expose the upper part of the Fort Union formation.

The Castle Mountains

The core of the Castle Mountains is a large mass of granitic rock that rose as molten magma about 50 million years ago. Part of the magma erupted to form lava flows and volcanic ash, some of which still remain north of the mountains. The rest of the magma crystallized at extremely shallow depth to form the granite. Like many granites that crystallized near the surface, that in the Castle Mountains contains blocky crystals of feldspar set in a matrix composed of crystals too small to see without a microscope.

The Castle Mountains granite tends to weather into a softly rounded landscape punctuated by occasional tall pinnacles of rock that rise through the deep mantle of soil. In dim light, and with the help of an active imagination, it is possible to see those rocky towers as ruined battlements left from some fabled age. That, apparently, is what gave the Castle Mountains their name.

Gordon Butte, the Northernmost Crazy Mountain

Except for a few dikes, Gordon Butte, just southwest of Martinsdale, is the northernmost outpost of the Crazy Mountains. It is shaped like a giant hockey puck, an almost perfectly circular intrusion more than a mile in diameter and about 700 feet thick. Like many other igneous intrusions in central Montana, Gordon Butte is a laccolith that formed as a blister of magma injected into sedimentary rocks, in this case the Fort Union formation. Rocks beneath the laccolith sank under the weight of the magma, so the center of Gordon Butte sags in a broad basin.

The igneous rock of Gordon Butte is the peculiar shonkinite of the northern Crazy Mountains, which differs from that of other igneous centers in central Montana in being very rich in sodium. The cooling magma lost much of its sodium to the enclosing sandstone and shale, converting them to a bright green rock called fenite—very rare stuff.

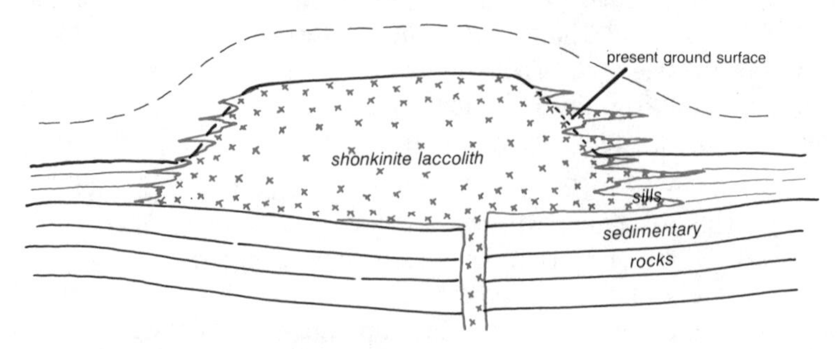

Section through Gordon Butte. Its base sags and its margins feather into the Fort Union formation in a swarm of sills.

A typical outcrop of sandstone in the upper part of the Fort Union formation.
—Montana Bureau of Mines and Geology photo by S. Fitch

Fort Union Formation

About halfway between Lavina and Roundup, the Musselshell River cuts deeply into the massive Tongue River sandstones in the upper part of the Fort Union formation. In the vicinity of Roundup, the river flows beneath bold cliffs of the yellowish sandstone, which absorb enough water to support a flourishing growth of pine trees.

As elsewhere, the sandstone in the upper part of the Fort Union formation contains good coal seams. Large mines operated until the late 1950s in Roundup and in the forested Bull Mountains south of the road. The coal seams are too deep for strip mining, so the mines worked underground. Even with growing energy shortages, it seems unlikely that such high cost mining will soon revive.

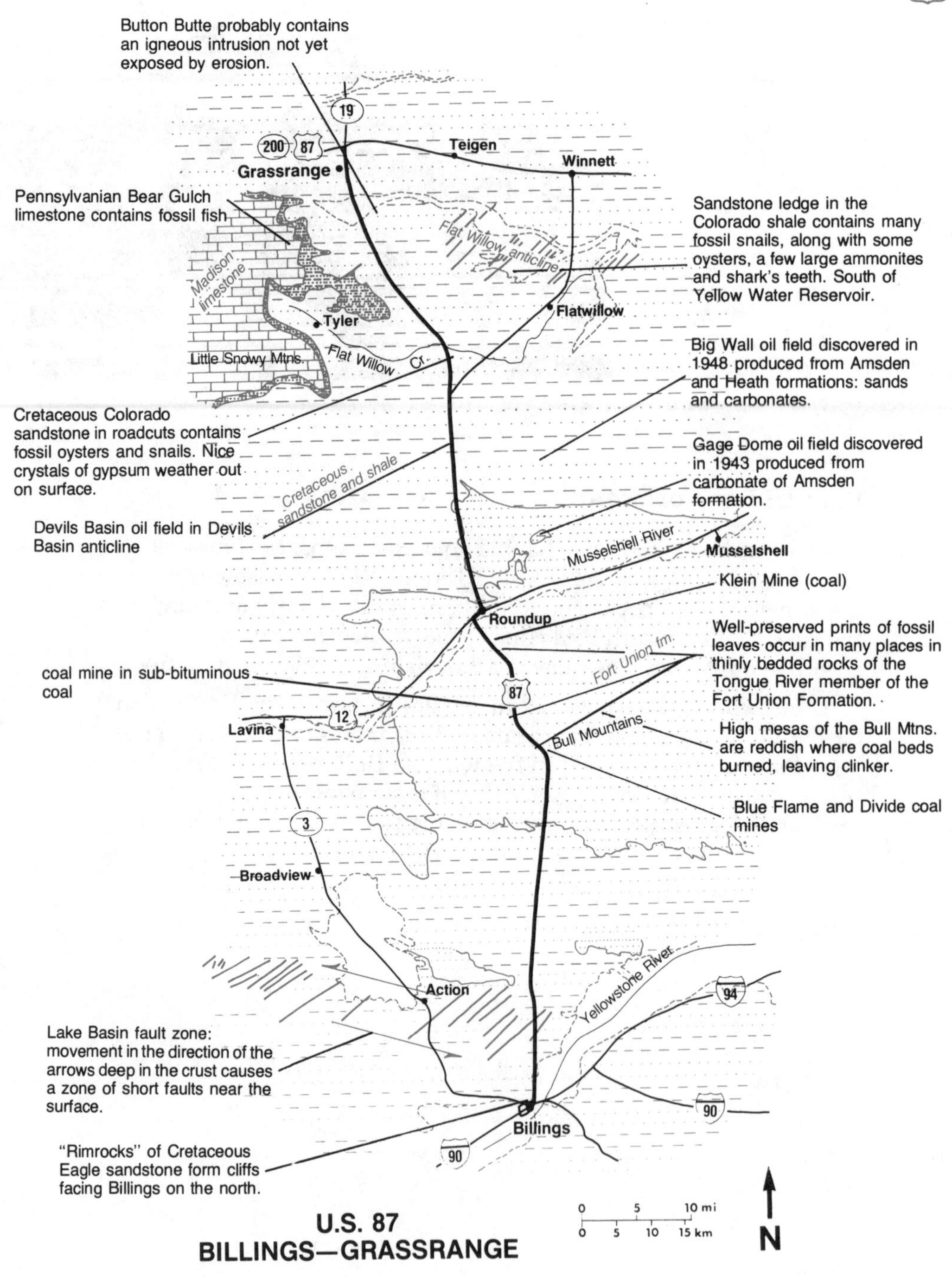
Button Butte probably contains an igneous intrusion not yet exposed by erosion.
19
200
87
Teigen
Winnett
Grassrange
Pennsylvanian Bear Gulch limestone contains fossil fish
Madison limestone
Flat Willow anticline
Sandstone ledge in the Colorado shale contains many fossil snails, along with some oysters, a few large ammonites and shark's teeth. South of Yellow Water Reservoir.
Tyler
Flatwillow
Little Snowy Mtns.
Flat Willow Cr.
Big Wall oil field discovered in 1948 produced from Amsden and Heath formations: sands and carbonates.
Cretaceous Colorado sandstone in roadcuts contains fossil oysters and snails. Nice crystals of gypsum weather out on surface.
Cretaceous sandstone and shale
Gage Dome oil field discovered in 1943 produced from carbonate of Amsden formation.
Devils Basin oil field in Devils Basin anticline
Musselshell River
Musselshell
Klein Mine (coal)
Roundup
Well-preserved prints of fossil leaves occur in many places in thinly bedded rocks of the Tongue River member of the Fort Union Formation.
Fort Union fm.
coal mine in sub-bituminous coal
87
12
Lavina
Bull Mountains
High mesas of the Bull Mtns. are reddish where coal beds burned, leaving clinker.
Blue Flame and Divide coal mines
3
Broadview
Yellowstone River
Action
94
Lake Basin fault zone: movement in the direction of the arrows deep in the crust causes a zone of short faults near the surface.
90
Billings
"Rimrocks" of Cretaceous Eagle sandstone form cliffs facing Billings on the north.
90
0 5 10 mi
0 5 10 15 km
N

U.S. 87
BILLINGS—GRASSRANGE

US 87: Billings—Grassrange
93 mi./149 km.

Bedrock between about ten miles north of Roundup and Grassrange is Cretaceous sedimentary formations, sandstone and shale. Some were deposited when this area was slightly below sea level and shallowly flooded, the rest on land along the coast. Exposures of shale are rare, but the more resistant sandstones appear in ledges in the valley walls, and in slabby outcrops capping some of the hills.

Most of those outcrops expose sedimentary layers still in their original horizontal position, but some reveal tilted layers, most noticeably along the south side of the valley of Flatwillow Creek, about 20 miles north of Roundup. Seeing nearly horizontal sedimentary beds in most exposures, steeply tilted beds more locally, leads one to suspect that the folded rocks are probably draped over deep faults.

Much of the landscape between Roundup and Grassrange consists essentially of an old and nearly level high plains surface in which the modern streams have cut their valleys. The road alternates between crossing broad expanses of nearly flat upland, the old plains surface, and the modern stream valleys eroded through it. Stream rounded pebbles, the Flaxville gravel, appear scattered in the wheat fields on the high, flat uplands, and roadcuts into that old surface expose thick beds of gravel.

Bull Mountains

For about 30 miles south of Roundup, the highway passes through the Bull Mountains, a tract of low, but rugged, hills well covered with a forest of bull pines. Rocks in this area are warped gently down into the broad Bull Mountain basin. Thick Tongue River sandstones of the upper part of the Fort Union formation are exposed throughout the Bull Mountains, along with some shale and numerous coal seams. Many of the coal seams burned, baking the rocks above them into

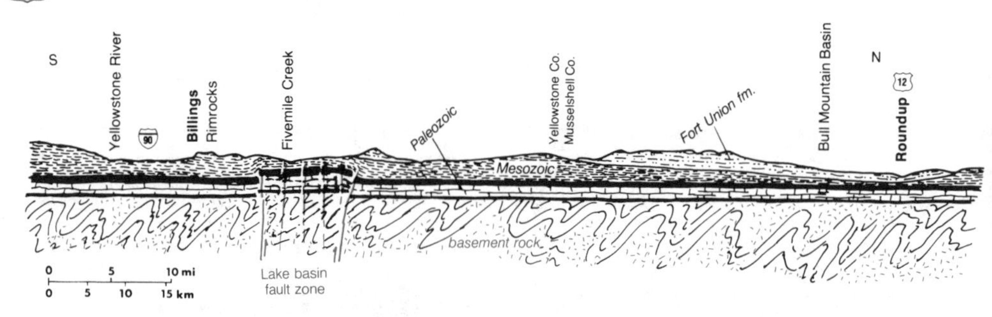

Section along the line of the road between Billings and Roundup.

zones of red clinker. The hard fired clinker and some of the sandstones resist erosion well enough to maintain the rugged hills. Watch for red caps of clinker on the crests of the ridges, broad bands of red on their flanks.

Coal mining began in the Bull Mountains in 1906, mostly to supply steam engines on the railroad, and continued through the 1950s. The coal came from at least 26 seams, all worked underground. Very little of the coal is shallow enough or in thick enough seams to permit strip mining, so it will be many years before large scale mining resumes in the Bull Mountain coal field.

The Devil's Basin Oil Field

About 15 miles north of Roundup, look for tilted layers of sandstone in the lower Cretaceous Kootenai formation. Geologists mapping those layers early in this century discovered them wrapping around a tightly folded anticlinal arch. A wildcat well drilled into the crest of that anticline in 1919 found oil, the Devil's Basin field, at a depth of 1185 feet. Of more than 25 wells drilled in the field during the early days, only three struck oil, and none of those produced very much. The field was shut down in 1937, then brought back into production about 20 years later, still not much good. Several wells on either side of the road still produce small quantities of oil.

Despite its dismal record, the Devil's Basin oil field is significant for its historic interest as the first oil discovery in central Montana. That first well inspired a rash of wildcat drilling elsewhere, which soon led to discovery of a number of good oil fields.

Button Butte

Button Butte, southeast of Grassrange, is an almost perfectly circular structural dome about three miles in diameter arched in the rocks beneath the plains, apparently not associated with any other folding. Several wildcat wells drilled into it many years ago found no oil, but did reveal that the rocks in the crest of the dome appear to have been baked. Although no igneous rocks have been found, geologists generally assume that the Button Butte dome arches over a deep igneous intrusion. Temperatures high enough to bake the rocks would certainly destroy any oil they might contain.

US 87: Great Falls—Grassrange

136 mi./218 km.

Between Great Falls and Grassrange, US 89 skirts around several mountain ranges full of a wide variety of rocks. But the only rocks the road actually crosses are sedimentary formations deposited in shallow sea water during Jurassic and Cretaceous time, between about 150 and 70 million years ago. Very few of those roadside rocks resist erosion well enough to make good outcrops.

The Highwood Mountains sprawl along the northern horizon between Belt and Geyser. On clear days, the geologically similar Bearpaw Mountains rise beyond them on the farthest horizon. Both ranges are volcanic centers that erupted shonkinite and other rocks about 50 million years ago. The Little Belt Mountains lie south of the road between Belt and Stanford. Near Lewistown, the road passes south of the Judith and Moccasin mountains, north of the Big and Little Snowy mountains.

Round Butte and Square Butte stand east of the main mass of the Highwood Mountains in the far distance north of Geyser. They are intrusions of shonkinite.

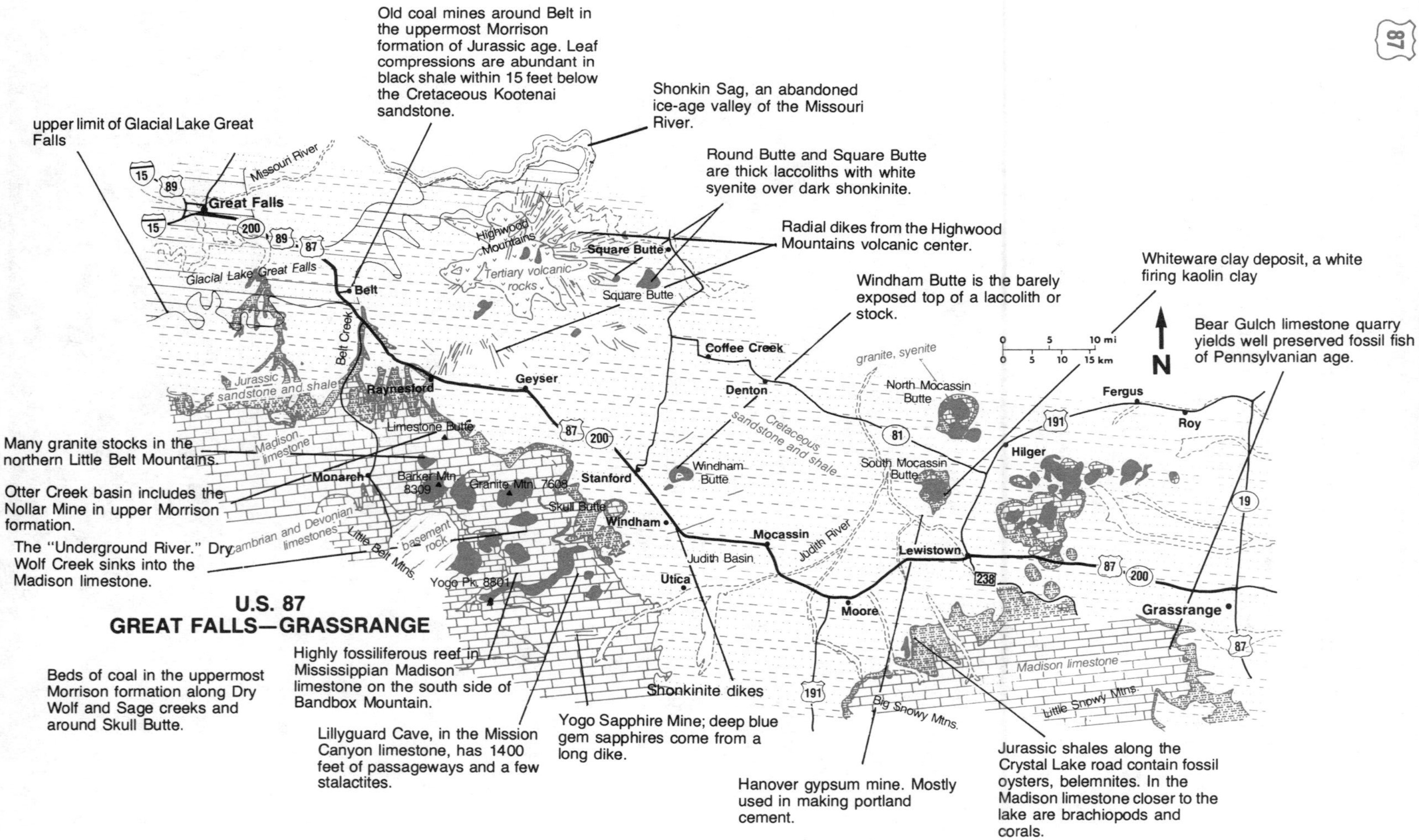

U.S. 87 GREAT FALLS—GRASSRANGE

upper limit of Glacial Lake Great Falls

Old coal mines around Belt in the uppermost Morrison formation of Jurassic age. Leaf compressions are abundant in black shale within 15 feet below the Cretaceous Kootenai sandstone.

Shonkin Sag, an abandoned ice-age valley of the Missouri River.

Round Butte and Square Butte are thick laccoliths with white syenite over dark shonkinite.

Radial dikes from the Highwood Mountains volcanic center.

Windham Butte is the barely exposed top of a laccolith or stock.

Whiteware clay deposit, a white firing kaolin clay

Bear Gulch limestone quarry yields well preserved fossil fish of Pennsylvanian age.

Many granite stocks in the northern Little Belt Mountains.

Otter Creek basin includes the Nollar Mine in upper Morrison formation.

The "Underground River." Dry Wolf Creek sinks into the Madison limestone.

Beds of coal in the uppermost Morrison formation along Dry Wolf and Sage creeks and around Skull Butte.

Highly fossiliferous reef in Mississippian Madison limestone on the south side of Bandbox Mountain.

Lillyguard Cave, in the Mission Canyon limestone, has 1400 feet of passageways and a few stalactites.

Yogo Sapphire Mine; deep blue gem sapphires come from a long dike.

Shonkinite dikes

Hanover gypsum mine. Mostly used in making portland cement.

Jurassic shales along the Crystal Lake road contain fossil oysters, belemnites. In the Madison limestone closer to the lake are brachiopods and corals.

Missouri River
Great Falls
15
89
200
87
Glacial Lake Great Falls
Belt
Belt Creek
Highwood Mountains
Tertiary volcanic rocks
Square Butte
Square Butte
Coffee Creek
Denton
Geyser
Raynesford
Jurassic sandstone and shale
Limestone Butte
Madison limestone
Monarch
Barker Mtn 8309
Granite Mtn 7608
Stanford
Skull Butte
Windham
Cambrian and Devonian limestones
Little Belt Mtns.
basement rock
Yogo Pk. 8801
Utica
Judith Basin
Mocassin
Judith River
Windham Butte
Cretaceous sandstone and shale
granite, syenite
North Mocassin Butte
South Mocassin Butte
81
Lewistown
238
Moore
191
Big Snowy Mtns.
Hilger
Fergus
Roy
19
Grassrange
Madison limestone
Little Snowy Mtns.
0 5 10 mi
0 5 10 15 km
N

An outcrop of coal near Belt.

The Great Falls-Lewistown Coal Field

Belt nestles in a canyon north of the highway almost in the shadow of Belt Butte, the prominent hill north of the road with a distinctive dark ledge of Cretaceous sandstone girdling its middle. Through a long sequence of connections, that ledge named Belt Butte, Belt, Belt Creek, the Belt and Little Belt mountains, even the Precambrian Belt rocks.

The first commercial coal mine in Montana opened in Belt in 1893, and the town mined coal for many years. A long decline began in 1930 when the Great Falls smelter switched to natural gas then accelerated toward final closure during the 1950s as the railroads switched to diesel engines. Remains of the old underground mines and their spoil heaps still form much of the Belt townscape.

The coal seams are in the upper part of the Morrison formation, which accumulated during Jurassic time, perhaps about 150 million years ago. To judge from the coal, the rocks associated with it, and the fossil plants they contain, this part of Montana was then a tropical and fairly arid coastal plain. The Rocky Mountains were beginning to rise in the west, and a shallow inland sea stretched away to the east. Beds of peat laid down in marshes became coal after they were buried under layers of sand and mud.

The mines at Belt were the largest workings in the Great Falls-Lewistown coal field, which also includes several areas along the north flanks of the Little Belt and Snowy mountains. Total production from the field during the years from 1885 to 1955 amounted to almost one quarter of all the coal produced in Montana. The entire coal field was moribund by 1955, and is likely to stay that way.

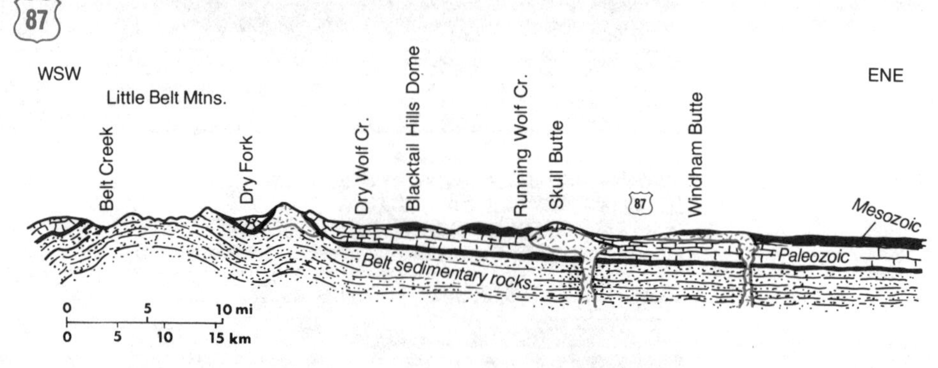

Section across the northern edge of the Little Belt Mountains near Stanford. Skull Butte apparently contains igneous rocks just exposed by erosion.

The Little Belt and Big Snowy Mountains

The Little Belt Mountains, a broad arch in the earth's crust, lie in the southern horizon along most of the route between Great Falls and Lewistown. The core of that range is an area of basement rock set within a mantle of Precambrian sedimentary rocks, Belt formations. Younger rock formations deposited during Paleozoic and Mesozoic time form the flanks of the arch. Like many of the broad folds in central Montana, the Little Belt Mountains also contain igneous rocks.

An array of igneous intrusions, laccoliths, forms a group of circular buttes in the northern part of the Little Belt Mountains. Other circular buttes in the same area probably contain igneous rocks still hidden beneath their cover of older sedimentary rocks. Windham Butte, near Stanford, is an outlying member of the group in which the igneous rocks peep out from under an uneroded cap of sedimentary rocks like the meat in a hamburger peeps out from beneath the bun. Although the igneous rocks vary somewhat from one intrusion to the next, all consist mostly of feldspar along with variable amounts of quartz and dark minerals—they are syenites and related rocks.

Another crustal arch, the broadly simple bulge of the Big Snowy Mountains, cuts a low profile in the horizon directly south of Lewistown. The Little Snowy Mountains are an eastward extension of the same structure. Rocks exposed in the middle of the Big Snowy uplift include Paleozoic formations, mainly the Madison limestone, deposited in shallow sea water around 300 million years ago. Elsewhere in this part of Montana, those rocks lie thousands of feet below the surface. Oil geologists find it most helpful to study such surface exposures of rocks they would otherwise see only in drill cuttings and cores.

Judith and Moccasin Mountains

The Judith Mountains north of Lewistown are a tight cluster of igneous intrusions, mostly granite and syenite, a pale igneous rock composed mostly of feldspar. The two isolated but basically similar large buttes of the Moccasin Mountains north of Hobson are syenite and related rocks. Some of those intrusions are known to be about 50 to 60 million years old, and the others are probably the same age. None of those centers include volcanic rock.

The intrusions invaded the older sedimentary rocks a few thousand feet below the surface. A number of other buttes in this part of Montana, including some along the road east of Lewistown, are domed sedimentary rocks that look suspiciously as though they may cover an igneous intrusion still not exposed by erosion.

Yogo Gulch Sapphires

Rumors of placer gold inspired a rush to Yogo Creek in 1879, but that excitement ended when it became clear that the reports lacked metallic substance. Another rumor of gold in 1894 actually led to construction of a big ditch, and once again the sluice boxes caught almost no gold. But the riffles did catch glassy blue pebbles that were finally recognized as sapphires in 1896. Mines that opened shortly thereafter produced large quantities of stones until 1929, when an assortment of legal and financial difficulties compounded by synthetic sapphires finally overwhelmed them.

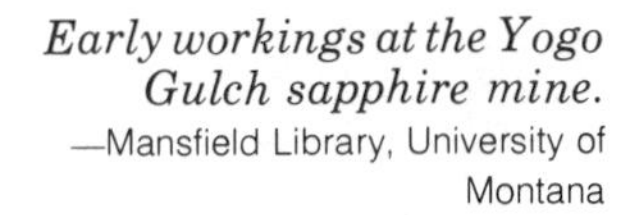

Early workings at the Yogo Gulch sapphire mine.
—Mansfield Library, University of Montana

For many years now, the Yogo Gulch mines have operated sporadically and generally at a loss, almost entirely as a source of gem rather than industrial sapphires. Yogo Gulch sapphires are greatly admired for their extraordinary color, an utterly distinctive deep blue untinged by the hint of green that blue sapphires from all other districts display. There is no other source for such stones.

Yogo Gulch sapphires come from a diatreme intrusion of greenish black rock melted directly from the earth's mantle. This diatreme is a dike about eight feet wide, and more than three miles long. Sapphire miners lay the freshly dug rock out to soften in the weather for several years, then break it up and wash the gems free. This is the only sapphire mine in the world that works hard rock; others wash the stones out of stream gravels or residual soil.

A dike of shonkinite stands like a giant ruined wall leading down to the Missouri River. A rind of baked sandstone clings to its sides. —Montana Department of Commerce photo

Giant Springs.
—Montana Tourist Bureau photo

US 87: Great Falls—Havre
113 mi./181 km.

Except for short stretches in the Marias and Missouri river valleys near Fort Benton, the route crosses high ground. Vast views of alternating wheat fields and summer fallow stretch into the far distance, a patterned floor under the sky. The Highwood Mountains east of Great Falls and the Bearpaw Mountains south of Havre are the deeply eroded remnants of old volcanic piles that also contain some masses of intrusive igneous rocks. They closely resemble each other in consisting largely of shonkinite erupted about 50 million years ago.

Bedrock near the road consists entirely of dark shale and brown sandstone, originally mud and sand deposited in shallow sea water during Cretaceous time. Outcrops are scarce because glacial deposits deeply mantle most of the landscape.

Giant Springs

Giant Springs is the centerpiece of a beautiful state park on the south bank of the Missouri River near the eastern edge of Great Falls. Enormous volumes of water well up through fractures in the Kootenai sandstone and pour into the river. The great flow of the springs poses a interesting problem because the sandstone does not

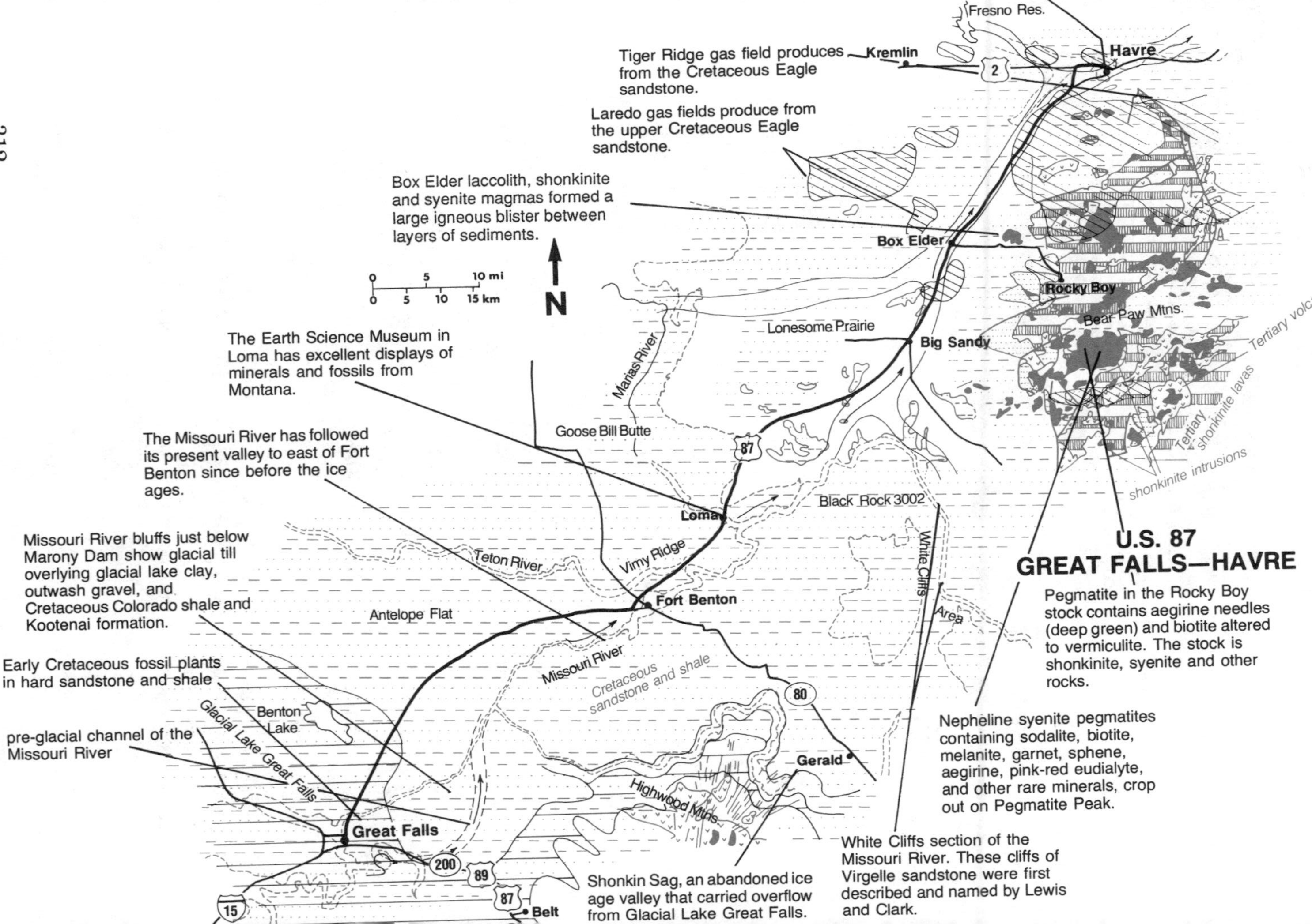
U.S. 87
GREAT FALLS—HAVRE
Tiger Ridge gas field produces from the Cretaceous Eagle sandstone.
Laredo gas fields produce from the upper Cretaceous Eagle sandstone.
Box Elder laccolith, shonkinite and syenite magmas formed a large igneous blister between layers of sediments.
The Earth Science Museum in Loma has excellent displays of minerals and fossils from Montana.
The Missouri River has followed its present valley to east of Fort Benton since before the ice ages.
Missouri River bluffs just below Marony Dam show glacial till overlying glacial lake clay, outwash gravel, and Cretaceous Colorado shale and Kootenai formation.
Early Cretaceous fossil plants in hard sandstone and shale
pre-glacial channel of the Missouri River
Pegmatite in the Rocky Boy stock contains aegirine needles (deep green) and biotite altered to vermiculite. The stock is shonkinite, syenite and other rocks.
Nepheline syenite pegmatites containing sodalite, biotite, melanite, garnet, sphene, aegirine, pink-red eudialyte, and other rare minerals, crop out on Pegmatite Peak.
White Cliffs section of the Missouri River. These cliffs of Virgelle sandstone were first described and named by Lewis and Clark.
Shonkin Sag, an abandoned ice age valley that carried overflow from Glacial Lake Great Falls.
Fresno Res.
Kremlin
2
Havre
Box Elder
Rocky Boy
Bear Paw Mtns.
Tertiary volcanics
Tertiary shonkinite lavas
shonkinite intrusions
Lonesome Prairie
Big Sandy
Marias River
Goose Bill Butte
87
Black Rock 3002
Loma
White Cliffs
Area
Teton River
Vimy Ridge
Fort Benton
Antelope Flat
Missouri River
Cretaceous sandstone and shale
80
Gerald
Benton Lake
Glacial Lake Great Falls
Highwood Mtns.
Great Falls
200
89
87
Belt
15
N
0 5 10 mi
0 5 10 15 km

contain nearly enough open pore space to produce water in such quantity.

It seems likely that the water discharging from Giant Springs must be rising from the Madison limestone, which lies several hundred feet below the surface. Limestones are the usual source of such large springs because they commonly contain caverns capable of conducting enormous flows. If the Madison limestone is indeed the source of Giant Springs, then the water probably comes from the Little Belt Mountains about 35 miles southeast of Great Falls. That is the closest area where the Madison limestone comes to the surface.

Look for leaves nicely preserved as black impressions on the bedding surfaces of the Kootenai sandstone around Giant Springs. The sandstone was deposited in early Cretaceous time, perhaps about 100 million years ago. Although that was about when the first flowering plants appeared, none are preserved in the Kootenai sandstone. The abundance of plant fossils and the complete absence of fossils of animals that lived in sea water show that the sandstone was deposited on land.

Glacial Debris

Except in the Great Falls area, the entire region along the line of the highway lay beneath glacial ice when the glaciers of the great Bull Lake ice age were at their maximum. Erratic boulders that sparsely litter a few of the fields beside the road are the most obvious evidence of glaciation. Many are rocks such as granite and gneiss, hard rocks full of glittering crystals that do not even slightly resemble the local bedrock. They are continental basement rocks that must have ridden in the ice all the way from central Canada, the closest place where the glacier could have picked them off the surface. They still lie precisely where the melting glacier dropped them.

A Game of Rivers and Valleys

Between Great Falls and Loma, the Missouri River flows in a broad valley appropriate to such a large stream. That part of the river is still in its original valley. At Loma, the Missouri River leaves its old valley to follow a narrow canyon east to Fort Peck Reservoir. The river probably started to flow in that course after a glacier blocked its former valley while the Bull Lake ice age was at its maximum sometime between 70,000 and 130,000 years ago. Water then emptying from Glacial Lake Great Falls eroded a channel, through the northern Highwood Mountains, that was abandoned after the lake drained. Elsewhere, water flowing near the edge of the glacier eroded such a deep channel that the river could not return to its old

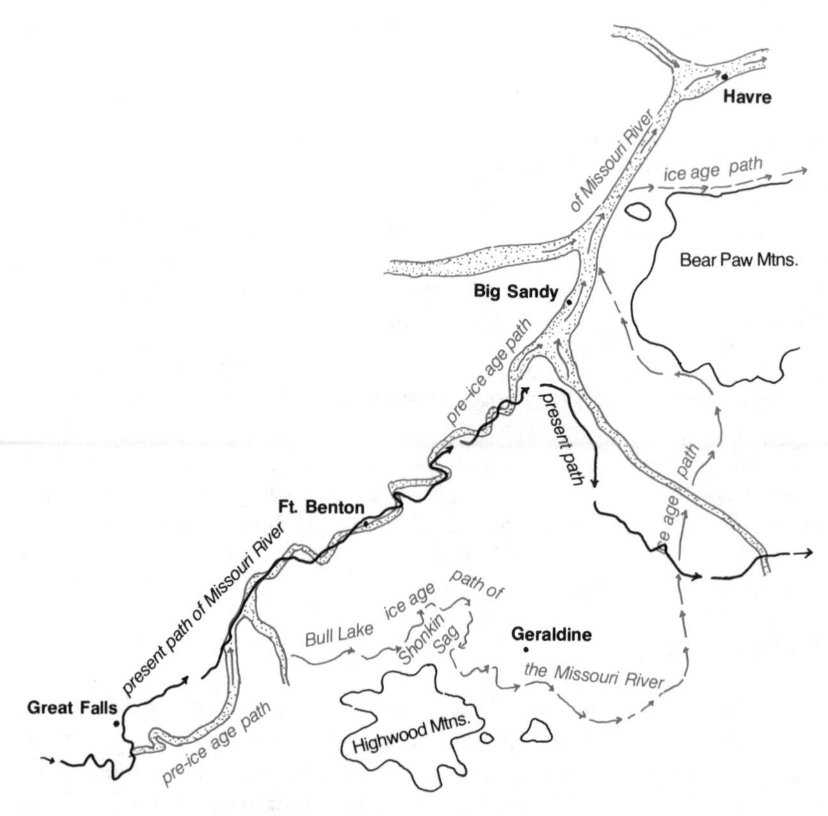

Preglacial, glacial, and present courses of the Missouri River.

valley after the ice melted. In any case, the melting ice left large parts of the old valley blocked with sediment.

Several stretches of highway between Loma and Havre follow Big Sandy Creek, which is much too small to have eroded its wide valley. In fact, the valley of Big Sandy Creek between Loma and Havre is just the right size to accommodate the Missouri River. It seems likely that the Missouri River did flow through that valley before the ice age glaciers pushed it into its present course. The outlines of the old valley are hard to see from the ground because it contains deep deposits of glacial sediment.

Bearpaw Mountains

The rugged Bearpaw Mountains rise east of the highway between Havre and Big Sandy. Fifty million years ago, an isolated cluster of volcanoes busily erupted large volumes of shonkinite, among other rocks, in that area. The Bearpaw Mountains also include a number of igneous intrusions, some of which contain extremely peculiar rocks. The Rocky Boy stock, for example, contains masses of carbonatite, an extremely rare igneous rock composed largely of the mineral calcite, calcium carbonate.

The big butte directly east of Box Elder is a laccolith that formed as a large blister of shonkinite magma pooled between layers of sedimentary rock. The upper part of the laccolith is sparkling white syenite composed almost entirely of potassium feldspar. As in several other such laccoliths, syenite magma appears to have separated from the molten shonkinite as oil separates from water, and floated to the top of the intrusion to form the white cap.

Like several other igneous centers in central Montana, the Bearpaw Mountains stand on the crest of a broad arch in the earth's crust, the Bearpaw uplift. That arch also trapped natural gas in the Eagle sandstone, a thick formation that accumulated along the shores of a shallow sea during Cretaceous time. The Tiger Ridge gas field, by far the largest in Montana, covers a vast area between Havre and the Bearpaw Mountains. About 200 wells there produce gas from as many square miles.

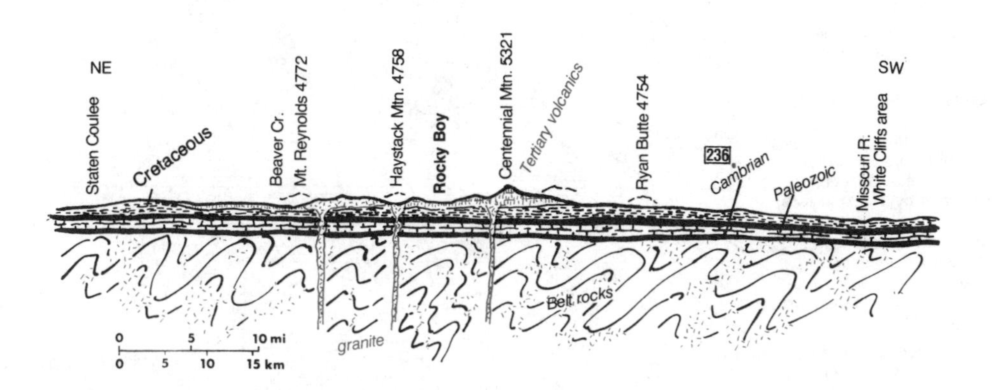

Section southeast of US 87 through the Bearpaw Mountains.

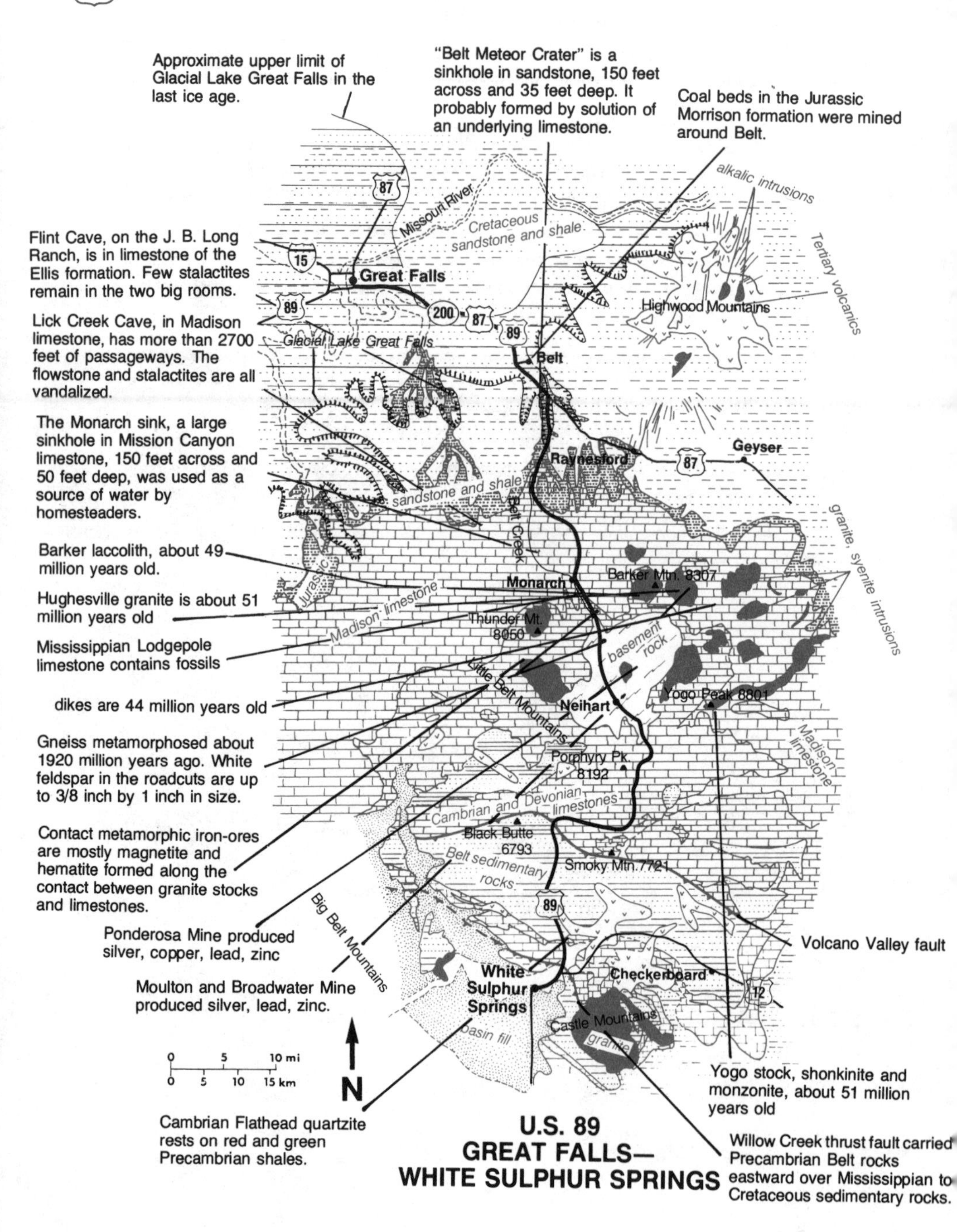
Approximate upper limit of Glacial Lake Great Falls in the last ice age.
"Belt Meteor Crater" is a sinkhole in sandstone, 150 feet across and 35 feet deep. It probably formed by solution of an underlying limestone.
Coal beds in the Jurassic Morrison formation were mined around Belt.
Flint Cave, on the J. B. Long Ranch, is in limestone of the Ellis formation. Few stalactites remain in the two big rooms.
Lick Creek Cave, in Madison limestone, has more than 2700 feet of passageways. The flowstone and stalactites are all vandalized.
The Monarch sink, a large sinkhole in Mission Canyon limestone, 150 feet across and 50 feet deep, was used as a source of water by homesteaders.
Barker laccolith, about 49 million years old.
Hughesville granite is about 51 million years old
Mississippian Lodgepole limestone contains fossils
dikes are 44 million years old
Gneiss metamorphosed about 1920 million years ago. White feldspar in the roadcuts are up to 3/8 inch by 1 inch in size.
Contact metamorphic iron-ores are mostly magnetite and hematite formed along the contact between granite stocks and limestones.
Ponderosa Mine produced silver, copper, lead, zinc
Moulton and Broadwater Mine produced silver, lead, zinc.
Cambrian Flathead quartzite rests on red and green Precambrian shales.
alkalic intrusions
Tertiary volcanics
Missouri River
Cretaceous sandstone and shale
Great Falls
Glacial Lake Great Falls
Highwood Mountains
Belt
Raynesford
Geyser
sandstone and shale
Belt Creek
Jurassic
Madison limestone
Monarch
Barker Mtn. 8307
Thunder Mt. 8050
granite, syenite intrusions
basement rock
Little Belt Mountains
Neihart
Yogo Peak 8801
Madison limestone
Porphyry Pk. 8192
Cambrian and Devonian limestones
Black Butte 6793
Belt sedimentary rocks
Smoky Mtn. 7721
Big Belt Mountains
Volcano Valley fault
White Sulphur Springs
Checkerboard
Castle Mountains
granite
basin fill
Yogo stock, shonkinite and monzonite, about 51 million years old
Willow Creek thrust fault carried Precambrian Belt rocks eastward over Mississippian to Cretaceous sedimentary rocks.
0 5 10 mi
0 5 10 15 km
N
15
87
89
200
12

U.S. 89
GREAT FALLS—
WHITE SULPHUR SPRINGS

89

Cylindrical fragments of crinoid stems litter this bit of Mississippian sea floor preserved in the Madison limestone in the Little Belt Mountains. The fossil in the center that looks almost like a clam is a brachiopod.
—Larry French photo

US 89: Great Falls—White Sulphur Springs
94 mi./150 km.

Between Great Falls and the area a few miles south of Belt, the highway crosses Jurassic and Cretaceous sedimentary rocks. Those don't resist weathering well enough to make many good exposures. Between Belt and White Sulphur Springs, the road passes through the Little Belt Mountains. Marvelous exposures of much older and more resistant rocks exist along that part of the route.

Highwood Mountains

The Highwood Mountains lie low along the horizon directly east of Great Falls, and north of the highway near Belt. They are the eroded remnants of a volcanic pile erupted about 50 million years ago. The volcanoes began by erupting a thick sequence of fairly normal volcanic rocks, medium gray stuff full of quartz grains. Then they changed their ways to produce vast quantities of shonkinite, very dark rocks greatly enriched in potassium and, like most dark igneous rocks, totally without quartz. The Highwoods also contain several interesting shonkinite intrusions.

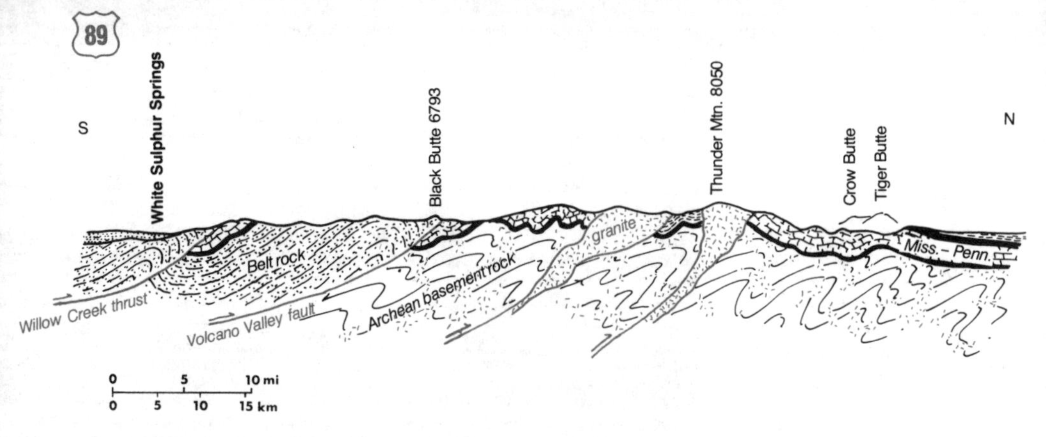

A faulted crustal arch brings basement rocks to the surface in the core of the Little Belt Mountains. Precambrian, Paleozoic, and Mesozoic formations flank the range. Section line just west of US 89.

Little Belt Mountains

Between Monarch and the area a few miles north of White Sulphur Springs, the highway winds through the Little Belt Mountains, following the valley of Belt Creek much of the way. Several miles of road just north of Neihart cross Precambrian basement rocks, the continental crust itself. This, an area in the Bridger Range, and the center of the Little Rocky Mountains are the only places in central Montana where those rocks appear at the surface. Several roadcuts north of Neihart expose a peculiar basement rock, the Pinto diorite, which contains large blocky crystals of pink feldspar floating in a sickly green matrix composed mostly of smaller grains of hornblende and plagioclase.

Big outcrops of pale gray Madison limestone along Belt Creek in the northern Little Belt Mountains.

Wu Tang Laundry and Drugs in Neihart, one of the last surviving buildings from the big mining boom of the last century.

In 1881, prospectors discovered silver and lead in Precambrian basement rocks north of Neihart, near the center of the Little Belt Mountains. Mining began the next year, and expanded rapidly, as did the towns of Neihart and Monarch, until the bottom dropped out of the silver market in 1893. That fairly effectively stopped mining in the Little Belt Mountains until 1917, when the price of silver rose. Mining ceased again, apparently permanently, in 1927 with the next big drop in the price of silver. The years since have brought occasional revivals of interest, but no real revival. Total production from the district amounts to something more than 17 million dollars, mostly in silver.

Laccoliths and their feeder dikes in the northern Little Belt Mountains east of US 89. Granite and finer-grained equivalents in color; syenite in black.

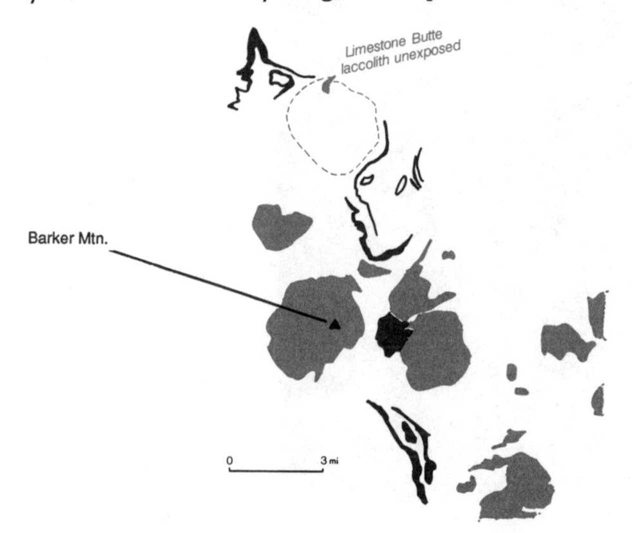

Like many of the crustal arches in central Montana, the Little Belt Mountains also contain large quantities of rocks intruded about 50 million years ago. Those rocks are in a swarm of laccoliths in the northern edge of the range, mostly east of the highway. Several suspicious looking circular buttes in the same area probably contain igneous intrusions not yet exposed by erosion. The laccoliths formed as blisters of molten magma injected between layers of sedimentary rock. The conduits that fed magma into the laccoliths survive as long dikes that radiate from a common center and connect with the laccoliths. Igneous rocks in the dikes and laccoliths closely resemble many of those in the Sweetgrass Hills and Judith and Moccasin mountains.

The Willow Creek and Volcano Thrust Faults

Several large thrust faults near and north of White Sulphur Springs carried a thick pile of sedimentary rocks that includes Precambrian Belt formations northeastward. Now they lie on top of Paleozoic formations that lie directly on basement rock, no Belt formations in between. It seems likely that those faults moved about 50 million years ago because the magma that became the Castle Mountains granite moved up along one of them. That granite is certainly 50 million years old. It seems likely that the hot water in White Sulphur Springs was heated as it circulated deep into the crust along the Willow Creek thrust fault.

Old mining debris near Neihart.

US 89: White Sulphur Springs—Livingston

69 mi./110 km.

Just south of White Sulphur Springs, the road passes west of the Castle Mountains. Between the Castle Mountains and Livingston, it crosses the western part of the Crazy Mountain Basin. Bedrock exposed near the road along the entire route is sandstone and shale, soft rocks that do not resist erosion well enough to form many good outcrops. Mountains on either side of the road contain a wide variety of bedrock types.

Hot Sulphur Springs

A city park near the west end of White Sulphur Springs contains a hot spring that does indeed smell faintly of hydrogen sulfide. The water surfaces at a temperature of 115 degrees Fahrenheit, a bit higher than most hot springs in the region. Although the details are not clear, it seems reasonable to suppose that the hot springs are producing water that has circulated deep along fractures, perhaps to the Willow Creek thrust fault. Rocks at a depth of several thousand feet are hot enough to heat water to the temperature of the hot springs. The hydrogen sulfide probably comes from pyrite, a common mineral in many kinds of rocks.

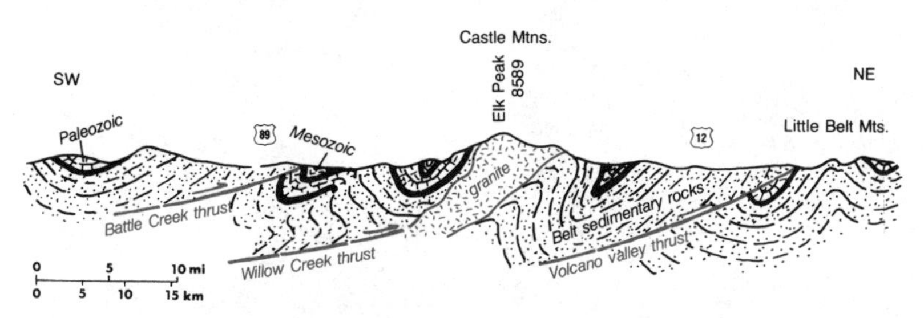

Section across US 89 near the junction with US 12.

The Castle Mountains

South of White Sulphur Springs, the highway passes the Castle Mountains. They look on the geologic map like a bullseye of granite set within crudely concentric rings of sedimentary rock that become younger outwards. The mass of granitic magma rose along the Willow Creek thrust fault, bulging the already folded sedimentary formations that lie beneath the plains to form a dome, now largely planed off by erosion. White rocks visible near the road in the west

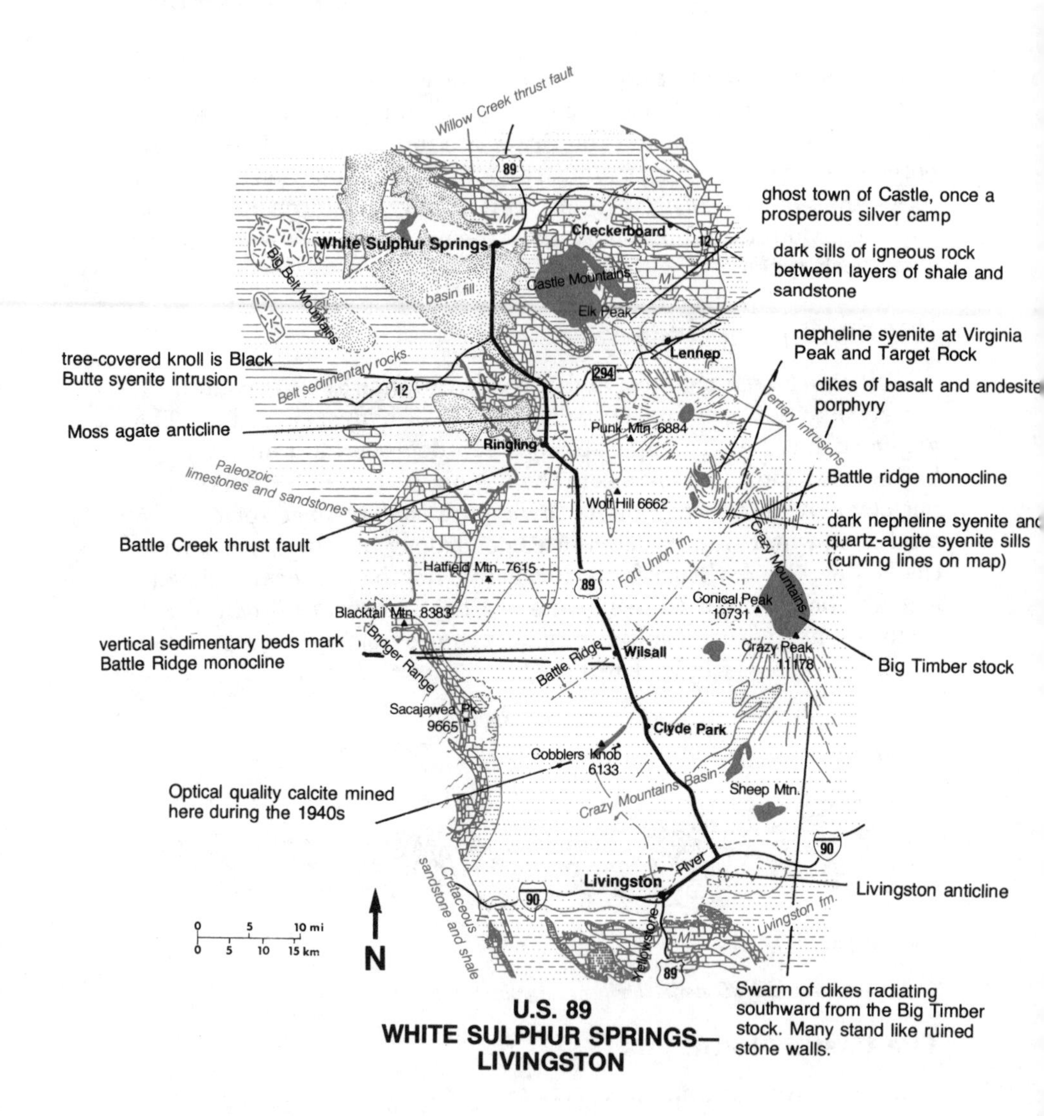
Willow Creek thrust fault
89
ghost town of Castle, once a prosperous silver camp
Checkerboard
12
White Sulphur Springs
dark sills of igneous rock between layers of shale and sandstone
Big Belt Mountains
Castle Mountains
M
basin fill
Elk Peak
nepheline syenite at Virginia Peak and Target Rock
Lennep
tree-covered knoll is Black Butte syenite intrusion
Belt sedimentary rocks
294
12
dikes of basalt and andesite porphyry
Tertiary intrusions
Moss agate anticline
Punk Mtn. 6884
Ringling
Paleozoic limestones and sandstones
Battle ridge monocline
Wolf Hill 6662
dark nepheline syenite and quartz-augite syenite sills (curving lines on map)
Battle Creek thrust fault
Fort Union fm.
Crazy Mountains
Hatfield Mtn. 7615
89
Conical Peak 10731
Blacktail Mtn. 8383
vertical sedimentary beds mark Battle Ridge monocline
Bridger Range
Battle Ridge
Wilsall
Crazy Peak 11178
Big Timber stock
Sacajawea Pk. 9665
Clyde Park
Cobblers Knob 6133
Optical quality calcite mined here during the 1940s
Crazy Mountains Basin
Sheep Mtn.
90
Livingston
River
Livingston anticline
Cretaceous sandstone and shale
90
Livingston fm.
0 5 10 mi
0 5 10 15 km
N
Yellowstone
89
Swarm of dikes radiating southward from the Big Timber stock. Many stand like ruined stone walls.

U.S. 89
WHITE SULPHUR SPRINGS—
LIVINGSTON

The lonely relics of Castle.

flank of the Castle Mountains are Madison limestone; red rocks farther back are the much older Spokane shale, Precambrian Belt rocks. Spires of granite rise on the skyline, near the crest of the mountains.

The abandoned town of Castle, on the southeast side of the Castle Mountains, was an active mining camp late in the last century and early in this. Tradition has it that a prospector found a hunk of ore in the Musselshell River, then tracked it 50 miles upstream by looking for similar pieces, to its source on the flank of the Castle Mountains. Mines began to produce lead, zinc, and silver in 1885, and reached their peak in 1892, when three smelters were in operation. The monetary panic of 1893 sent the mines into a steep decline from which they revived for a while, then relapsed into hard times in 1898. A few small mines limped along sporadically for a while longer; all died before 1920.

Crazy Mountain Basin

From near Ringling to Livingston, the highway follows the western side of the Crazy Mountain Basin. The Bridger Range defines the western margin of the basin well west of the highway, and the Crazy Mountains rise in its center east of the highway. Near Ringling, the road crosses Cretaceous sedimentary rocks. Farther south, it crosses the Fort Union formation, which accumulated during latest Cretaceous and early Tertiary time.

The Crazy Mountain Basin is a deep trough buckled down in the earth's crust with a number of smaller folds within it. Those lesser folds are obvious from the air, hard to see from the road. The Crazy Mountain Basin and the folds within it have greatly disappointed those who hoped to find oil there.

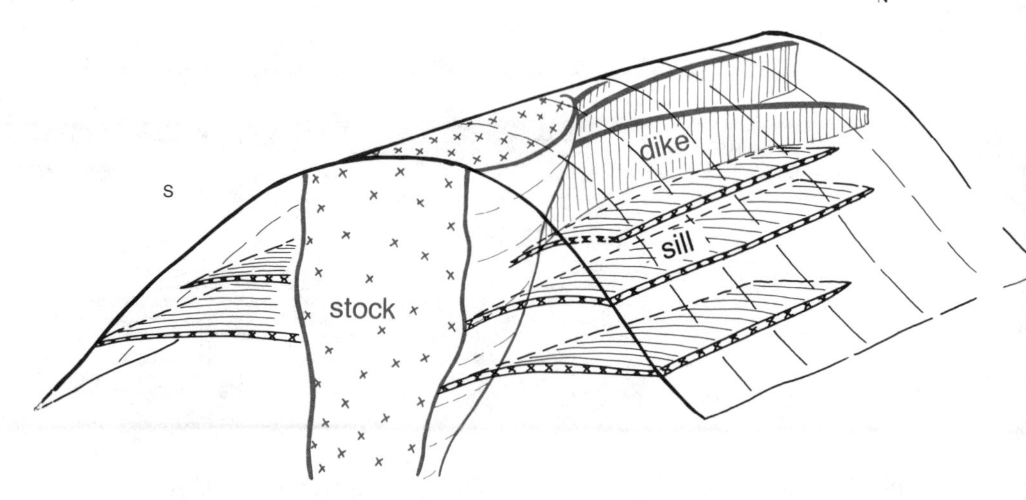

Schematic diagram of the structure in the southern Crazy Mountains.

The usual first step in looking for oil is to find a sedimentary basin, a persistently low area of the earth's crust that holds an uncommonly thick section of sedimentary formations deposited over a long period of time. Although the name sounds right, the Crazy Mountain Basin doesn't quite fit that pattern. It formed late, with the crustal movements that made the Rocky Mountains, so it does not contain a thick section of the older sedimentary formations. Perhaps it is not too surprising that attempts to find oil there have failed.

Crazy Mountains

The rather low north end of the Crazy Mountains contains a swarm of small intrusions, most of them the peculiar Crazy Mountain shonkinite with its large sodium content. The assortment of intrusions includes sills injected between layers of sedimentary rocks, vertical dikes formed as magma-injected fractures, and laccoliths formed where blisters of magma pooled under a strong layer of sedimentary rock. The Shields River Road heads east from Wilsall into that part of the range.

A lone road on the east side of the range reaches the much higher and more rugged southern end of the Crazy Mountains where the angular peaks and sharp ridges tell of enormous ice age glaciers in the big valleys. The core of that part of the range is a large igneous intrusion, the Big Timber stock, emplaced in the arch of an anticline. Spectacular dikes that radiate north and south from that intrusion follow the axis of the anticline, and swarms of sills follow the sedimentary layers on the east and west flanks of the fold.

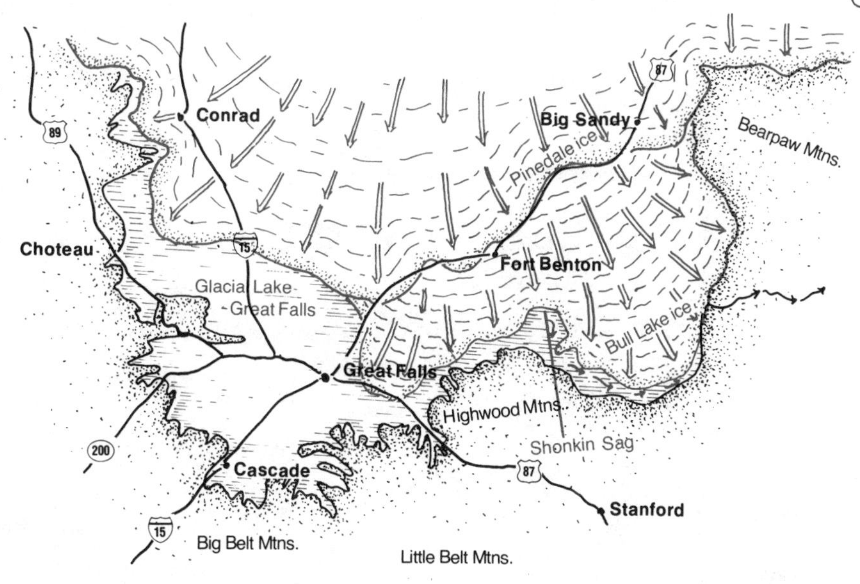

US 89, Glacial Lake Great Falls, and the approximate margin of the ice.

US 89: Great Falls—Browning

110 mi./176 km.

Glacial deposits of various kinds mantle so much of the ground along this road that bedrock outcrops are few. The rare outcrops that do exist expose sedimentary formations deposited on land along the coastal plain or in shallow sea water near the coast. Most were laid down during the latter part of Cretaceous time, sometime between 100 and 65 million years ago.

Continental Glaciers and Glacial Lakes

During the Bull Lake ice age of sometime between 70,000 and 130,000 years ago, Glacial Lake Great Falls flooded the area between the great continental ice sheet that lay east and north of the highway and higher ground to the south and west. Most of the route between Great Falls and Choteau crosses the old lake floor. The smaller glacier of the last ice age impounded a smaller version of Glacial Lake Great Falls. Old shorelines at elevations of about 3900 and 3600 feet probably date from the two glaciations. They are hard to see.

The glacier carried large boulders from central Canada to the edge

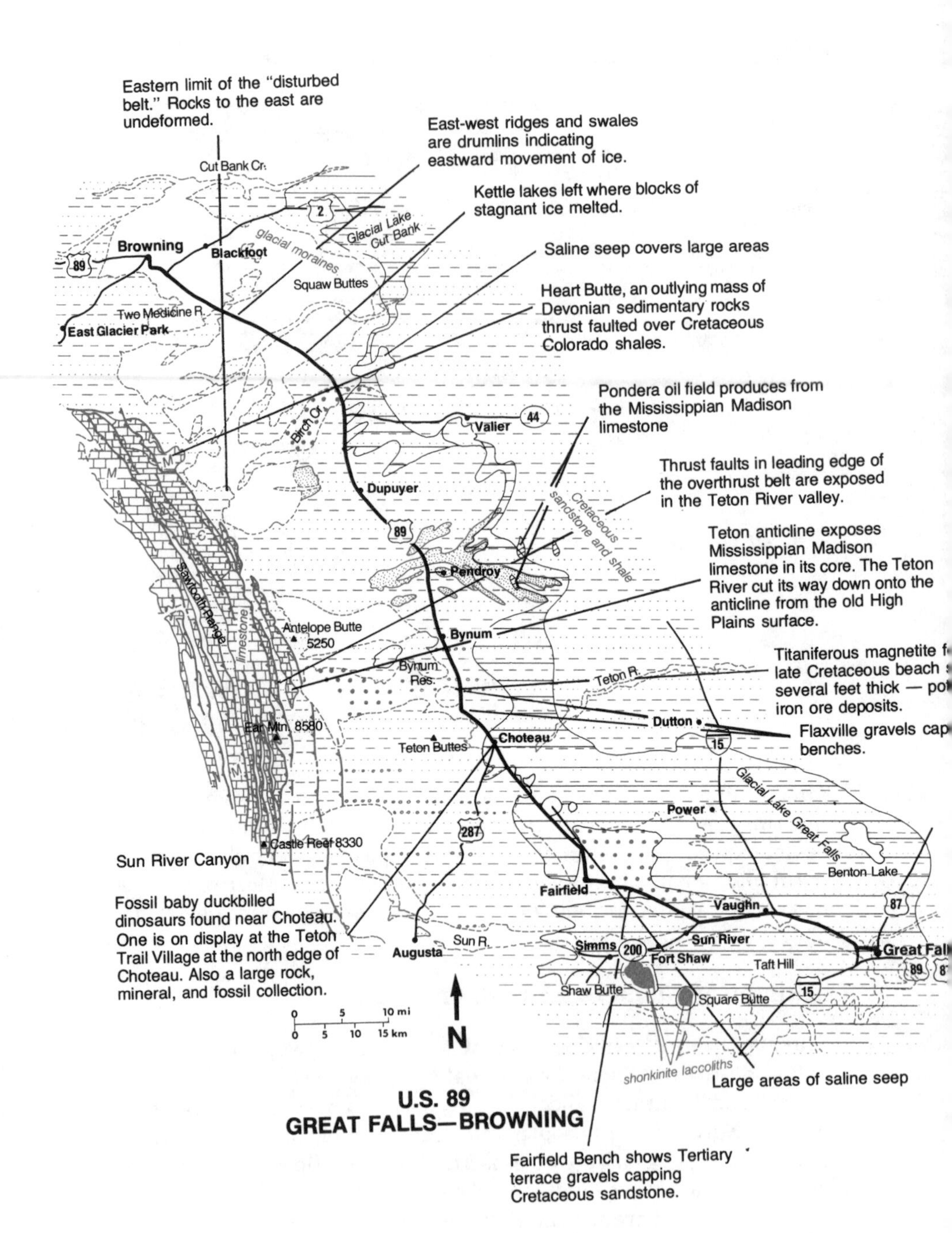
Eastern limit of the "disturbed belt." Rocks to the east are undeformed.
East-west ridges and swales are drumlins indicating eastward movement of ice.
Kettle lakes left where blocks of stagnant ice melted.
Saline seep covers large areas
Heart Butte, an outlying mass of Devonian sedimentary rocks thrust faulted over Cretaceous Colorado shales.
Pondera oil field produces from the Mississippian Madison limestone
Thrust faults in leading edge of the overthrust belt are exposed in the Teton River valley.
Teton anticline exposes Mississippian Madison limestone in its core. The Teton River cut its way down onto the anticline from the old High Plains surface.
Flaxville gravels cap benches.
Sun River Canyon
Fossil baby duckbilled dinosaurs found near Choteau. One is on display at the Teton Trail Village at the north edge of Choteau. Also a large rock, mineral, and fossil collection.
Large areas of saline seep
Fairfield Bench shows Tertiary terrace gravels capping Cretaceous sandstone.
Cut Bank Cr.
Glacial Lake Cut Bank
glacial moraines
Browning
Blackfoot
Squaw Buttes
Two Medicine R.
East Glacier Park
Birch Cr.
Valier
Dupuyer
Cretaceous sandstone and shale
Pendroy
Sawtooth Range
limestone
Antelope Butte 5250
Bynum
Bynum Res.
Teton R.
Dutton
Ear Mtn. 8580
Teton Buttes
Choteau
Glacial Lake Great Falls
Power
Castle Reef 8330
Benton Lake
Fairfield
Vaughn
Simms
Sun River
Fort Shaw
Great Fal
Taft Hill
Augusta
Sun R.
Shaw Butte
Square Butte
shonkinite laccoliths
0 5 10 mi
0 5 10 15 km
N
U.S. 89
GREAT FALLS—BROWNING

of the ice, which doubtless floated in Glacial Lake Great Falls. Then large icebergs broke off the margin of the glacier and sailed out across the lake carrying the rocks embedded in them to the end of their long journey. Now we see those rocks strewn across the fields along the road between Great Falls and Choteau. Isolated boulders may have simply dropped from a melting iceberg as it drifted along; groups of boulders probably mark places where grounded icebergs melted in one place.

Through about 20 miles south of Browning, the road crosses hummocky glacial moraines dropped from ice that came from the Rocky Mountains. Large glaciers pouring out of the mountains ponded on the high plains to form piedmont glaciers, vast expanses of nearly stagnant ice. When the piedmont glaciers melted, they left great tracts of morainal landscape full of little hills and ponds all strewn with erratic boulders carried out of the Rocky Mountains.

The Great Flood of June, 1964

Most highway bridges between Choteau and Browning date from 1964, the year of the great flood. Heavy snow lingered in the Sawtooth Mountains until well into June that year. Then torrential rains fell for several days. Catastrophic floods of rain and snowmelt water rushed down all the streams between Choteau and Browning, inflicting heavy damage. Watch especially in the area north and south of Dupuyer for flood rafted wood still littering large areas well away from the streams, and for the remains of old bridges east of the highway, downstream.

The Overthrust Belt

Section across the Sawtooth Range and US 89 at Pendroy. The strongly deformed, scrambled rocks of the Rocky Mountains are the same formations that lie flat beneath the plains.

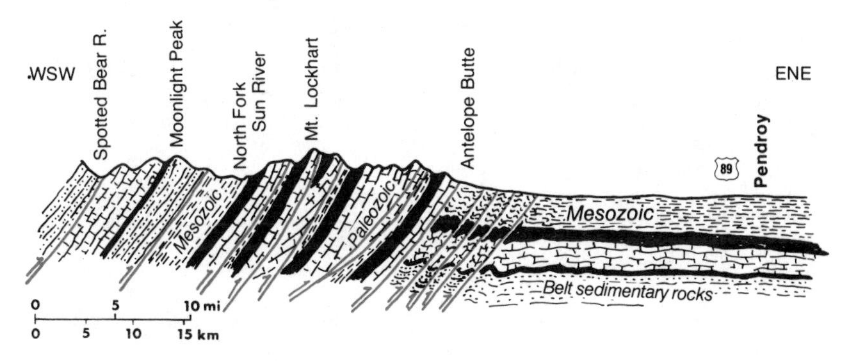

Overthrust slabs of limestone rear above Lake Lavale, in the Sawtooth Mountains about 15 miles past the end of the road that goes west from Choteau.
—U.S. Forest Service photo by G. Roskie

The Sawtooth Range on the skyline west of the highway is the northern Montana overthrust belt. Those mountains are big slabs of rock that slid east from western Montana, then piled on each other. All that probably happened about 70 million years ago. Since then, erosion has stripped out the layers of softer rock, leaving the edges of the hard layers standing high as ridges.

Heart Butte stands boldly east of the main front of the Rocky Mountains a few miles north of Dupuyer. It is a mass of Devonian sedimentary rock some 350 million years old resting on Cretaceous sedimentary rock about 80 million years old. Those rocks are stacked in the wrong order. It seems that the Devonian rock is part of the overthrust belt that moved a bit farther east than the rest.

Greenish volcanic sandstones of the Livingston formation in an outcrop north of Big Timber.

US 191: Big Timber—Lewistown

101 mi./162 km.

The road crosses the plains of central Montana passing but not entering several mountain ranges. Bedrock along the road consists almost entirely of sandstone and mudstone, sedimentary rock deposited in shallow sea water during Cretaceous time. Those formations weather so easily that outcrops are rare along most of the route.

Many of the original sediments were laid down in the shallow sea that flooded most of Montana during long intervals of Cretaceous time. Others accumulated on land, probably not far from the coast. During late Cretaceous time, 70 to 80 million years ago, large volcanoes were erupting from the Boulder batholith between Butte and Helena. Enormous volumes of volcanic debris went eastward into the Crazy Mountain Basin to become the Livingston formation.

The road passes east of the Crazy Mountains between Big Timber and Harlowton, east of the Little Belt Mountains and west of the Big Snowy Mountains between Harlowton and the junction with Montana 87. Lewistown lies southwest of the beautiful forested mass of the Judith Mountains.

Massive beds of sandstone in the Fort Union formation resist weathering and erosion to make the rugged Cayuse Hills in the area about midway between Big Timber and Harlowton. The sandstone makes distinctively picturesque bouldery outcrops, and the pore spaces between sand grains hold enough water to support a good growth of pine trees.

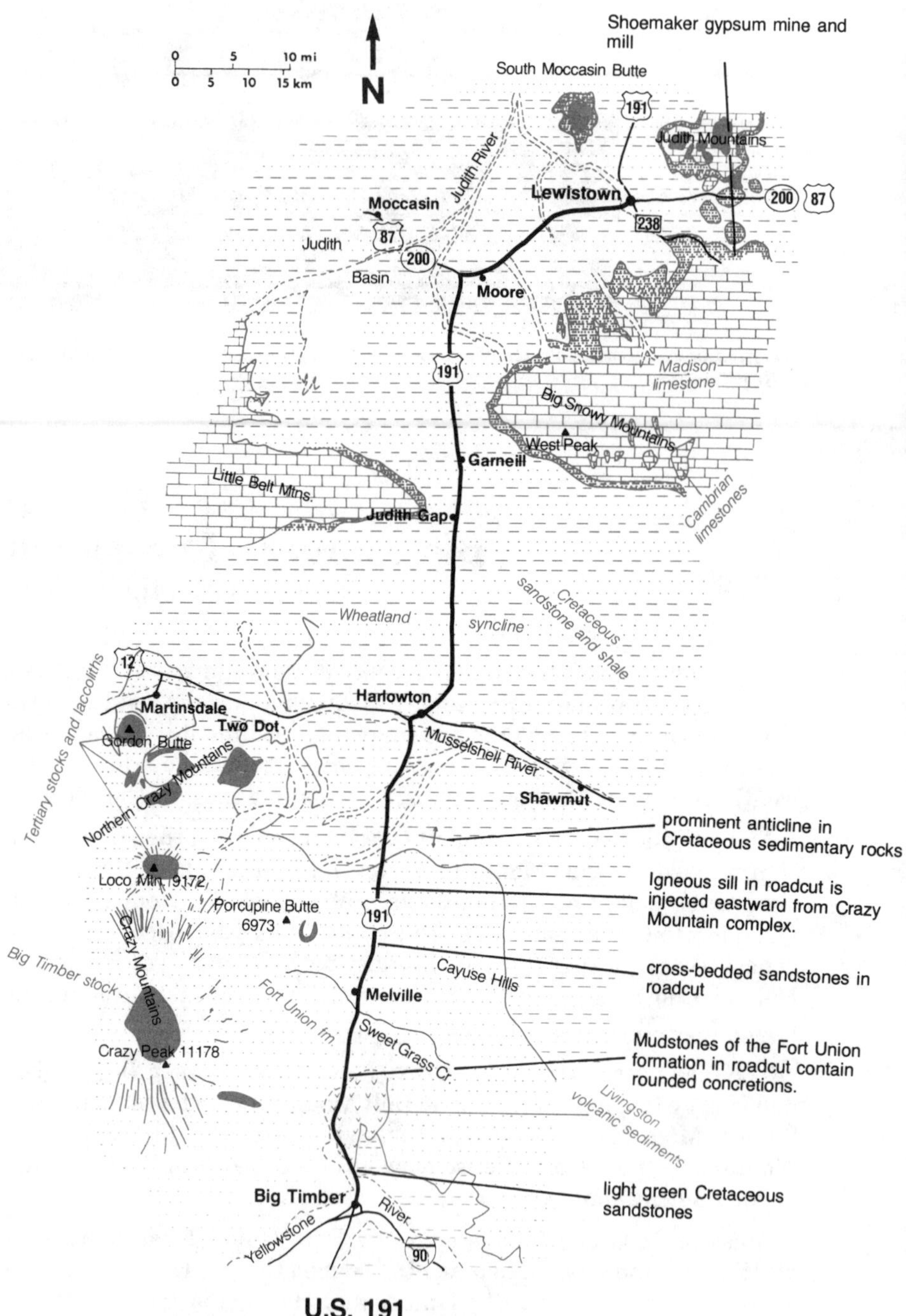

Shoemaker gypsum mine and mill
South Moccasin Butte
0 5 10 mi
0 5 10 15 km
N
191
Judith Mountains
Judith River
Lewistown
200
87
Moccasin
238
87
Judith
200
Basin
Moore
191
Madison limestone
Big Snowy Mountains
West Peak
Garneill
Little Belt Mtns.
Judith Gap
Cambrian limestones
Cretaceous sandstone and shale
Wheatland
syncline
12
Tertiary stocks and laccoliths
Harlowton
Martinsdale
Two Dot
Gordon Butte
Musselshell River
Northern Crazy Mountains
Shawmut
prominent anticline in Cretaceous sedimentary rocks
Loco Mtn. 9172
Igneous sill in roadcut is injected eastward from Crazy Mountain complex.
Porcupine Butte 6973
191
Big Timber stock
Crazy Mountains
Cayuse Hills
cross-bedded sandstones in roadcut
Melville
Fort Union fm
Sweet Grass Cr.
Crazy Peak 11178
Mudstones of the Fort Union formation in roadcut contain rounded concretions.
Livingston volcanic sediments
light green Cretaceous sandstones
Big Timber
River
Yellowstone
90

U.S. 191
BIG TIMBER—LEWISTOWN

Crazy Peak, in the southern Crazy Mountains. —Montana Bureau of Mines and Geology photo by H. L. James

The Crazy Mountains

The high and deeply glaciated southern part of the Crazy Mountains forms an alpine backdrop on the horizon northwest of Big Timber, east of the road along most of the way between Big Timber and Harlowton. That part of the range contains a large igneous intrusion, the Big Timber stock. The Fort Union formation was still freshly deposited sand and mud, still soft and full of water, when that big mass of white hot magma invaded it about 50 million years ago. Heat and mineral matter escaping from the magma baked and altered the soft sediments as much as a mile from the contact, converting them into extremely hard rocks.

Consider that 50 million years ago the soft sediments of the Fort Union formation must have extended across this entire countryside to an elevation about as high as the Crazy Mountains. Erosion has since removed most of that blanket of soft sedimentary rock, leaving the resistant igneous rock and the hard baked sedimentary rocks around its contacts standing in high erosional relief. That is why the Crazy Mountains are mountains.

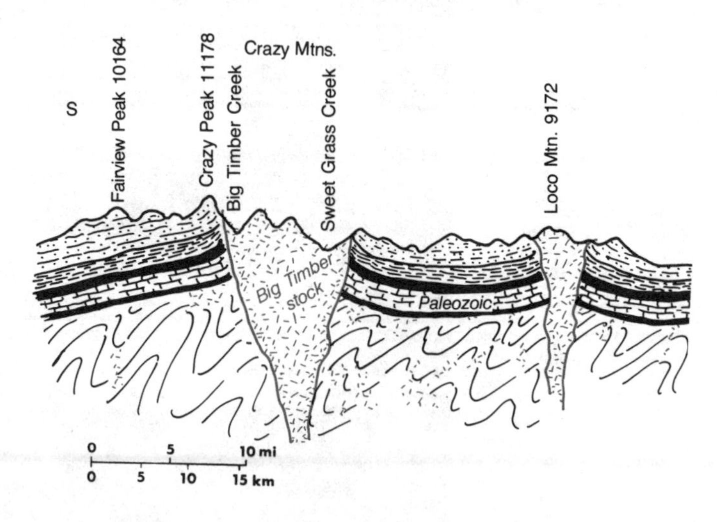

A long section west of US 191 between Big Timber and north of Harlowton

Arches in the Plains

The Little Belt Mountains lie along the distant western horizon between Harlowton and the junction with Montana 87. They are a broad arch that exposes Precambrian basement rocks in its core. The low profiles of the Big Snowy Mountains, and of the Little Snowy Mountains at their eastern end, make them look exactly like what they are: a simple, broad bulge in the plains. That bulge brings Cambrian rocks to the surface near the southern margin of the Big Snowy Mountains. A little more erosion there will expose the Precambrian basement rocks.

The Little Belt, Big and Little Snowy, and Judith mountains all escaped glaciation during the ice ages because those ranges are too low to catch the enormous quantities of snow that big glaciers need for their nourishment. Contrast the relatively rounded skyline profiles of those ranges to the jagged crest of the ice-carved southern Crazy Mountains.

The Judith and Moccasin Mountains

The Judith Mountains consist essentially of a group of igneous intrusions with two outliers in the Moccasin Mountains, another in Black Butte, northeast of the main cluster of intrusions. Geologically, they all belong together. Large masses of magma rose, one after the other, into this area about 50 million years ago, pushing through the rock formations beneath the plains like great pistons. Now we find those older formations tightly wrapped around the intrusive masses.

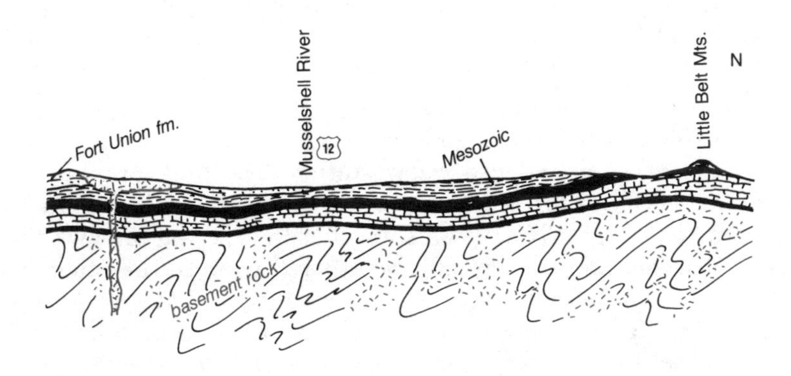

showing arches of the Little Belt and Big Snowy mountains.

Large exposures of pale gray Madison limestone, rock normally several thousand feet down in this part of Montana, are especially conspicuous.

The rising masses of magma popped up the sedimentary formations above them as though they were trapdoors hinged at one edge. Some of the igneous intrusions have now lost their sedimentary lids to erosion, and appear on the map as round masses of igneous rock. The lids of other intrusions are only partially eroded off, so the igneous rock peeps out from beneath one side of its sedimentary cover. Still other igneous intrusions remain covered, revealing themselves only as domes bulged in the sedimentary rock.

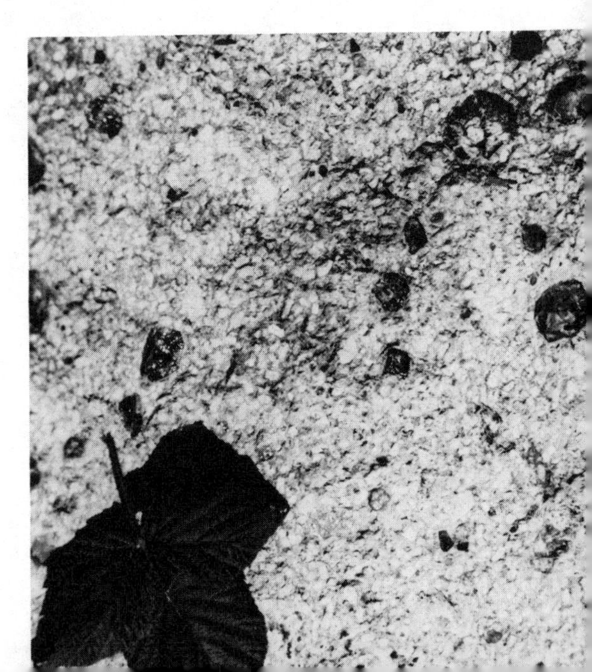

Quartz phenocrysts in granitic matrix from the top of Judith Peak.

All the igneous rocks in the Judith Mountains crystallized below the surface as intrusions. Some are fairly normal granite, others are peculiar rocks rich in the alkali elements, potassium and sodium. Igneous rocks in the Judith Mountains closely resemble those in the Sweetgrass Hills and Little Rocky Mountains.

One good place to look at some of those rocks is at the top of Judith Peak, near the abandoned radar station. The pink granite there contains perfectly formed quartz crystals about the size of the last joint of your finger. The crystals weather free, so you can pick them out of the soil and the partially weathered rock. In the same area, there are small exposures of a spectacular alkalic rock called tinguaite that contains blocky crystals of white or pink orthoclase feldspar set in a fine-grained groundmass of dark green rock.

North Moccasin Mountain from near Hilger. —Montana Bureau of Mines and Geology photo by H. L. James

US 191: Lewistown—Malta
133 mi./213 km.

The long road between Lewistown and Malta passes an amazing variety of geologic features: the Judith, Moccasin, and Little Rocky mountains, the canyon of the Missouri River, and an assortment of glacial landforms. Bedrock along the road consists almost entirely of sedimentary formations deposited in and near the shallow sea that flooded this part of Montana during much of Cretaceous time, between 135 and 65 million years ago. Bedrock in the mountains includes all the sedimentary formations that lie beneath the plains, as well as some very unusual igneous rocks.

The Judith and Moccasin Mountains

North of Lewistown, the road winds past the west and north sides of the Judith Mountains, a tight cluster of igneous intrusions emplaced about 50 million years ago. The side road east from the highway into the valley of Warm Springs Creek invites a visit. Most of the igneous rocks are syenites that consist almost entirely of feldspar, but there are some granites and related rocks that contain both feldspar and quartz. The rising magmas punched through some of the older sedimentary rocks of the plains and arched others into domes that wrap around the intrusions.

The two large buttes of the Moccasin Mountains, two large masses of syenite wrapped in the older sedimentary rocks they intruded, rise west of the road. Think of them as outposts of the Judiths. Another

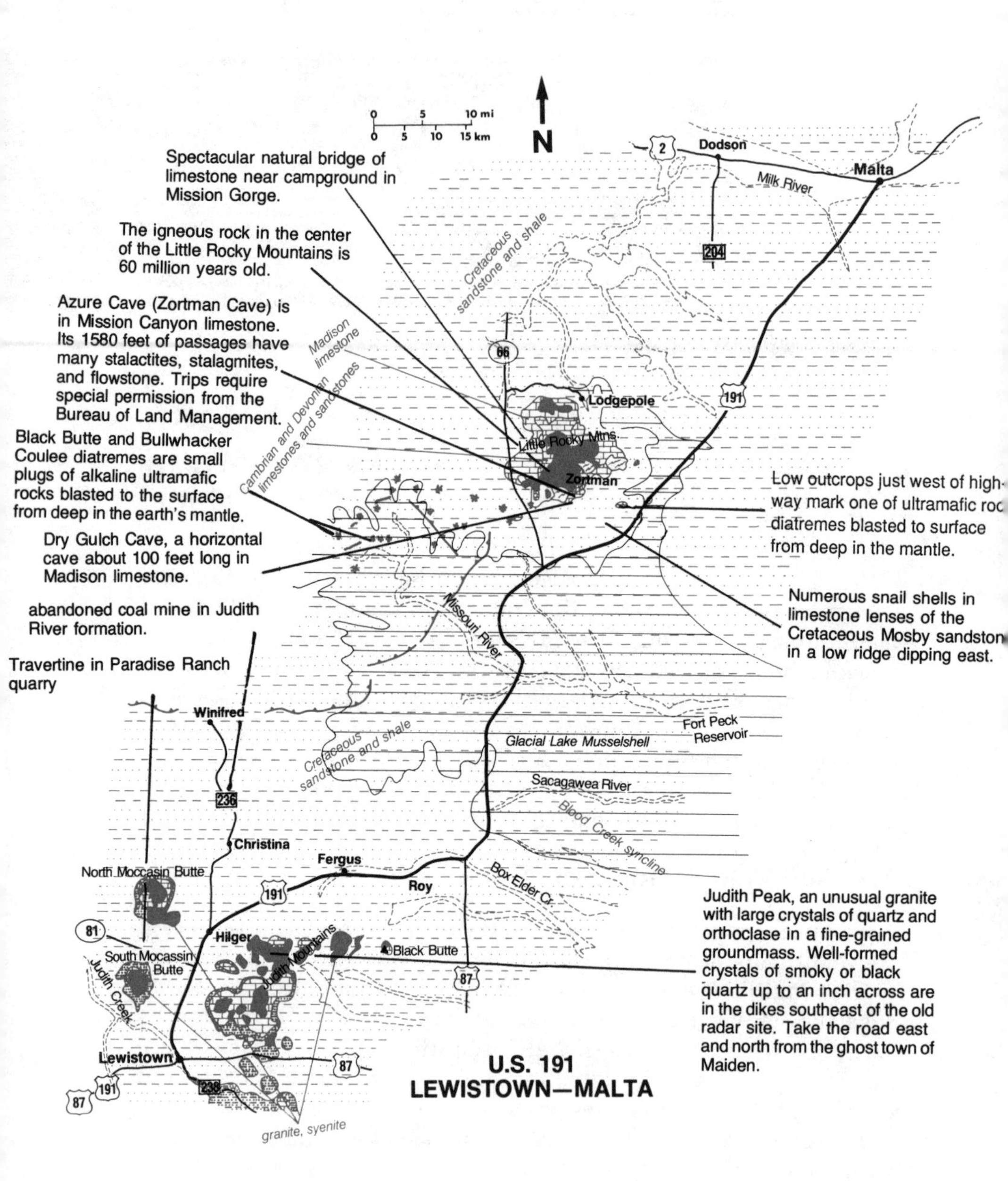

N
0 5 10 mi
0 5 10 15 km
Spectacular natural bridge of limestone near campground in Mission Gorge.
The igneous rock in the center of the Little Rocky Mountains is 60 million years old.
Azure Cave (Zortman Cave) is in Mission Canyon limestone. Its 1580 feet of passages have many stalactites, stalagmites, and flowstone. Trips require special permission from the Bureau of Land Management.
Black Butte and Bullwhacker Coulee diatremes are small plugs of alkaline ultramafic rocks blasted to the surface from deep in the earth's mantle.
Dry Gulch Cave, a horizontal cave about 100 feet long in Madison limestone.
abandoned coal mine in Judith River formation.
Travertine in Paradise Ranch quarry
Low outcrops just west of highway mark one of ultramafic roc diatremes blasted to surface from deep in the mantle.
Numerous snail shells in limestone lenses of the Cretaceous Mosby sandston in a low ridge dipping east.
Judith Peak, an unusual granite with large crystals of quartz and orthoclase in a fine-grained groundmass. Well-formed crystals of smoky or black quartz up to an inch across are in the dikes southeast of the old radar site. Take the road east and north from the ghost town of Maiden.
2
Dodson
Malta
Milk River
204
Cretaceous sandstone and shale
Madison limestone
Cambrian and Devonian limestones and sandstones
66
Lodgepole
191
Little Rocky Mtns.
Zortman
Missouri River
Winifred
236
Christina
Fergus
Roy
Fort Peck Reservoir
Glacial Lake Musselshell
Sacagawea River
Blood Creek syncline
Box Elder Cr.
North Moccasin Butte
81
South Mocassin Butte
Judith Creek
Hilger
Judith Mountains
Black Butte
87
Lewistown
granite, syenite
U.S. 191
LEWISTOWN—MALTA

such outpost, the isolated mass of Black Butte, stands as a sentinel northeast of the main body of the Judiths. The igneous rock there is more nearly related to granite. Heat and steam from that mass of magma created a spectacular zone of metamorphism in the older sedimentary rocks within a few hundred feet of the contact.

Prospectors found placer gold in Warm Springs Creek in the spring of 1880, lode deposits in quartz veins around the old town of Maiden later that same year. As usual, the placer deposits were soon worked out. Underground mines that worked lode deposits thrived mightily until 1885, then went into rapid decline. A few mines stayed marginally active until about 1920 even though the district was long past its peak by 1895. Like most gold quartz vein deposits, those around Maiden contain occasional pockets of inspiring bonanza ore isolated within vast barrens of profitless quartz.

The Little Rocky Mountains

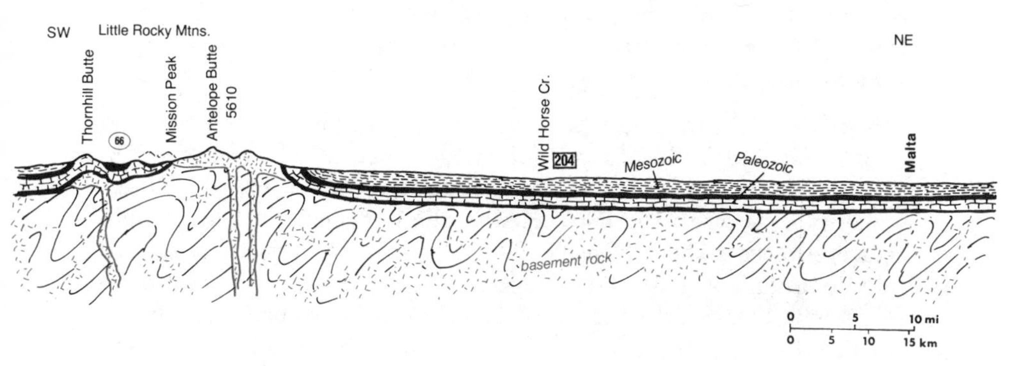

Section from the Little Rocky Mountains to Malta.

For more than 50 miles south of Malta, the Little Rocky Mountains rise west of the highway. Like several other isolated ranges of central Montana, the Little Rocky Mountains are a dome sharply punched in the plains. Large quantities of igneous rocks, pale syenites rich in sodium, form the core of the uplift, along with some large exposures of Precambrian basement rock.

The Madison limestone stands steeply tilted as it wraps around the Little Rocky Mountains. The thick limestone resists weathering in this dry climate, and so stands up in erosional relief to form a steep wall that encloses the range. From a good vantage point, and in the raking light of early morning or late evening, it is easy to imagine that wall of limestone as the ruined fortifications around a medieval city.

The syenite intrusions in the Little Rocky Mountains like those elsewhere in central Montana, brought gold with them. Although the Indians apparently knew of gold in the Little Rocky Mountains, and there are accounts of a discovery in 1868, mining did not begin until after prospectors found placer gold in Alder Gulch in 1884. Several thousand of the free floating miners of those days rushed to the district, quickly cleaned out the easy placer deposits, and left within a few months.

Ice Ages and Glacial Lakes

Prospectors found extremely rich gold quartz ore in bedrock in 1890, and a new rush to the district began. Landusky sprang up in 1894 near the southern margin of the igneous intrusion. Zortman started at about the same time near another bedrock deposit that had been discovered in 1868, and then ignored for more than 20 years. Acres of spoil heaps and tailings in Ruby Gulch north of Zortman testify to the enormous amount of rock mined and fed through cyanide mills, most of it between 1894 and 1939.

North and south of the Missouri River, the highway crosses the floor of Glacial Lake Musselshell, which formed when ice moving southwest out of central Canada dammed the Missouri River. Icebergs floating in the lake dropped the big rocks that lay widely scattered across the landscape. Those, and some deposits of glacial lake sediments are all that remain to tell us the story. Glacial Lake Musselshell existed during one of the earlier glacial periods, probably the Bull Lake ice age of sometime between 70,000 and 130,000 years ago, when the ice advanced considerably farther than during the last glaciation.

The road to Zortman in the southern part of the Little Rocky Mountains.
—Montana Bureau of Mines and Geology photo by H. L. James

Last continental glacial maximum and small ice-marginal lakes north of the Little Rocky Mountains.

When the glaciers of the last ice age were at their maximum sometime around 15,000 years ago, ice reached south of Malta, almost to the northern edge of the Little Rocky Mountains. That glacier impounded a much smaller glacial lake in the area north of the mountains. The road crosses the moraine near the turnoff to Lodgepole.

About midway between Lewistown and Malta, the highway crosses the Missouri River Canyon, another relic of one of the earlier ice ages. This crossing near the downstream end of the canyon gives no hint of the spectacular gorge to the west. The secondary road that connects the highway to Winifred crosses a number of high vantage points near the canyon rim that provide marvelous views of the colorful canyon walls.

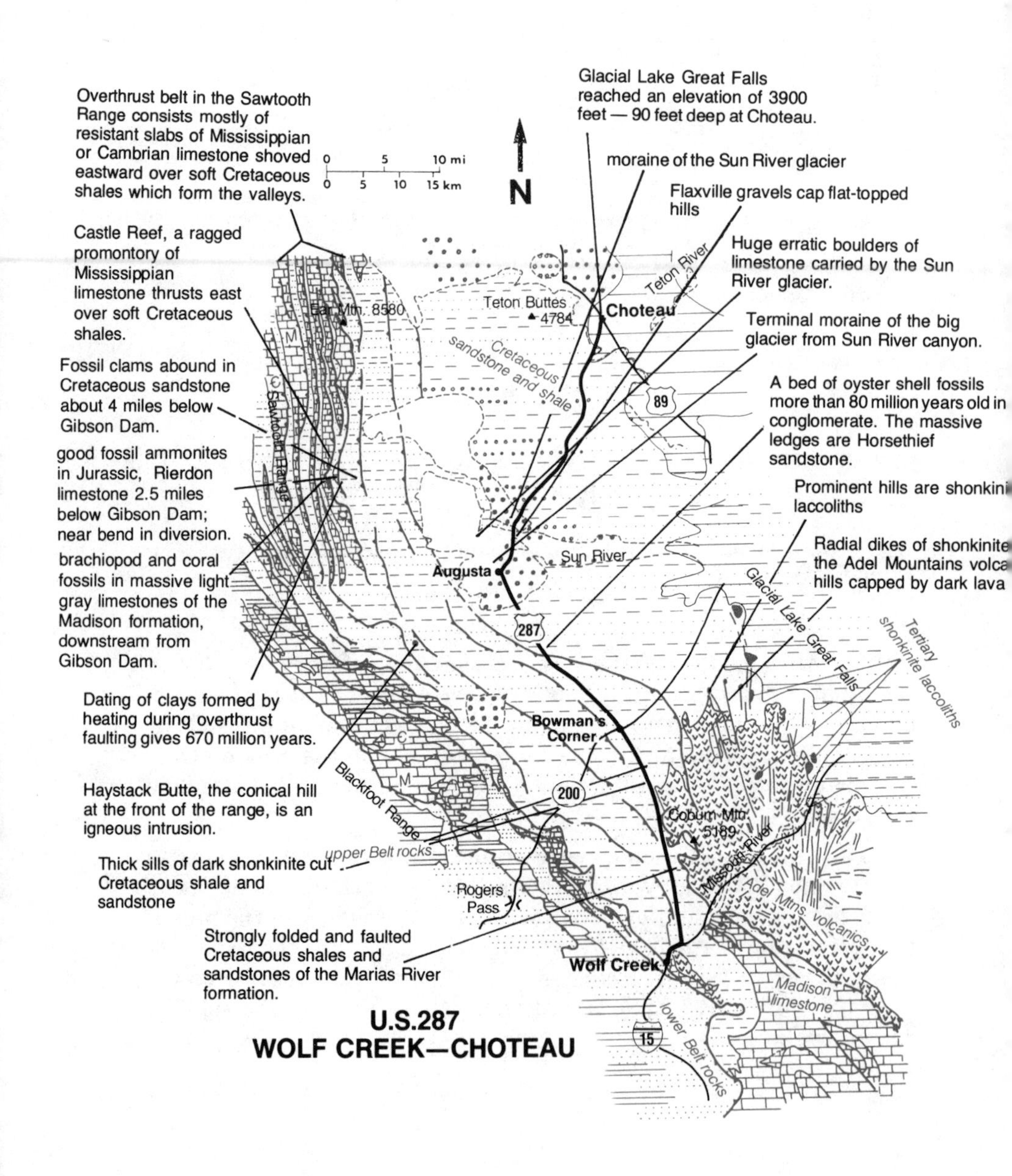
Overthrust belt in the Sawtooth Range consists mostly of resistant slabs of Mississippian or Cambrian limestone shoved eastward over soft Cretaceous shales which form the valleys.
Castle Reef, a ragged promontory of Mississippian limestone thrusts east over soft Cretaceous shales.
Fossil clams abound in Cretaceous sandstone about 4 miles below Gibson Dam.
good fossil ammonites in Jurassic, Rierdon limestone 2.5 miles below Gibson Dam; near bend in diversion.
brachiopod and coral fossils in massive light gray limestones of the Madison formation, downstream from Gibson Dam.
Dating of clays formed by heating during overthrust faulting gives 670 million years.
Haystack Butte, the conical hill at the front of the range, is an igneous intrusion.
Thick sills of dark shonkinite cut Cretaceous shale and sandstone
Strongly folded and faulted Cretaceous shales and sandstones of the Marias River formation.
Glacial Lake Great Falls reached an elevation of 3900 feet — 90 feet deep at Choteau.
moraine of the Sun River glacier
Flaxville gravels cap flat-topped hills
Huge erratic boulders of limestone carried by the Sun River glacier.
Terminal moraine of the big glacier from Sun River canyon.
A bed of oyster shell fossils more than 80 million years old in conglomerate. The massive ledges are Horsethief sandstone.
Prominent hills are shonkini laccoliths
Radial dikes of shonkinite the Adel Mountains volca hills capped by dark lava
0 5 10 mi
0 5 10 15 km
N
Ear Mtn. 8580
Teton Buttes 4784
Choteau
Teton River
Cretaceous sandstone and shale
89
Sawtooth Range
M
C
Augusta
Sun River
Glacial Lake Great Falls
Tertiary shonkinite laccoliths
287
Bowman's Corner
Blackfoot Range
200
Cobum Mtn. 5189
Missouri River
upper Belt rocks
Rogers Pass
Adel Mtns. volcanics
Wolf Creek
Madison limestone
15
lower Belt rocks

U.S.287 WOLF CREEK—CHOTEAU

The dark mass of Haystack Butte forms a pointed peak just east of the Sawtooth Range.

US 287: Wolf Creek—Choteau

65 mi./104 km.

The entire route follows the disturbed belt, a zone in which the rocks are deformed like those in the overthrust belt, but do not withstand weathering and erosion well enough to stand up as high mountains. Ragged peaks of the deeply glaciated Sawtooth Range, the overthrust belt, form the western skyline along the entire route. Most of those visible from the highway are slabs of Madison limestone that slid to where we now see them along large overthrust faults. In many places, the Madison limestone lies on much younger formations.

The mostly forested hills east of the road between Augusta and Wolf Creek are the Adel Mountains—many people call them the Birdtail Mountains. They are the remains of a volcanic pile that probably erupted about 50 million years ago, part of the central Montana alkalic province. Most of the rocks are shonkinite.

Haystack Butte stands slightly but distinctly east of the mountain front near Augusta. Except that it is an igneous intrusion, no one seems to know much about Haystack Butte. In the absence of more definite knowledge, it seems reasonable to assume that it is probably part of the central Montana igneous activity of about 50 million years ago.

The View of the Overthrust Belt in Sun River Canyon

The best place in Montana to get a close roadside view of geologic structures in the overthrust belt is along the road that goes west from Augusta to Gibson Dam. Just outside of Augusta, the road to Gibson Dam crosses a large moraine, its hummocky surface littered with erratic boulders. The moraine marks the edge of a glacier that poured out of the canyon of the Sun River, and spread sluggishly across the plains east of the mountain front as a piedmont ice sheet.

The road enters the Sun River Canyon through a narrow gorge eroded through the slab of Madison limestone that forms the steep front of the Sawtooth Range. It is the easternmost overthrust slab, and lies on top of the Cretaceous sedimentary rocks exposed at the mouth of the canyon. The valley widens in the Wagner Basin west of the first slab of Madison limestone where less resistant younger rocks are exposed. Then the valley narrows again to pass through a second slab of Madison limestone that slides over the younger formations on another overthrust fault. Another wide place in the valley, Hannan Gulch, marks the second outcrop of the same set of less resistant younger formations exposed in Wagner Basin. A third narrow gorge in the third overthrust slab of Madison limestone holds Gibson Dam.

The Bob Marshall Wilderness lies beyond, so the road goes no farther than Gibson Dam. But the succession of overthrust slabs marches another 80 miles west to the Mission Range, with the rocks becoming generally older westward. Throughout the northern Montana overthrust belt, the ridges trend generally from north to south along the upturned edges of resistant slabs, and the long valleys follow the less resistant rocks between them.

Disturbed Belt

Rocks near the road in the disturbed belt consist mostly of

Large erratic boulder along the road to Gibson Dam, the Sawtooth Range in the background.

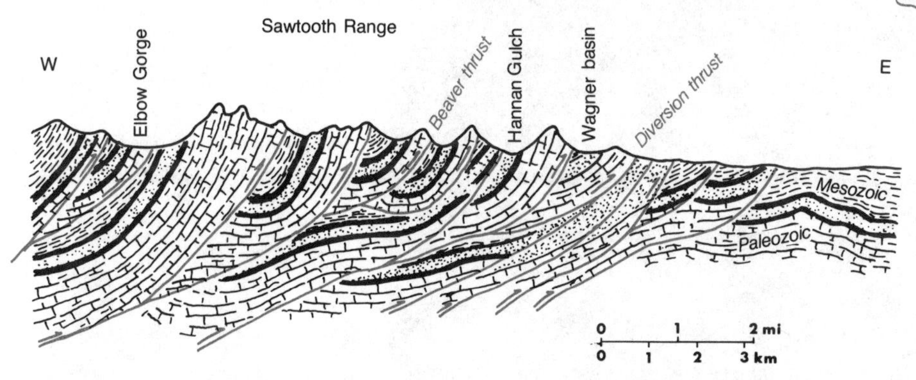

Section along the line of the road to Gibson Dam showing multiple slabs of Madison limestone stacked on each other with slices of younger Mesozoic formations sandwiched between them. Only overthrust faulting on many surfaces can create such a bizarre arrangement of sedimentary layers.

sedimentary strata laid down during Jurassic and Cretaceous time, between 100 and 65 million years ago, in round numbers. Most of the formations accumulated in shallow sea water that covered this part of Montana from time to time during those years. In many places between Choteau and Wolf Creek, conspicuous ledges of resistant sandstone mark the trend of the sedimentary layers, most of which are too weak to form good outcrops. Watch those ledges to see how in places they wind across the countryside in sinuous hairpin curves, evidence that the rocks are tightly folded.

One conspicuous anticlinal arch, the Robertson Dome, lies east of the highway for several miles between Augusta and the junction with Montana 200. Virtually all such anticlines exposed at the surface had wildcat wells drilled into them during the early years of oil exploration, in most cases before 1925. If you don't see producing wells and storage tanks on them now, you can safely assume that the early wildcat wells were dry.

High Plains Surface and Flaxville Gravel

About ten miles north of Augusta, the road crosses a small remnant of the High Plains surface. It is a low hill with a conspicuously flat top. Roadcuts at the edge of the hill expose beds of the Flaxville gravel, and stream-rounded pebbles litter the wheat fields on the flat top. Clearly, the flat top of the hill is part of the High Plains surface, a remnant of the desert plain that stretched continuously from the Rocky Mountains to the central Dakotas during late Miocene and Pliocene time, as recently as three million or so years ago.

Nothing protects a land surface against erosion quite as effectively

Pebbles of the Flaxville gravel in a wheat field on the High Plains surface north of Augusta.

as a cap of gravel, which absorbs surface water that would otherwise erode gullies. Perhaps an old alluvial fan made the Flaxville gravel thicker here than elsewhere in the area, and so preserved this remnant of the High Plains surface by preventing any streams from starting to flow across it.

Dinosaurs

One of the world's most spectacular dinosaur discoveries came from a small area of badlands a few miles west of Choteau, in particular, from a low hill now nicknamed "Egg Mountain." There the Two Medicine formation, which accumulated on land during late Cretaceous time, contains the bones of numerous duck billed dinosaurs, along with their nests full of eggs and baby dinosaurs. Volcanic ash falls buried the dinosaurs. The ash presumably erupted from the Elkhorn Mountains volcanoes that were then in the full

Model of a baby duck billed dinosaur–the pencil indicates its size. —Jack Horner photo

vigor of their activity in the area of the Boulder batholith. After the discovery on Egg Mountain, similar sites were found in other exposures of the Two Medicine formation elsewhere in this part of Montana. Those nests show that dinosaurs were not quite so reptilian as many people have imagined them.

Some of the nests contain only eggs, others broods of babies and partly grown dinosaurs. Some of the young dinosaurs lived long enough before clouds of volcanic ash overwhelmed and buried them to put distinct signs of wear on their teeth. Evidently, their parents nurtured them in the nest, brought food to them, for some months after they hatched. Duck billed dinosaurs must have cared for their young in much the same way that birds do.

A nest of fossil dinosaur eggs west of Choteau. —Jack Horner photo

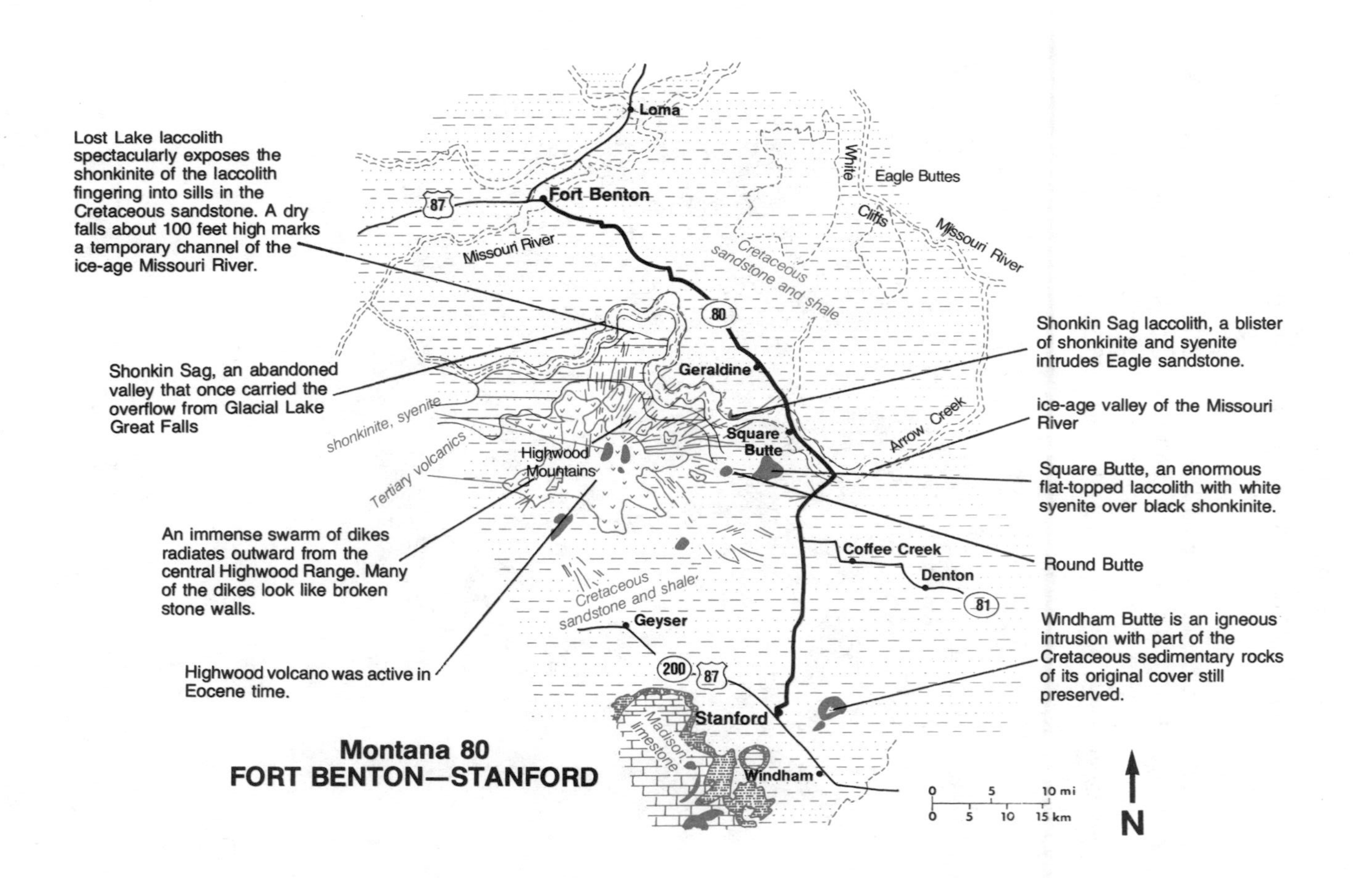
Lost Lake laccolith spectacularly exposes the shonkinite of the laccolith fingering into sills in the Cretaceous sandstone. A dry falls about 100 feet high marks a temporary channel of the ice-age Missouri River.
Shonkin Sag, an abandoned valley that once carried the overflow from Glacial Lake Great Falls
An immense swarm of dikes radiates outward from the central Highwood Range. Many of the dikes look like broken stone walls.
Highwood volcano was active in Eocene time.
Shonkin Sag laccolith, a blister of shonkinite and syenite intrudes Eagle sandstone.
ice-age valley of the Missouri River
Square Butte, an enormous flat-topped laccolith with white syenite over black shonkinite.
Round Butte
Windham Butte is an igneous intrusion with part of the Cretaceous sedimentary rocks of its original cover still preserved.
Loma
Fort Benton
87
Missouri River
White
Eagle Buttes
Cliffs
Missouri River
Cretaceous sandstone and shale
80
Geraldine
Arrow Creek
Square Butte
shonkinite, syenite
Tertiary volcanics
Highwood Mountains
Coffee Creek
Denton
81
Cretaceous sandstone and shale
Geyser
200
87
Stanford
Madison limestone
Windham
0 5 10 mi
0 5 10 15 km
N

Montana 80
FORT BENTON—STANFORD

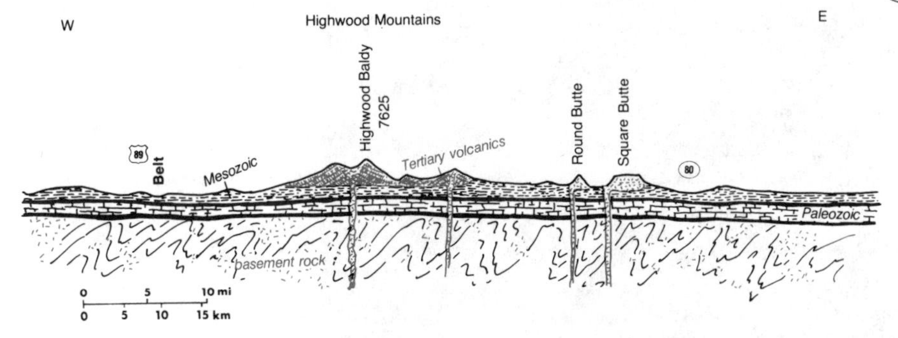

Section across the Highwood Mountains. Although not shown here, Precambrian Belt rocks may exist between the Paleozoic formations and the older basement.

Montana 80: Fort Benton—Stanford

65 mi./104 km.

The road between Fort Benton and Stanford traces a route along the northern and eastern sides of the Highwood Mountains, past several of the most fascinating rocks and landscapes of Montana. Some of these rocks and landscapes are even famous, at least among geologists.

The Highwood Mountains

The Highwood Mountains rise south and west of the road along the entire route. They are the deeply eroded remains of a volcanic center that erupted large volumes of shonkinite and other rocks about 50 million years ago. Many of the dark shonkinite intrusions in the Highwood Mountains wear a glistening white cap of syenite, composed mostly of potassium feldspar.

For miles north and south of Geraldine, the bulky mass of Square Butte with its flat top and the smaller mass of Round Butte with its pointed top rise as prominent sentinels west of the road. Both are very large intrusions of shonkinite with syenite caps conspicuous enough to be easily visible from the road.

Many people contend that this Square Butte near Geraldine, not the one in the Adel Mountains west of Great Falls, is the mountain that appears on the skyline in so many of C. M. Russell's paintings. In

Square Butte south of Geraldine. The dark lower part is shonkinite; the light upper part is syenite.

fact, the distinctive profiles of both mountains appear in those paintings. Both are shonkinite intrusions. Maybe the artist liked shonkinite.

For many years, most geologists thought that growing crystals of feldspar floated to the top of the still molten shonkinite intrusion, and accumulated there to form the white syenite cap. But that theory fails to explain several things, most notably the presence of numerous round globs of syenite about the size of small oranges suspended in the dark shonkinite beneath the cap. It now seems more reasonable to suppose that syenite magma separated from the molten shonkinite in much the same way that oil separates from water. Round blobs of syenite magma floated up through the molten shonkinite like drops of oil rising in a cruet of salad dressing, and coalesced at the top to form the syenite cap. A few stragglers caught in the shonkinite as it crystallized tell the story.

The Shonkin Sag

During the big ice age of Bull Lake time, perhaps sometime around 70,000 to 130,000 years ago, the continental glacier moving southwest from central Canada crossed the line of this road, and lapped onto the north end of the Highwood Mountains. That glacier was thin and slow moving in this area near its farthest edge, and it left little mark on the landscape beyond scattered boulders of granite and gneiss, rounded chunks of northern Manitoba. Nevertheless, the ice was thick enough to block the Missouri River drainage and impound Glacial Lake Great Falls. That vast inland sea overflowed through a narrow channel that closely followed the ice front through the northern part of the Highwood Mountains.

Enough water rushed through the meltwater channel in the northern Highwood Mountains to erode a valley large enough to carry a river at least the size of the Missouri. Water spilled through that channel, and then down the valley of Arrow Creek, which the road crosses about five miles south of the community of Square Butte.

As the ice age ended, the glacier melted back to the line of the Missouri River, draining Glacial Lake Great Falls. That left the meltwater channel through the Highwoods a dry valley with a string of shallow lakes along its floor, the Shonkin Sag. Secondary roads west of Geraldine follow the Shonkin Sag through the little communities of Shonkin and Highwood. It is one of the two or three most spectacular glacial melt water channels in the country.

Looking across the Shonkin Sag, an old channel that drained Glacial Lake Great Falls. In the cliff, Shonkin Sag laccolith fingers to the right into paler Eagle sandstone.

The Shonkin Sag slices neatly through the Shonkin Sag laccolith, cutting it almost exactly in half to expose a marvelous cross section in the valley wall. The big mass of dark shonkinite resting on pale Eagle sandstone is conspicuous from the secondary road that follows the floor of the abandoned valley. That magnificent exposure has made the Shonkin Sag laccolith the most thoroughly studied and best known igneous intrusion of its type in the country, perhaps in the world.

Virgelle Sandstone

South of Square Butte, the highway passes areas of rugged badlands eroded in the Virgelle sandstone, a formation deposited in shallow sea water during Cretaceous time. Most of the Virgelle sandstone is white, but it contains near its top a layer of rock rich in the iron mineral magnetite, which weathers to the red iron oxide mineral hematite. So the formation tends to erode into picturesque toadstools with dark reddish caps standing on stems of white sandstone.

The base of the Shonkin Sag laccolith. Black rock in the upper part of the cliff is shonkinite lying on pale Eagle sandstone, a Cretaceous formation. The valley in the background is the Shonkin Sag.

The view west from Montana 200 reveals ridges of the Sawtooth Range receding into the distance. Haystack Butte punctuates the right-hand edge of the photo.

Montana 200: Bowman's Corner—Great Falls

37 mi./59 km.

Between Bowman's Corner and Simms, the road crosses the disturbed belt, a zone of tightly deformed Cretaceous sedimentary formations along the eastern margin of the overthrust belt. Even though the land is not mountainous, that is geologically the easternmost edge of the northern Rocky Mountains. Except for a few roadcuts that expose tilted beds of rock and several places where folded beds of sandstone make ledges that wind around in the hills, little evidence of the disturbed belt is visible from the road.

East of Simms, the bedrock, still Cretaceous sedimentary formations, lies in a broadly arching fold so gentle that the subtle tilt of the sedimentary layers is hardly noticeable. And the road passes north of three conspicuous buttes. They are the northernmost outposts of the Adel Mountains, a volcanic center that was probably active sometime around 50 million years ago, but that age may be questionable.

According to some geologic maps, several of the faults in the

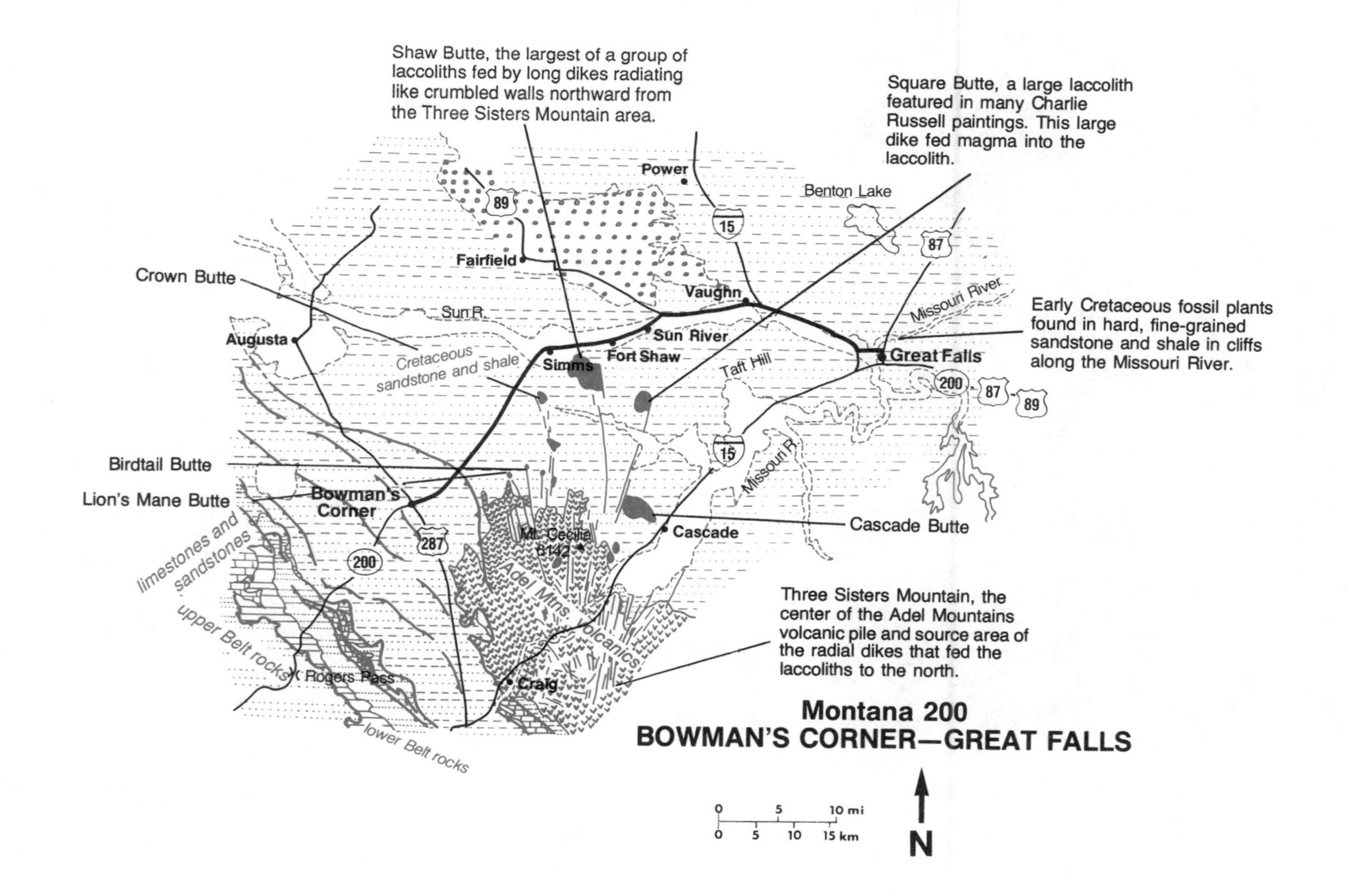
Shaw Butte, the largest of a group of laccoliths fed by long dikes radiating like crumbled walls northward from the Three Sisters Mountain area.
Square Butte, a large laccolith featured in many Charlie Russell paintings. This large dike fed magma into the laccolith.
Early Cretaceous fossil plants found in hard, fine-grained sandstone and shale in cliffs along the Missouri River.
Three Sisters Mountain, the center of the Adel Mountains volcanic pile and source area of the radial dikes that fed the laccoliths to the north.
Crown Butte
Birdtail Butte
Lion's Mane Butte
Cascade Butte
Power
Benton Lake
Fairfield
Vaughn
Sun R.
Sun River
Augusta
Simms
Fort Shaw
Great Falls
Missouri River
Taft Hill
Cretaceous sandstone and shale
Bowman's Corner
Mt. Cecilia 6142
Cascade
Missouri R.
Adel Mtns. volcanics
limestones and sandstones
upper Belt rocks
Rogers Pass
Craig
lower Belt rocks
89
15
87
200
287
0 5 10 mi
0 5 10 15 km
N

Montana 200
BOWMAN'S CORNER—GREAT FALLS

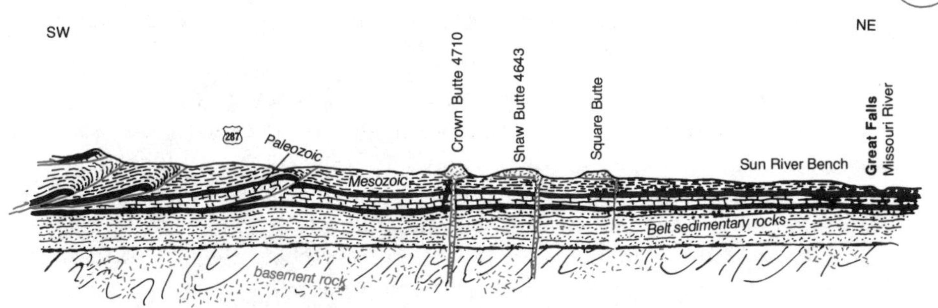

Section near the line of Montana 200 from west of Bowman's Corner to Great Falls showing the complexly thrust faulted front of the Rocky Mountains fading into the undisturbed rocks beneath the High Plains.

easternmost edge of the disturbed belt break the volcanic rocks of the Adel Mountains. If those maps are correct, then the volcanic rocks are almost certainly more than 50 million years old. If so, that would make it difficult to fit the rocks in the Adel Mountains into the picture with their cousins in the Highwood and Bearpaw mountains. But it is very hard to trace faults through volcanic rocks, so those maps could easily be wrong. A few age dates on rocks from the Adel Mountains would resolve the question. Whatever their age, the Adel Mountains are the main geologic attraction along this road.

A Parade of Laccoliths

Crown, Shaw, and Square buttes, from west to east, all bear a thick cap of shonkinite, dark igneous rock densely studded with blocky crystals of glossy black augite. The shonkinite intruded as blisters of magma that swelled beneath the Eagle sandstone, a formation of late Cretaceous sedimentary rock, laccoliths. Now, erosion has stripped

Section through an idealized laccolith like those south of Montana 200 between Simms and Great Falls.

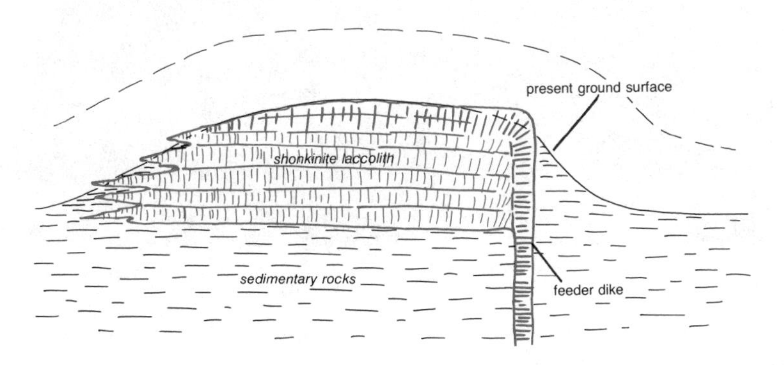

away the original sedimentary cover, leaving the more resistant shonkinite standing in high erosional relief.

The relationship between laccoliths and whatever conduit may have fed the magma into them has been a puzzle because the conduit is not ordinarily exposed. Most geologists have imagined laccoliths standing above feeder conduits like mushrooms stand on their stems. The northern Adel Mountains are one of the few places where the plumbing connection between laccoliths and their feeders is well enough exposed to leave little to the imagination.

Long dikes, ancient fissures about 20 or 30 feet wide filled with shonkinite, strike southward from one edge of each laccolith, not from beneath. The dikes clearly connect to the laccoliths, and are equally clearly the conduits that fed shonkinite magma into them. All the dikes radiate from a common center, a volcanic neck about 20 miles south of Montana 200 and just east of Interstate 15. They look on a map or from the air like the broken spokes of an enormous wagon wheel.

Although Shaw Butte is larger than Crown Butte, it is less imposing because it has not so completely lost its original cover of sedimentary rocks. A large dike, obviously the feeder conduit, enters Shaw Butte near its southeastern edge. That dike continues more than ten miles south as a great rock wall in the fields.

View west toward Crown Butte. The steep embattlements of its cliffs protect one of the best preserved remnants of the original prairie vegetation on the top of the butte.

The low ridge in the middle ground is the shonkinite dike that runs into the eastern side of Square Butte.

Dark hills visible along the southeastern horizon from high vantage points between Simms and Bowman's Corner are the eroded remnants of the old Adel Mountains volcano, the center of the activity that produced the laccoliths south of the road. Those distant hills consist almost entirely of volcanic rocks, mostly shonkinite, probably erupted about 50 million years ago. Several of the closer, more prominent peaks are laccoliths similar in many respects to those near Montana 200.

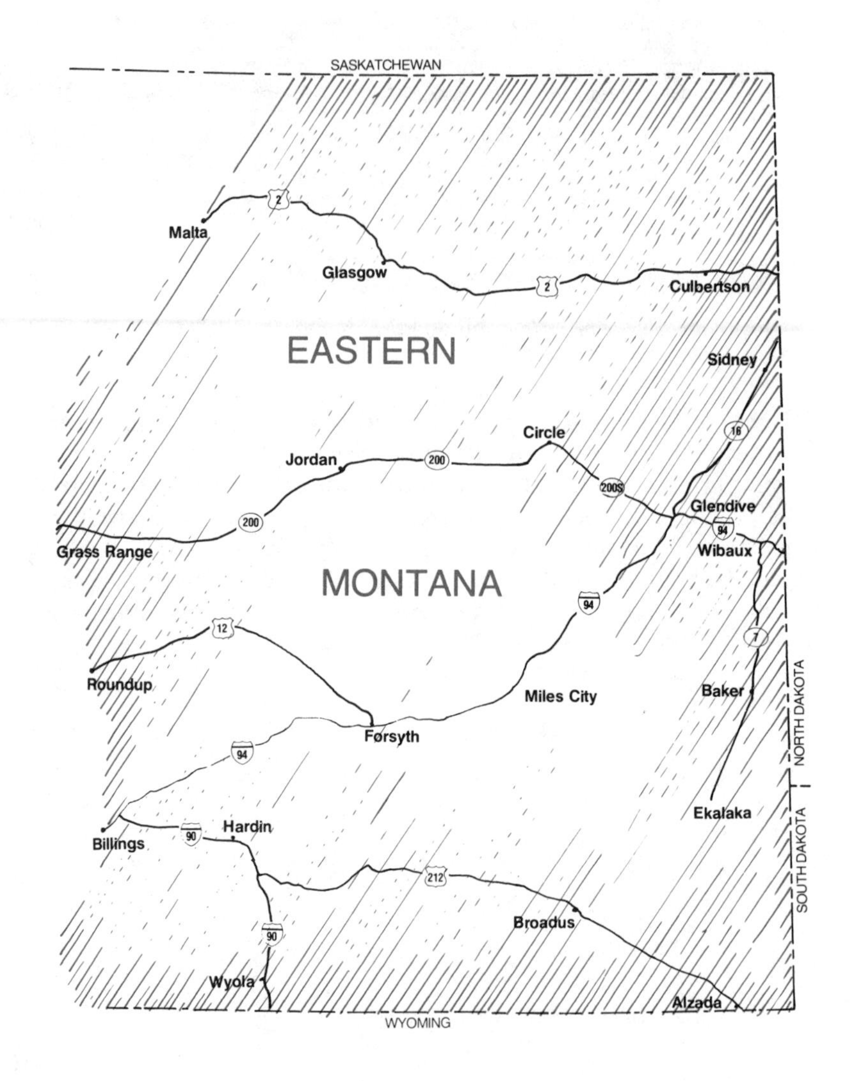

Roads covered in this section.

The heavy counterweight on this large pumping jack balances the weight of a long string of sucker rods in a deep oil well–the size of the counterweight reflects the depth of the well.

Eastern Montana—
The Far Horizons

Everything in eastern Montana stretches out to the farthest horizon. The rocks seem as endless as the landscapes. All the rocks exposed at the surface in eastern Montana are sedimentary formations laid down rather late in the long geologic history of the state. In most of the region, those layered rocks still lie as flat as they were the day they were laid down, so the roads of eastern Montana tend to stay on the same layer of rock all the way from here to there, and there may be halfway to infinity. That gives the roadside geology a certain unavoidable sameness from one road to the next. But the rocks are interesting. In fact, from an economic point of view the rocks of eastern Montana are the most interesting in the state—they contain the overwhelming bulk of Montana's oil and coal, most of its potential mineral wealth.

Oil tends to occur in places where the veneer of sedimentary rocks that covers the continental basement is abnormally thick. Geologists call those areas basins because they are low places in the continent that have accumulated large volumes of sediment. Throughout long intervals of geologic time, the continental basins were below sea level longer than other parts of the continent. Because oil typically occurs in sedimentary formations laid down in shallow sea water, these persistent inland seas created the best opportunities to look for oil. The Williston Basin is typical.

Williston Basin

Think of the Williston Basin as a broad depression in the continental crust, an enormous shallow saucer filled with sedimentary rocks. In western North Dakota, the deepest part of the Williston Basin, the continental basement rock lies at a depth as great as 13 thousand feet below sea level, almost three times its average depth in most of central and eastern Montana. Throughout most of Paleozoic time, starting at least 500 million years ago, and continuing well into Mesozoic time, the Williston Basin was a persistently low area. When shallow seas invaded the continent, they flooded the Williston Basin sooner, deeper, and longer than nearby areas. The southwestern part of that saucer is in northeastern Montana, and it contains a large fraction of the state's mineral wealth.

Sedimentary formations contsistently tend to thicken as we trace them from one well to another into the Williston Basin, and some of the formations there do not exist outside. Many formations in the Williston Basin contain thick beds of salt, evidently because there were long periods when the area was

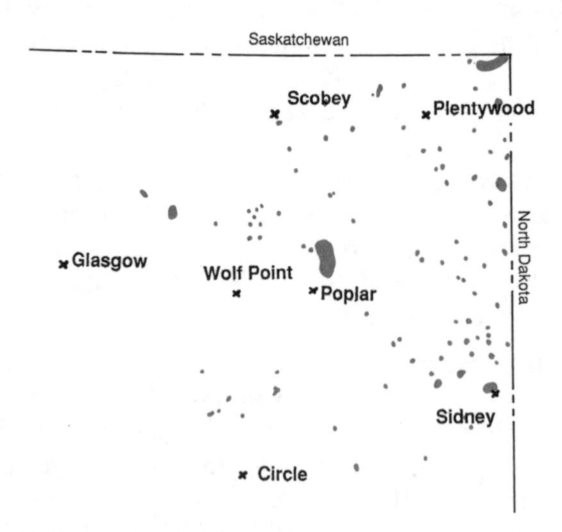

Oil fields in the Williston Basin.

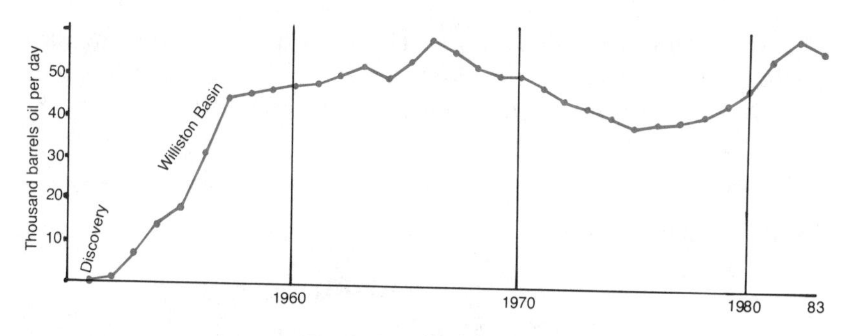

Oil production in the Williston Basin.

an isolated arm of an inland sea where evaporating sea water dumped its load of salt.

Oil and gas typically form in sedimentary rocks deposited in shallow sea water. They are especially likely to exist in places where those rocks contain beds of salt. The greater the thickness of such rocks, the greater the likelihood that the region will contain oil and gas. The first step in looking for oil and gas is to find a sedimentary basin. The next problem is to locate places within the basin where oil and gas are likely to be trapped as they move through the rocks. The Williston Basin was among the last discovered in North America, and the search for good oil traps within it continues.

Wildcat drillers began working in the Williston Basin in the years after the Second World War, and brought in the first oil well in the Montana portion in 1951. Production climbed rapidly, peaked in 1966, then began the inevitable long downward trend. Vigorous exploration drilling during the oil shortage years of the late 1970s and early 1980s discovered a number of new oil fields that greatly slowed, but did not quite reverse, the long downward trend in production.

Exploration for new oil fields is extremely difficult in the Williston Basin because the oil there tends to occur in stratigraphic traps. Those most typically form where layers of permeable rock such as sandstone grade into impermeable rock such as shale. Oil migrating upwards through the permeable sandstone finds its path blocked where the sandstone turns into shale. The oil accumulates there because it cannot continue to migrate upwards. Stratigraphic traps are so difficult to find because they depend upon subtle changes in the rock that are invisible to seismic profiling and other techniques geologists use to locate hidden folds or faults. It is easy to imagine that more large oil fields could wait discovery in the Williston Basin.

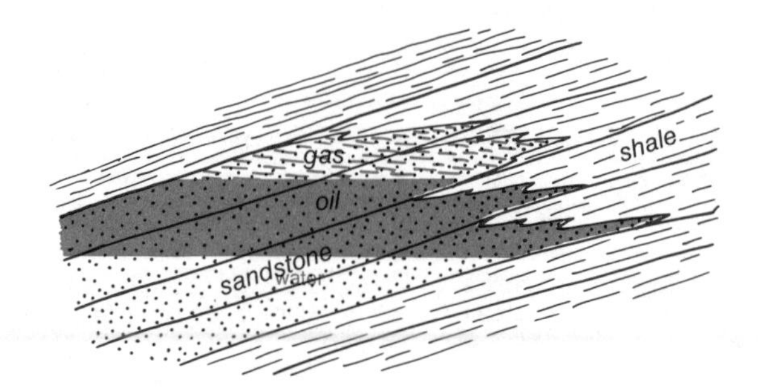

How a stratigraphic trap works.

At the end of 1983, the Montana part of the Williston Basin contained a total of 1446 active wells, which produced almost 21 million barrels of oil that year. That is more than twice the combined production of all the other oil wells in Montana.

Tricone drilliing bit—18 inches across, used in drilling oil wells.

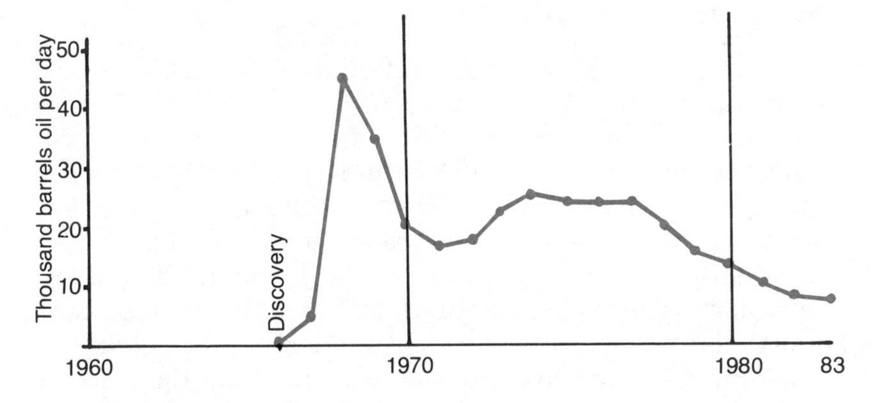

Oil production in the Montana portion of the Powder River Basin.

The Powder River Basin

Most of the Powder River Basin is in Wyoming, generally in the region between the Bighorn Range and the Black Hills. The Powder River Basin owes its existence more to deformation of the earth's crust during development of the Rocky Mountains than to a long history of crustal subsidence like that of the Williston Basin. In its deepest part, the basement rock beneath the Powder River Basin lies at an elevation of about 10,000 feet below sea level. The north end of the Powder River Basin in Montana is much shallower, but still deep enough to contain a very thick pile of sedimentary rocks.

Oil actually seeps to the surface in the Powder River Basin, so drilling began early. Early wildcatters discovered a shallow oil field in 1887 by drilling near a seep, another in 1889. Several important oil fields were found early in this century in Wyoming by drilling on the crests of folds exposed at the surface. That is so easy to do that all surface folds had been drilled before 1925.

The first big oil find in the Montana part of the Powder River Basin came in 1966 with the discovery of the giant Bell Creek field. At the end of 1983, that was still the only significant discovery in the Montana portion of the Powder River Basin with about 98 percent of the total production. Like many other oil fields in the Powder River Basin, the Bell Creek field produces from a stratigraphic trap.

The Cedar Creek Anticline

The Cedar Creek anticline is a sharp fold arched in the sedimentary rocks of extreme eastern Montana, a failed attempt at a mountain range. The folded rocks we see at the surface are probably sharply draped over a fault block in the basement. The Black Hills of South Dakota are on a parallel fault trend that moved farther, brought more resistant rocks, including basement rocks, to the surface, and did indeed become a mountain range. The Black Hills contain some igneous rocks that formed about 50 million years ago, probably at the same time the range rose. So it seems likely that the parallel Cedar Creek anticline formed about 50 million years ago, too.

Striking as it is on the geologic map of eastern Montana, the Cedar Creek anticline is not much of a spectacle on the ground. None of the rock formations it brings to the surface resist erosion well enough to stand up as prominent ridges. The most obvious clues to its existence come in occasional glimpses of tilted layers of rock.

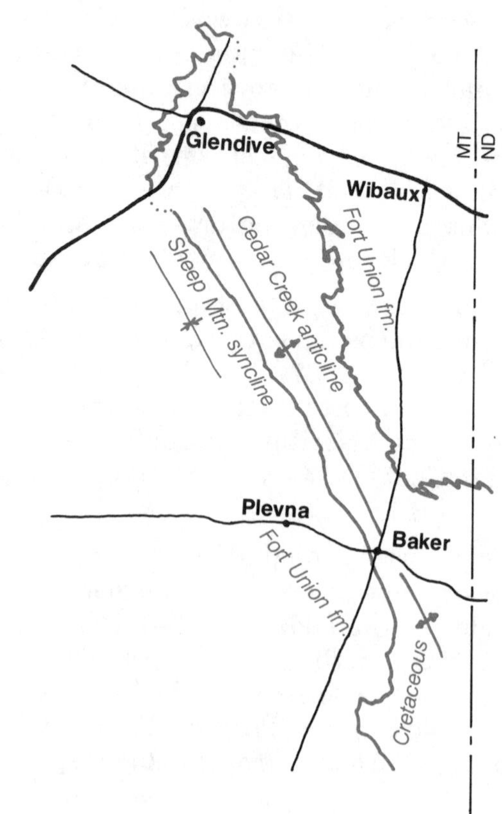

The Cedar Creek anticline.

The earliest geologists in eastern Montana noticed those tilted beds and mapped the Cedar Creek anticline by about 1910. They immediately recognized it as a perfect example of the kind of fold that quite commonly traps oil and gas in its crest. A successful wildcat well found natural gas in the crest of the Cedar Creek anticline in 1912. Within a few years, an almost continuous line of oil and gas fields was producing along the crest of the fold.

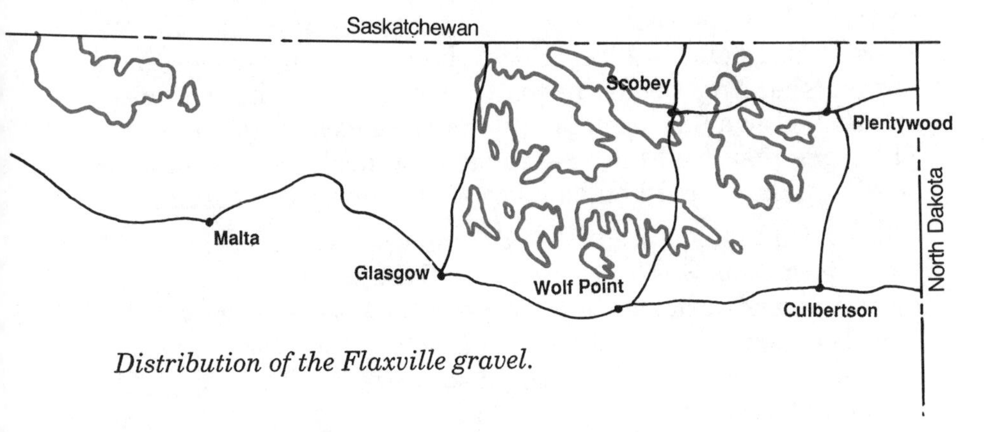

Distribution of the Flaxville gravel.

The Fort Union formation

As Cretaceous time drew to a close, the inland sea retreated east for the last time. Sediments continued to accumulate in the shallow sea water offshore, by now mostly in North Dakota and Saskatchewan, and in a vast area west of the coastline that was slightly above sea level. Those sediments continued to accumulate through the end of Cretaceous time and into the early part of Tertiary time, until about 55 million years ago.

Although the sediments deposited in latest Cretaceous and early Tertiary time look exactly the same, geologists name them differently because only the Cretaceous rocks contain dinosaur fossils. Those are called the Lance and Hell Creek formations. The sediments deposited after the end of Cretaceous time, those that lack the remains of dinosaurs, are called the Fort Union formation. It covers vast areas of eastern and central Montana.

Nearly everywhere, the Fort Union formation consists mostly of pale yellowish mudstone and sandstone still so soft that it might better be called mud and sand. Geologists working with those rocks often find a shovel more useful than a hammer. That softness of the Fort Union formation makes it so easily erodible that it rarely forms prominent outcrops, and tends to erode into vast and nearly featureless plains.

The Fort Union formation also contains enormous, incredible, amounts of coal that originally accumulated as peat deposited in marshes along the old sea coast and in the floodplains of rivers that drained into it. Picture the scene that must have existed then: a shallow sea still flooded extreme eastern Montana and points east and north. Barrier islands stood slightly offshore, separating broad coastal lagoons and salt marshes from the main coast. Broad tracts of coastal dunes encroached on a flat coastal plain that extended far to the west. The scene may well have resembled the modern coast of the Gulf of Mexico from Louisiana to south Texas and Mexico.

It is hard to know what the climate was like when the Fort Union formation was laid down, but all the sediment that piled up on the landward side of the coast suggests that it may well have been arid. Sediments accumulate on land along modern arid coasts, including the coast of south Texas, but not along humid coasts such as those of Florida.

Even in arid regions, abundant vegetation may grow along the floodplains of rivers and in the coastal lagoons. If partially decayed vegetation accumulates in those places, it forms mucky deposits of peat that will turn into coal if more sediments bury and compress them. If changes in sea level or elevation of the land shift the position of the shoreline, the position of the peat deposits shifts with it. That happened again and again as the sediments of the Fort Union formation piled up across eastern Montana. Exploring for coal in eastern Montana is largely a matter of tracing old shorelines and the rivers that flowed into them.

The End of the Dinosaurs

As long as the Cretaceous period lasted, dinosaurs roamed the broad coastal plain in eastern Montana, frequently left their bones in the accumulating sediments of the Hell Creek and Lance formations. Then they suddenly vanished, along with many other important groups of animals that lived on land and in the sea. Some terrible catastrophe nearly annihilated animal life on earth.

Every place in the world where geologists can examine rocks that were being deposited when the dinosaurs vanished, they find a thin layer of dark rock that seems to mark the geologic moment of the great extinction. That layer generally contains about 30 parts per billion of iridium, along with comparable amounts of the other elements related to platinum. Those elements are so rare on the earth's surface that they hardly exist in normal sedimentary rocks, so that minute concentration is actually amazingly high. About the only thing that could spread iridium all over the earth's surface is fallout from the explosion of a giant meteorite, an asteroid.

Although not themselves explosive, asteroids course through space at speeds of 25 to 40 miles per second, fast enough to give them an energy of motion that is greater, pound for pound, than the energy contained in a hydrogen bomb. If they hit something solid, such as the earth, all that energy of motion suddenly converts into more than enough heat to vaporize them. When that happens, they detonate with a blast violent beyond anything in human experience. Sudden conversion of all that energy of motion into heat could have no other consequence. Imagine the results.

Most geologists envision a scenario of extinction that hinges on the dust cloud, rather than on the explosion itself. Imagine the high altitude winds spreading that dust cloud around the earth in a matter of a few days. Then it hung in the atmosphere for weeks or months, casting a pall of darkness and cold that stopped growth of green plants. First, the animals that ate plants starved, and then as the supply of flesh and carrion vanished, the predatory animals starved. As the dust finally settled, the sun shone again on a world reduced to plants surviving as roots and seeds, and animals that ate roots and seeds. Birds and small mammals lived through the disaster, but the dinosaurs were gone. That drew the curtain on Cretaceous time.

Even though the dinosaurs and many other animals were gone, deposition of mud and sand continued in eastern Montana as though nothing had happened. Except for the abrupt change in fossil content, the only evidence of the catastrophe within the sediments is the thin layer of dust. Geologists more commonly recognize the change by the disappearance of dinosaur bones.

Coal

Its astronomically large coal content makes the Fort Union formation the most valuable body of rock in the state, one of the most valuable in the entire world. The actual amount of coal is unknown, but certainly staggering, probably something more than 50 billion tons, an unbelievable reserve of fossil fuel.

Geologists divide the Fort Union formation into three zones, called members. The lowest and oldest is the Tullock member, which consists mostly of thinly bedded sandstone and mudstone. Next comes the Lebo shale. You rarely see much of that because it weathers too readily to make good outcrops; it erodes into vast plains. The Tongue River member at the top of the formation is mostly thick beds of sandstone that you do see in cliffs along rivers and as outcrops in the sides of rugged hills. Together, the Tullock, Lebo, and Tongue River members of the Fort Union formation generally total somewhere between two and three thousand feet in thickness.

All three members of the Fort Union formation contain coal, and all three have supported mines, but the overwhelming bulk of the Fort Union coal is near the top of the formation in the Tongue River sandstone. Most of the coal is underground where no one can see it, so estimating the exact amount is

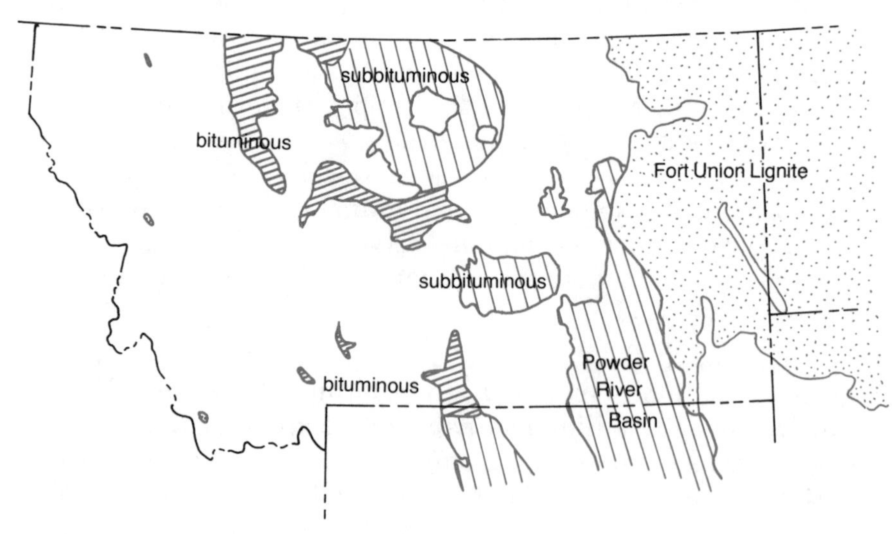

The coal fields of eastern Montana.

difficult. Geologists begin by carefully mapping the areas underlain by the Tongue River member of the Fort Union formation. Then they focus their attention on those areas because that is where the large coal deposits will be. In most cases, the next step is drilling to see whether the coal is actually there, and how thick it is.

The quality of coal depends largely upon how much heat it yields on burning, and that depends mostly upon how deeply it has been buried, how hard it has been compressed. In general, the quality of Fort Union coal improves to the west and south because the depth of former burial increases in those directions. The brown Fort Union coal of easternmost Montana is mostly lignite, soft crumbly stuff that tends to contain a lot of moisture and has relatively low heat content. The black sub-bituminous and bituminous coals farther west and south are denser, drier, hotter burning than the lignites, and yield far more heat per ton.

Sulfur in any form is an undesireable impurity in coal because it burns into sulfur dioxide, a gas that kills plants and causes acid rain. But you have to consider the heat and sulfur content of coal together because it is the combination of those two factors that determines how much pollution the coal will cause. For example, a lignite with low sulfur content may cause worse pollution than bituminous coal with medium sulfur content because you have to burn so much more of the lignite to get the same amount of heat.

Burned Coal and Red Hills

The scarcity of hard rock in areas where the Fort Union formation is at the surface creates real shortages of road material and construction aggregate in large parts of eastern Montana. Hills are in short supply too, unless the Tongue River sandstone is at the surface. Coal changes all that.

As streams cut down through a coal seam, the water drains out of it. Then sooner or later a lightning strike or a range fire ignites the dry coal, and it burns. Burning coal seams may smoke and smolder for centuries until the fire finally runs out of fuel, or burns right through to the other side of the hill. Coal burns hot, so a smoldering coal seam bakes the soft mudstone and sandstone above it into rock as red and hard as kiln-fired

A close view of clinker left from a burned out coal bed.
—Larry French photo

brick, clinker. A bed of coal five to ten feet thick generally produces ten to 30 feet of clinker as it burns; one 50 feet thick may cook as much as 200 feet of clinker. One way to prospect for coal seams is to trace thick clinker zones into low areas where the coal did not burn.

Coal fired Fort Union clinker is very hard, difficult to erode, and generally bright red or yellow, although some is black. Clinker provides much of the road metal and construction aggregate in eastern Montana—watch for the red roads. It also stands up in erosional relief to become hills, colors the tops of those hills red, and paints red stripes along their sides. So it is no accident that the coal fields tend to lie among rugged and colorful hills. The coal provides the scenery. Neither is it any accident that the big coal mines operate in the valleys where the coal is still too wet to burn.

Most clinker beds consist largely of broken rubble that collapsed into the space left as the coal burned. The large open spaces between the fragments of rock hold water, and the water

Collapsing ground over a burning coal bed.
—U.S. Forest Service photo by Phil South

Badlands near Ashland, Montana. —U.S. Forest Service photo by K. D. Swan

seeps easily from one space to another. That capacity to store water and let it move makes the clinker beds good aquifers that yield abundant water to a well. Watch the hillsides in the rugged coal fields to see trees and bushes growing largest and most abundantly along the red streaks of exposed clinker. That is where water seeps out of the ground. And in most cases a well drilled into the hill will hit water where it penetrates the clinker horizon.

Badlands

Here and there throughout eastern and central Montana, the rolling prairies give way to isolated patches of badlands, miniature deserts. They probably mark places where something, perhaps an intense range fire, destroyed the plant cover that normally shelters the soil. Once started, badlands tend to perpetuate themselves indefinitely by making it extremely difficult for the plants to return.

Normally, the natural plant cover in eastern Montana is continuous enough to cast an umbrella of leaves over every part of the ground surface. Raindrops strike a leaf first, then drop gently onto the soil. Meanwhile, the roots of the plants and the burrowing insects that live among them keep the soil full of open pores and burrows that absorb surface water. The general effect of the plant cover is to retard erosion by making it impossible for raindrops to disturb the soil surface, and by preventing surface runoff that might carve gullies.

If something destroys the plant cover, the sheltering umbrella of leaves is gone, and raindrops splatter directly onto the soil. Such raindrops spattering onto wet soil pound the surface tight, forming the impermeable crust that everyone has seen in gardens and fields during late summer. And the roots and insects that kept the soil open are gone, leaving nothing to destroy that crust so water can soak into the ground. The effect of all that is that rain falling on bare soil runs off the surface, eroding gullies and carrying away any particles of soil that raindrops splattered off the surface. The combination of rain splash erosion of the soil and increased surface runoff is erosion in the desert style. It is far more rapid than the kind of erosion that affects more humid areas with a good plant cover.

When rain water beings to run off the surface, instead of soaking into the ground, the soil dries out, and no longer contains enough moisture to support its natural plant cover. And rapid erosion tends to strip seeds and sprouts off the surface before they can gain a foothold. Once rainsplash and surface runoff erosion have well begun, it is extremely difficult to restore the plants. Badlands tend to be permanent little deserts even though they receive the same rainfall as nearby areas with a good plant cover.

Saline Seep

The spreading development of saline seeps is one of the biggest agricultural problems in large areas of central and eastern Montana. To a considerable extent, the problem is the result of farming practices.

Saline seeps are springs that produce water heavily laden with dissolved salts. The water varies in composition, but normally contains some combination of sodium, calcium and magnesium chlorides, carbonates, sulfates, and nitrates—an evil alkaline brine that tastes salty and bitter. The first sign of a developing saline seep is a patch of almost unnaturally dark green vegetation that turns out upon inspection to consist largely of sedges, reedy plants with tall stems that have a

Crystallizing salts creeping up from the soil are splintering the base of this fence post planted in a saline seep.

triangular cross section. Sedges like plenty of water and they don't mind a bit of salt, so they can thrive in a new saline seep.

Most of the seeping brine evaporates as it reaches the surface, leaving the dissolved salts behind. Within a few years, those salts accumulate in the soil to a concentration that kills the sedges, along with everything else. Then the seep becomes a patch of barren ground that turns into sticky mud during wet weather, and acquires a glaring white crust of bitter salts when the weather turns dry. Nothing grows on that ground.

The dry plains of Montana have always had a few saline seeps. Now there are a great many, and more appear every year. Most of the new ones are probably the result of dry farming practices that became widespread during the 1920s. Dry farming entails planting a crop on alternate strips of land while keeping those between fallow and bare of all plants. With no plants to use it, an entire season's worth of water accumulates in the fallow strips. The next crop is planted in the fallow strips while those last harvested are laid fallow. The effect is to use the moisture of two seasons to raise one crop. The problem is that the system catches a bit too much water.

Water accumulating in the fallow strips soaks down through the soil to the ground water reservoir, and raises the water table. Normally, little or no water passes all the way through the dry soils of eastern and central Montana and the ground water reservoir recharges mainly by seepage through the beds of streams. The new source of recharge would be good if it were not that the soils of those dry regions of Montana tend to contain rather large amounts of salts that dissolve in the water passing through them. So the water is salty and alkaline by the time it reaches the water table. Wells that had produced sweet water become bitter. As the water table continues to rise, it eventually intersects the ground surface, and that is where the new seeps appear.

There is no quick solution for a problem that developed over a period of more than 50 years. In the long range, changes in dry farming techniques can probably limit the future development of new saline seeps. The only cure for most of those that already exist is to let the rain leach the accumulated salts out of the soil. That will be an extremely slow process in a region as dry as central and eastern Montana.

Bighorn Reservoir in the Pryor Mountains.
Montana Department of Commerce photo

Interstate 90: Billings—Wyoming
101 mi./163 km.

The high and usually snowcapped mountains on the distant skyline southwest of Billings are the eastern end of the Beartooth Plateau, a block of basement rock that rose along faults about 50 million years ago. The lower and much broader rise in the skyline almost directly south of town is the Pryor Mountains, a broad arch that brings large areas of Madison limestone to the surface, along with older sedimentary formations exposed in deep canyons. Farther south, that low bulge rises to become the towering Bighorn Range that hovers snowcapped on the skyline southwest of the highway between Hardin and the Wyoming skyline.

Rocks along and near the highway between Billings and the Wyoming state line are all Cretaceous sandstones and shales that belong to several different formations. Only the thick sandstones make rugged hills and prominent outcrops. Hills with brilliant caps of red clinker, the usual sign of coal country, become conspicuous near the Wyoming border.

Bighorn Canyon

Yellowtail Dam west of Lodge Grass impounds the Bighorn River to flood its canyon and those of its branching tributaries in the Pryor Mountains, the low northern end of the high Bighorn Range. Hundreds of miles of high cliffs, canyon walls eroded mostly in

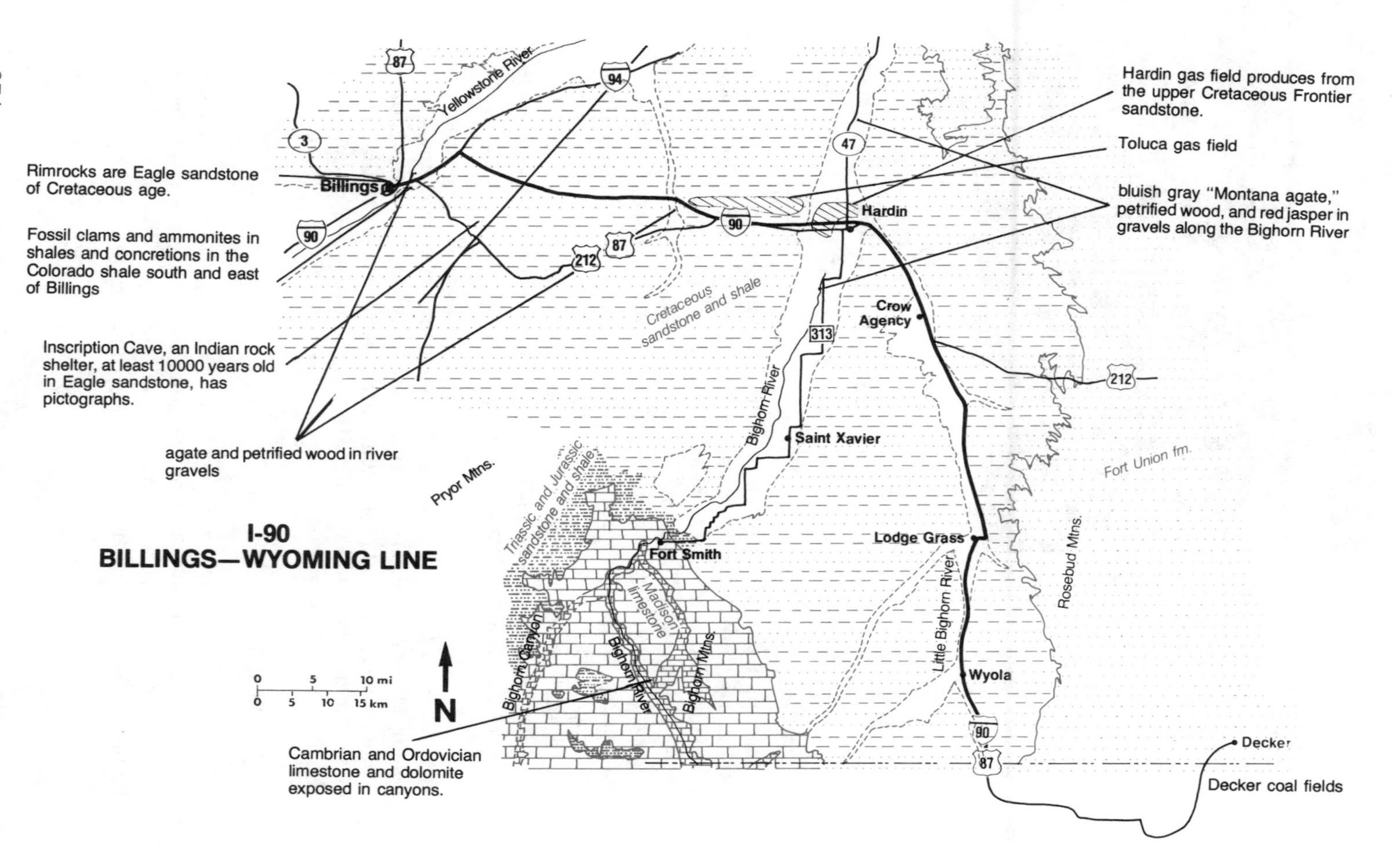

I-90
BILLINGS—WYOMING LINE
Rimrocks are Eagle sandstone of Cretaceous age.
Fossil clams and ammonites in shales and concretions in the Colorado shale south and east of Billings
Inscription Cave, an Indian rock shelter, at least 10000 years old in Eagle sandstone, has pictographs.
agate and petrified wood in river gravels
Hardin gas field produces from the upper Cretaceous Frontier sandstone.
Toluca gas field
bluish gray "Montana agate," petrified wood, and red jasper in gravels along the Bighorn River
Cambrian and Ordovician limestone and dolomite exposed in canyons.
Decker coal fields
Decker
Billings
Yellowstone River
Hardin
Crow Agency
Saint Xavier
Lodge Grass
Wyola
Fort Smith
Bighorn River
Little Bighorn River
Cretaceous sandstone and shale
Triassic and Jurassic sandstone and shale
Madison limestone
Bighorn Mtns.
Bighorn Canyon
Pryor Mtns.
Rosebud Mtns.
Fort Union fm.
N
0 5 10 mi
0 5 10 15 km
87
94
3
90
212
47
313

Buff-colored Cretaceous sandstone cliffs a few miles north of Lodge Grass

Madison limestone, rise above the reservoir. Some of the canyons bite through the Madison limestone into older formations, of which the Big Horn dolomite is the most conspicuous. It was deposited during Ordovician time.

The Sarpy Creek Coal Mine

A big strip mine began producing coal in the Sarpy Creek area about 30 miles east of Hardin in 1974. It works the Robinson seam, which is about 20 feet thick, and the Rosebud-McKay seam, which is about 30 feet thick. Both of those enormous energy reserves are in the massive Tongue River sandstones in the upper part of the Fort Union formation. The big strip mines at Colstrip produce from the same pair of oversized coal seams, which together contain a large part of the economic future of Montana.

The Rosebud-McKay seam lies between 50 and 120 feet above the Robinson seam, so the mine works it first, then strips off the intervening rock as it steps down to the lower deposit. Like other modern strip mines in eastern Montana, this one restores most of the original ground surface, soil, and plant cover after the coal is gone—state law now requires such thorough reclamation of worked coal lands.

Coal Mines at Decker

One of Montana's biggest coal producing areas is around Decker, about 30 miles southeast of Wyola on a road that enters from Sheridan, Wyoming. Three large coal seams in the Decker area range from about 12 to 37 feet thick, and lie at shallow depth beneath

Digging coal in the West Decker Mine.
—Western Energy Company photo by Richard Kehrwald

hundreds of square miles of the countryside. Together, those incredible seams contain more than two billion tons of unmined coal, by eminently sober and conservative estimate. That staggering reserve is ample to keep the district in business for many decades, to keep the whole country in coal for many years.

The Decker area produces bituminous coal, which releases about twice as much heat per ton as the lower grade brown coals from fields farther northeast in Montana. Coal from the Decker field is also prized for its low sulfur content, the main reason for its ability to compete in the national market. Low sulfur coals are in great demand because they minimize the need for investment in expensive equipment to scrub sulfur dioxide from stack gases.

Rosebud coal seam at Decker.
Montana Bureau of Mines and Geology photo by Gary Cole

Fossil leaves of the Dawn Redwood in sandstone of the Fort Union formation west of Miles City. Dawn Redwood trees thrive today, but not in places as cold as Montana. —Larry French photo

Interstate 94: Billings—Miles City
142 mi./227 km.

Between Billings and Miles City, the highway follows the south side of the Yellowstone River Valley; alternating between short stretches on the valley floor, and longer routes that cross the high ground slightly to the south. Long stretches of the road above the valley floor cross remnants of the High Plains surface, nearly level ground underlain by the Flaxville gravel. Watch for pebbles strewn through the fields and occasional roadcuts that expose beds of gravel.

Bedrock is late Cretaceous sedimentary formations along most of the route, Fort Union formation near Miles City. Near Billings, the late Cretaceous Eagle sandstone makes spectacular cliffs. It is a thick deposit of sand that probably accumulated as barrier islands with sand dunes on them along the coast of an inland sea. Try to imagine this area looking something like the modern coast of south Texas except that dinosaurs roamed the countryside instead of cows. Most of the other formations are too poorly exposed to attract much attention.

Pompey's Pillar is simply a monumental outcrop of Cretaceous sandstone that stands isolated from the main line of bluffs on the north side of the Yellowstone River. It is one of Montana's most widely celebrated rocks because Captain Clark of the Lewis and Clark expedition carved his name in the soft sandstone in 1806.

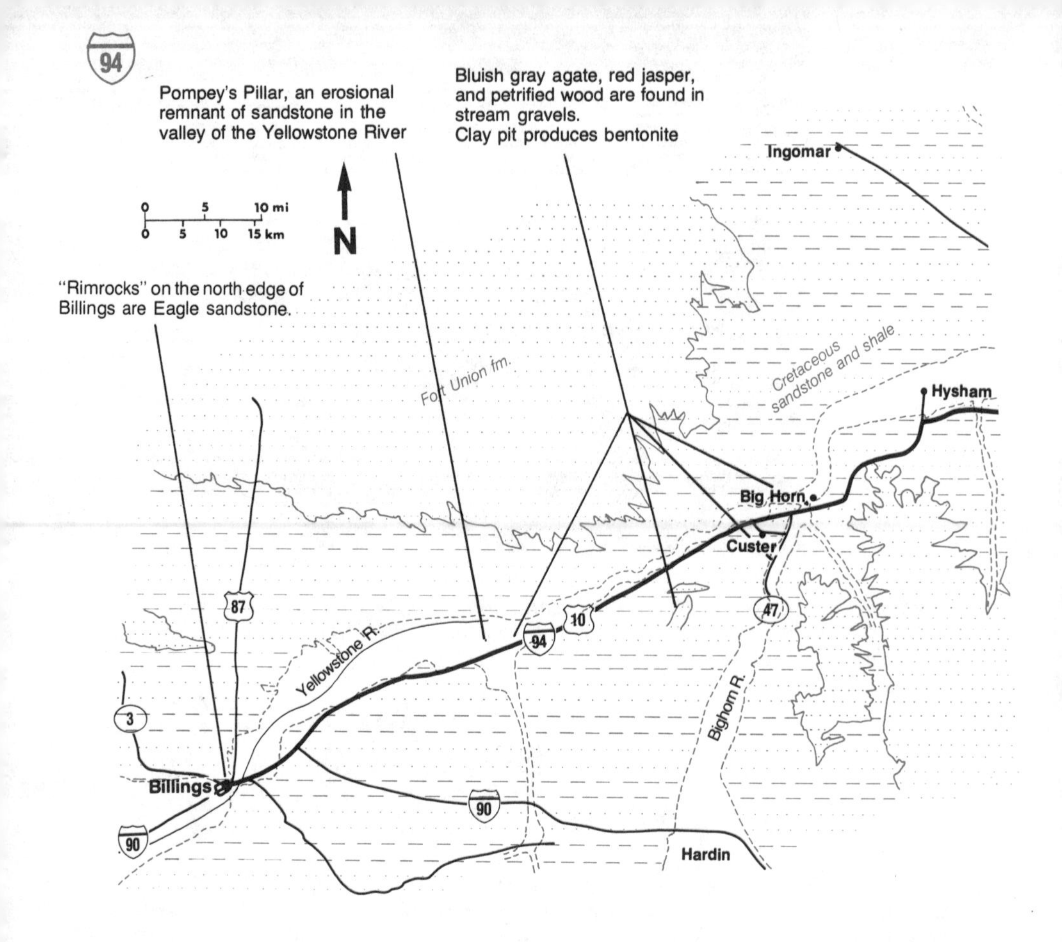

The Lake Basin Fault Zone

East of Billings, watch for many roadcuts and outcrops that expose tilted beds of tan Cretaceous sandstone. They are in the Lake Basin fault zone, one of the major structures of central and eastern Montana. Plotting those tilted beds on a map shows that the zone consists of a long row of small faults that align along a trend oriented slightly south of east even though the individual faults trend northwest. The block south of the zone rose.

One way to investigate structures like the Lake Basin fault zone is to make a model consisting of a loosely jointed rectangular wooden frame that holds a metal screen—that represents the continental crust. A thin layer of plaster of paris poured onto the screen can represent the sedimentary rocks that cover the continental crust, in this area to a depth of several thousand feet. Then it is possible to simulate movements of the continental crust by racking the wooden

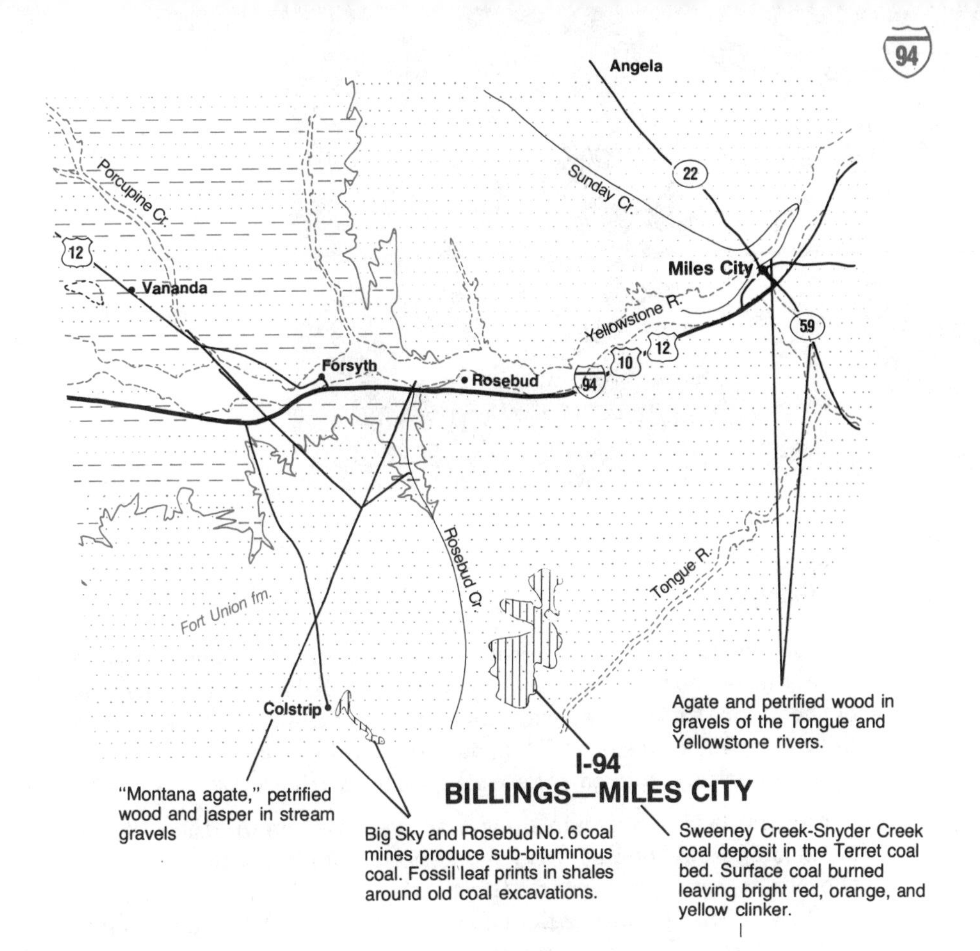
Angela
22
Sunday Cr.
Porcupine Cr.
12
Vananda
Miles City
Yellowstone R.
59
12
10
Forsyth
Rosebud
94
Rosebud Cr.
Tongue R.
Fort Union fm.
Colstrip
Agate and petrified wood in gravels of the Tongue and Yellowstone rivers.
I-94
BILLINGS—MILES CITY
"Montana agate," petrified wood and jasper in stream gravels
Big Sky and Rosebud No. 6 coal mines produce sub-bituminous coal. Fossil leaf prints in shales around old coal excavations.
Sweeney Creek-Snyder Creek coal deposit in the Terret coal bed. Surface coal burned leaving bright red, orange, and yellow clinker.

frame out of shape, and see how that deforms the plaster of paris. Such experiments show that the continental crust south of the Lake Basin fault zone probably moved southeast relative to that to the north, tearing the sedimentary rocks above into the pattern we see on the map.

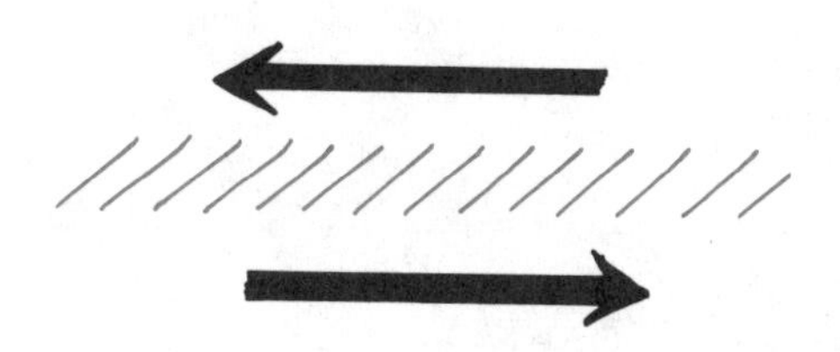

The Lake Basin fault zone. The arrows show the kind of crustal movement that probably caused the faulting.

The dark and resistant bed of red clinker that caps this ridge is responsible for its existence.
—Larry French photo

Colstrip

The side road to Colstrip crosses Fort Union formation all the way except for a short stretch just south of Forsyth, where it crosses some of the older Cretaceous formations. Much of the route crosses a coal field. Burning coal seams left beds of red clinker that resist erosion to make colorfully rugged hills well covered with trees.

Enormous transmission lines march across the landscape to feed electricity from the big coal fired generating plants at Colstrip into the power networks serving the northern High Plains and Pacific Northwest. Strip mining flourished mightily at Colstrip during the 1920s, then faded as competition from cheap oil and natural gas nearly drove northern High Plains coal from the market. Then increasing demand for power and diminishing reserves of oil and gas inspired a revival of strip mining during the late 1960s to the accompaniment of intense public debate. The major issues have been mine reclamation, maintenance of the ground water reservoir, and the effect of stack gases on the region downwind. All those issues ramify into geology.

Strip mining at Colstrip during the 1920s.
—Montana Historical Society photo

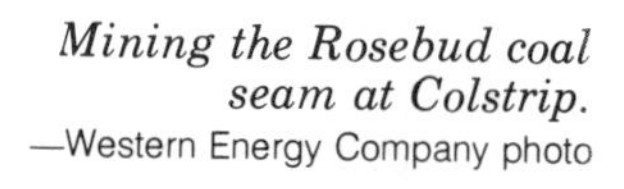
Mining the Rosebud coal seam at Colstrip.
—Western Energy Company photo

Many coal seams serve as aquifers that hold large quantities of water and supply it to wells, so the possible effects of mining on availability and quality of ground water is a serious issue. Pumping out a mine certainly does lower the water level in the immediately surrounding area, but that will restore itself naturally within a few years after mining and pumping cease. However, a return to the original water table does not necessarily restore the original situation because the aquifer has been replaced with waste rock from the mine.

In most cases, the broken rubble used to backfill mine pits during reclamation should contain enough open space to be a reasonably good aquifer, better in some cases than the original coal. But there is some question about the quality of the water. Mineral matter dissolving out of the freshly broken rock used to backfill the worked out mines tends to make the ground water both salty and alkaline, too much so to be useable in some cases. It is not yet possible to predict with confidence how far that effect will extend, or how many years must pass before the natural circulation will flush the aquifer fresh.

Early miners left an ugly legacy of enormous tracts of formerly beautiful and productive land furrowed as though by a gargantuan plow. Long trenches opened where the coal was stripped remain unfilled to this day; the parallel ridges of rock dumped as it was stripped off the coal remain unlevelled. The slow natural processes of weathering and soil formation cannot return such a devastated landscape to any semblance of productiviy for many thousands of years.

Montana law now requires strip miners to use their waste rock to refill the trenches, grade the landscape to its original form, and establish a productive plant cover. The art and science of mine reclamation are now so highly developed that the recently worked sites are visible only to a knowing and practiced eye.

The caps of hard sandstone protect the underlying soft mud from rainsplash erosion to create these pillars in the Makoshika badlands near Glendive.
—U.S. Forest Service photo by K. D. Swan

Interstate 94: Miles City—Wibaux
96 mi./154 km.

The road follows the south side of the Yellowstone River between Miles City and Fallon, the north side between Fallon and Glendive. Along most of the route, the road stays on remnants of the High Plains surface, overlooking the valley the river eroded into it. Watch for exposures of Flaxville gravel in the roadcuts, for pebbles in the fields.

Except near Glendive, bedrock along the entire route is Fort Union formation, mostly the Tullock and Lebo members that tend to erode into level plains. Around Glendive, the bedrock looks exactly the same, but is called the Hell Creek formation because it contains a few dinosaur bones. That stuff was deposited during latest Cretaceous time, shortly before the great extinction.

East of Glendive, many patches of badlands north and south of the road provide good exposures of the lower Fort Union formation, the Tullock member. The rock is pale gray mudstone that erodes into rugged little buttes covered with rills between ledges of brown sandstone that jut from the slopes. Farther east, the badlands become

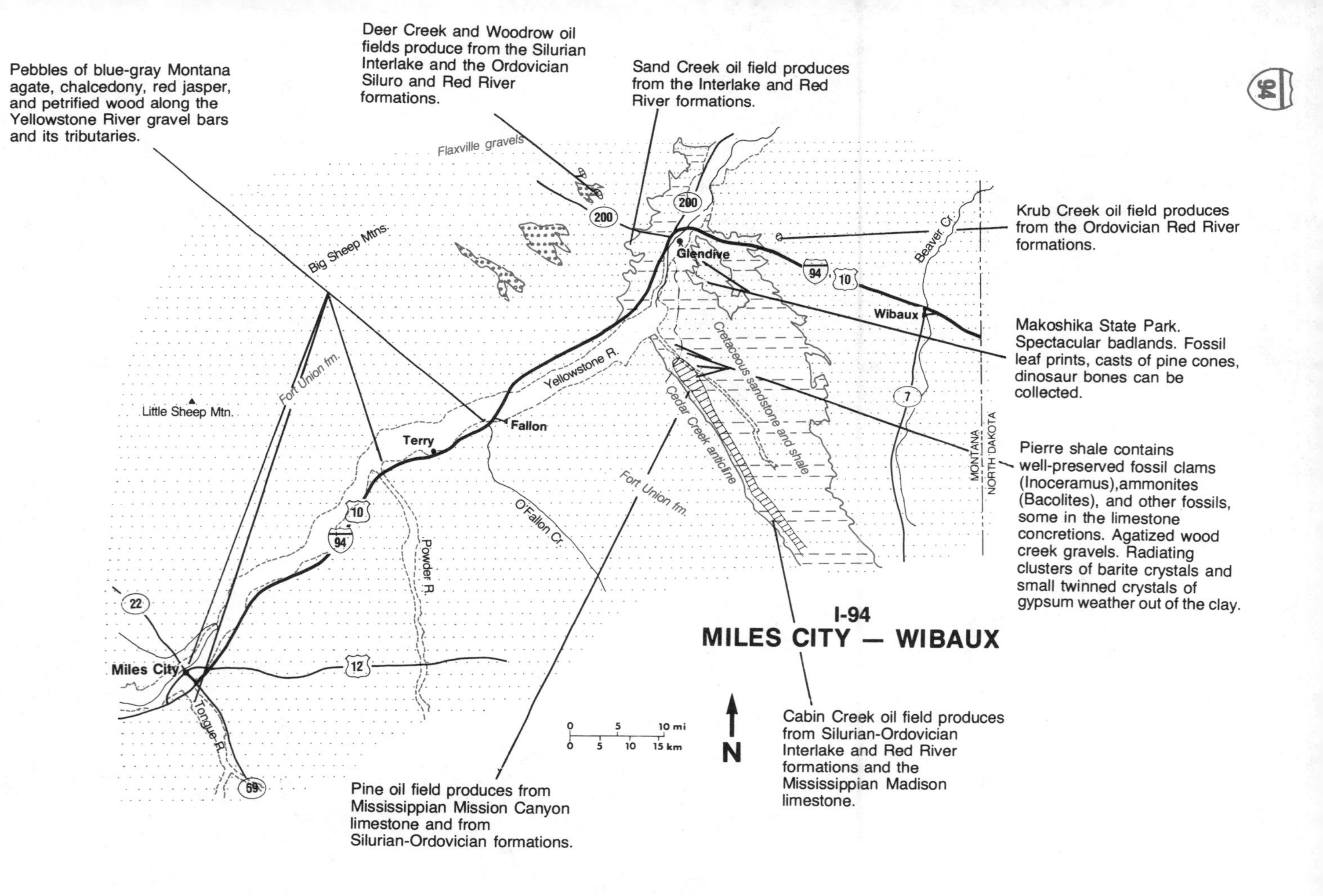
Pebbles of blue-gray Montana agate, chalcedony, red jasper, and petrified wood along the Yellowstone River gravel bars and its tributaries.
Deer Creek and Woodrow oil fields produce from the Silurian Interlake and the Ordovician Siluro and Red River formations.
Sand Creek oil field produces from the Interlake and Red River formations.
Krub Creek oil field produces from the Ordovician Red River formations.
Makoshika State Park. Spectacular badlands. Fossil leaf prints, casts of pine cones, dinosaur bones can be collected.
Pierre shale contains well-preserved fossil clams (Inoceramus),ammonites (Bacolites), and other fossils, some in the limestone concretions. Agatized wood creek gravels. Radiating clusters of barite crystals and small twinned crystals of gypsum weather out of the clay.
Cabin Creek oil field produces from Silurian-Ordovician Interlake and Red River formations and the Mississippian Madison limestone.
Pine oil field produces from Mississippian Mission Canyon limestone and from Silurian-Ordovician formations.
Flaxville gravels
Big Sheep Mtns.
Little Sheep Mtn.
Fort Union fm.
Yellowstone R.
Glendive
Wibaux
Fallon
Terry
Miles City
Powder R.
O'Fallon Cr.
Tongue R.
Beaver Cr.
Cretaceous sandstone and shale
Cedar Creek anticline
MONTANA
NORTH DAKOTA
200
94
10
7
12
22
59
0 5 10 mi
0 5 10 15 km
N

I-94
MILES CITY — WIBAUX

less numerous, and then disappear as the road passes onto the Lebo shale, and the country flattens into a grassy prairie that stretches out to the horizon. Near Wibaux, the Tongue River sandstone in the upper part of the Fort Union formation contains beds of coal that may someday support large coal mines.

The Cedar Creek Anticline

Glendive is on the northern end of the Cedar Creek anticline, a sharply arched fold that extends southeast into South Dakota. Although it is the most striking geologic structure in eastern Montana, the rocks visible from the road give no hint that the Cedar Creek anticline exists. The fold warps the Fort Union formation, so it must have formed sometime after the last of those sediments accumulated about 55 million years ago.

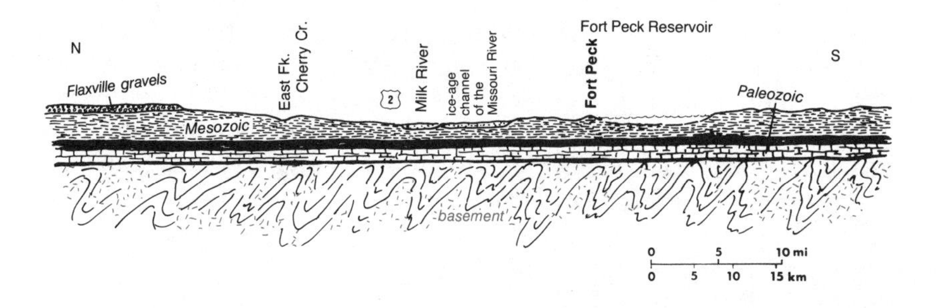

Section across US 2 between Glasgow and Nashua. The layers of sedimentary rock beneath this part of Montana still lie nearly flat. Flaxville gravel floors the broad upland surfaces, remnants of the High Plains surface that formed during Pliocene time, when this region was a desert.

Makoshika Badlands

The eastern suburbs of Glendive trail into the edge of the Makoshika badlands, one of the most spectacular in the state. Rainsplash and surface runoff erosion has carved soft mudstones and sandstone into fantastic shapes that suggest hundreds of fairyland castles fallen into ruins. Evidently something, perhaps an intense range fire, destroyed the plant cover in this area centuries ago, permitting splashing raindrops to attack the rocks.

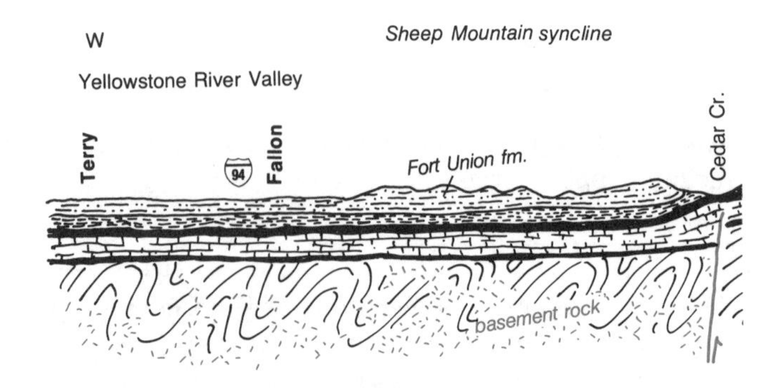

Section south of the highway through the Cedar Creek anticline. The fold

Most of the rocks exposed in the Makoshika badlands are the Cretaceous Hell Creek shale. The lower part of the early Tertiary Fort Union formation caps some of the hills in the eastern part of the area. The two formations look alike except that the Hell Creek shale contains a few widely scattered fragments of dinosaur bone, and the Fort Union formation does not. The catastrophe that exterminated the dinosaurs happened while these sediments were accumulating near the margin of the shallow inland sea that still covered much of North Dakota and Saskatchewan.

View into part of the Makoshika badlands, a miniature desert landscape.
—Montana Bureau of Mines and Geology photo by H. L. James

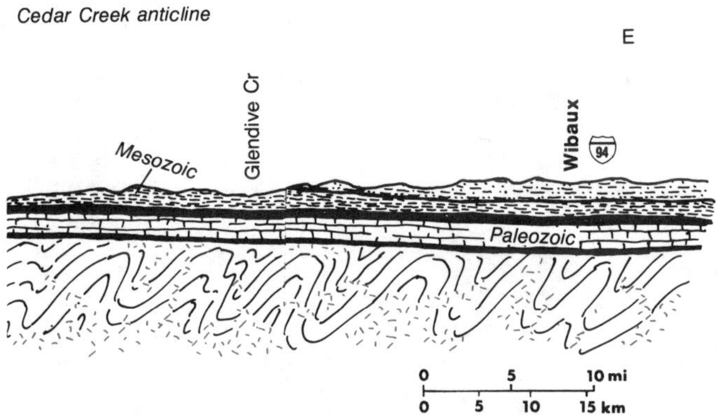

probably formed as a fault in the basement rocks at depth buckled the sedimentary formations above.

US 2: Malta—Wolf Point
117 mi./187 km.

Widely scattered exposures show that all the bedrock between Malta and Wolf Point is dark shales and brown sandstone deposited in shallow sea water during Cretaceous time. Very little of that bedrock is exposed, largely because it is too weak to make bold outcrops. In any case, the deep blanket of glacial debris that the great continental ice sheet spread over most of this part of Montana buries the bedrock in most areas.

Glaciation

During the earlier and larger of the great ice ages, the one that corresponds to the Bull Lake ice age in the Rocky Mountains, a continental glacier reached far south of the line of US 2. The smaller continental glacier of the last ice age advanced south of US 2 only in the area west of Hinsdale. Both of those glaciers were near their southernmost reach, moving slowly, and rapidly thinning when they got into northeastern Montana. Neither was capable of eroding the landscape in this area so far from their source, so the chief evidence of their passage is in deposits of glacial sediments.

Erratic boulders, transported rocks that lie on a different kind of bedrock, are the most obvious sign of glaciation. Watch for rounded rocks that litter the fields nearly everywhere, sparsely in most areas, densely in a few places. Farmers have picked a good many of them out of their fields, and piled them along the fences. Although most erratic

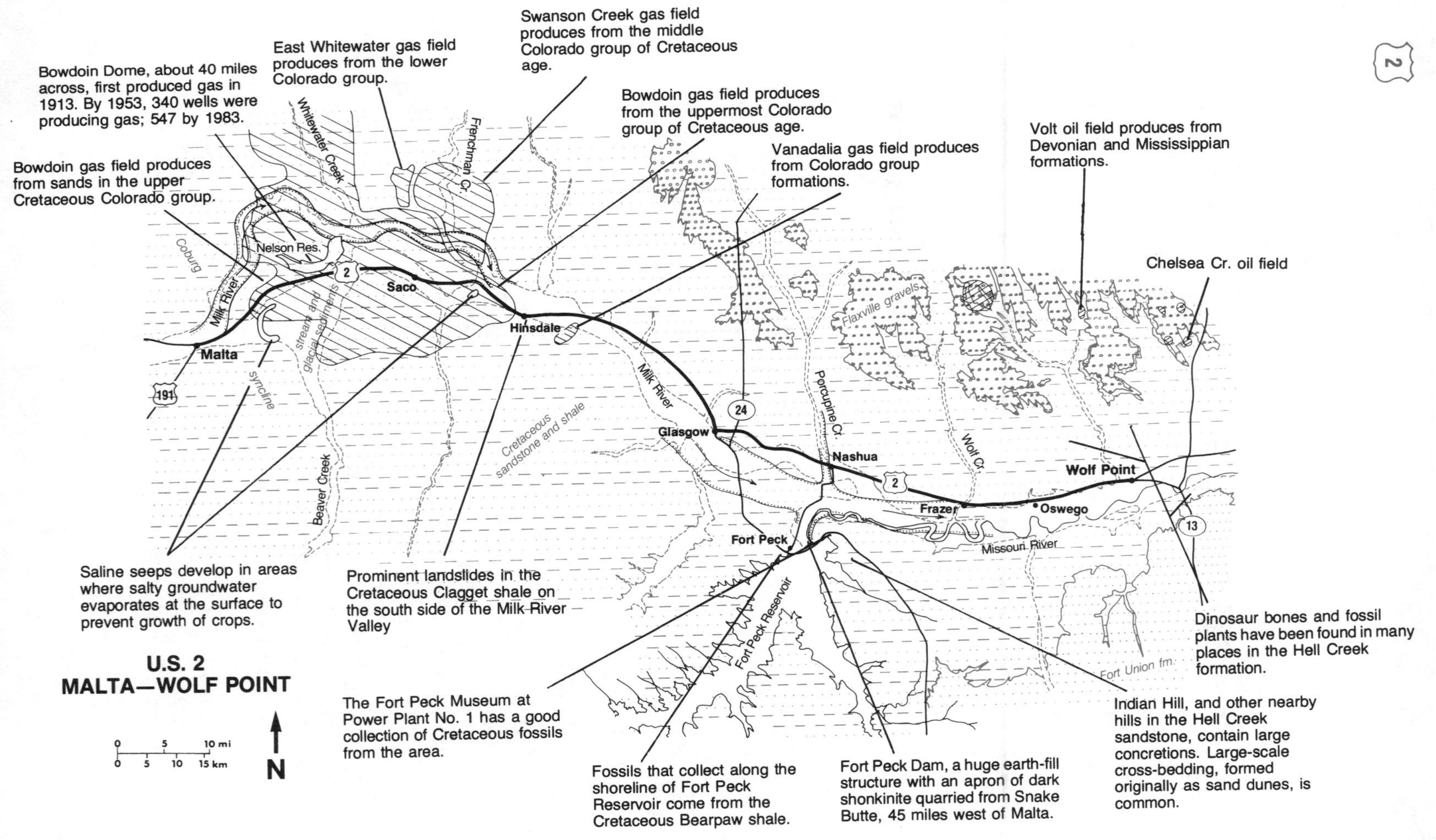

U.S. 2
MALTA—WOLF POINT

Bowdoin Dome, about 40 miles across, first produced gas in 1913. By 1953, 340 wells were producing gas; 547 by 1983.

Bowdoin gas field produces from sands in the upper Cretaceous Colorado group.

East Whitewater gas field produces from the lower Colorado group.

Swanson Creek gas field produces from the middle Colorado group of Cretaceous age.

Bowdoin gas field produces from the uppermost Colorado group of Cretaceous age.

Vanadalia gas field produces from Colorado group formations.

Volt oil field produces from Devonian and Mississippian formations.

Chelsea Cr. oil field

Saline seeps develop in areas where salty groundwater evaporates at the surface to prevent growth of crops.

Prominent landslides in the Cretaceous Clagget shale on the south side of the Milk River Valley

The Fort Peck Museum at Power Plant No. 1 has a good collection of Cretaceous fossils from the area.

Fossils that collect along the shoreline of Fort Peck Reservoir come from the Cretaceous Bearpaw shale.

Fort Peck Dam, a huge earth-fill structure with an apron of dark shonkinite quarried from Snake Butte, 45 miles west of Malta.

Indian Hill, and other nearby hills in the Hell Creek sandstone, contain large concretions. Large-scale cross-bedding, formed originally as sand dunes, is common.

Dinosaur bones and fossil plants have been found in many places in the Hell Creek formation.

Malta
Saco
Hinsdale
Glasgow
Nashua
Frazer
Oswego
Wolf Point
Fort Peck
Nelson Res.
Milk River
Whitewater Creek
Frenchman Cr.
Beaver Creek
Porcupine Cr.
Wolf Cr.
Missouri River
Fort Peck Reservoir
Coburg
syncline
stream and glacial sediments
Cretaceous sandstone and shale
Flaxville gravels
Fort Union fm.
191
2
24
13

0 5 10 mi
0 5 10 15 km
N

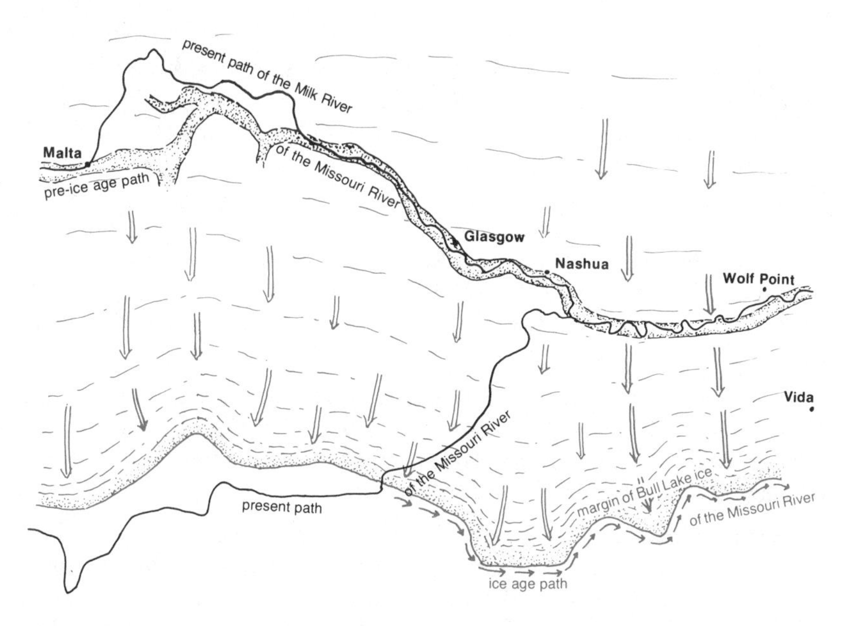

Courses of the Missouri River before, during, and after the Bull Lake ice age. The later Pinedale glacier reached a few miles south of Malta.

boulders are smaller than a wash tub, a few stand waist high or more and must weigh several tons.

Many of the boulders in the fields are yellow and brown Cretaceous sandstone that came from some local source, but some are a long way from home, very erratic indeed. Watch for big chunks of red and gray granite, streaky red and black gneiss, and various other coarsely crystalline rocks. Those are basement rocks, pieces of the continental crust. They must have come from northern Manitoba, the nearest area where the glacier could have picked them up on its way to Montana.

This cobble was once embedded in the sole of a glacier, and acquired its pattern of scratches as the moving ice dragged it across the bedrock. Nothing in nature but a glacier can do this to a rock.

The coming and going of the great ice-age glaciers shifted the Milk and Missouri rivers from one valley to another. Between Malta and the area a few miles west of Hinsdale, the highway winds in and out of the broad swale of the old valley of the Missouri River. In that area, the Milk River is in its old valley a few miles north of the highway. A few miles west of Hinsdale, the little Milk River enters the former valley of the Missouri River, and wanders nearly lost in its wide expanse to its mouth near Fort Peck. Between Fort Peck and Wolf Point, the road follows the Missouri River, in this reach still in its original valley.

Just west of Hinsdale, US 2 passes a moraine, the tract of low hummocky hills heavily littered with rocks in the valley floor south of the highway. It helps explain the roughly broken valley wall, a series of large landslides, directly south of Hinsdale.

Evidently, the edge of the glacier lay here for a while late in the last ice age, and the melting ice dumped large quantities of debris to build the moraine. Meanwhile, glacial meltwater poured between the edge of the ice and the south wall of the Milk River Valley. Rocks in that valley wall consist of soft shale overlain by a thick layer of hard, brown sandstone. The torrential rush of meltwater eroded the shale, thus undercutting the sandstone and causing the landslides that created the rough topography.

The Bowdoin Gas Field

Between Malta and Saco, the road passes through the Bowdoin gas field, which was discovered in 1913 when a water well began to produce natural gas from a depth of 640 feet. Deliberate drilling for gas for local use began in 1916, and the area developed into a large producing gas field after a pipeline was laid to Malta and Glasgow in 1929. Now the field connects to a regional pipeline net.

The crest of a large dome warped in the rocks at the westernmost edge of the Williston Basin trapped the gas in two layers of Cretaceous sandstone, one a little less than 1000 feet below the surface, the other a few hundred feet deeper. At the end of 1983, the Bowdoin field was still the second largest gas producer in the state with more than 300 active wells.

Wheat fields and summer fallow pattern the High Plains surface.
—U.S. Geological Survey photo by C. W. Balsley

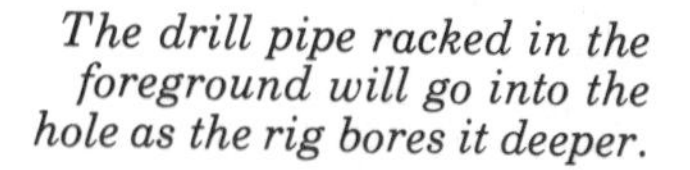

The drill pipe racked in the foreground will go into the hole as the rig bores it deeper.

US 2: Wolf Point—North Dakota
77 mi./123 km.

Bedrock between Wolf Point and Brockton is soft sandstone and mudstone deposited during late Cretaceous time. From Brockton east, bedrock is exactly similar sandstones and mudstones of the Fort Union formation deposited during early Tertiary time. Glacial deposits effectively cover that bedrock along almost the entire route.

The Williston Basin

East of Wolf Point, oil wells begin to dot the landscape as the road gets into the Williston Basin. Scattered clusters of pumps peck rhythmically at the ground like giant birds as they patiently suck their daily bit of oil out of the earth. But that slow production does add up to fortunes in oil as it continues year after year.

The East Poplar oil field and its smaller neighbor the Poplar Northwest field are the biggest in this area. Both were discovered in 1952; both produce from sedimentary formations deposited during Mississippian time. Most of the wells are between 5,000 and 7,000

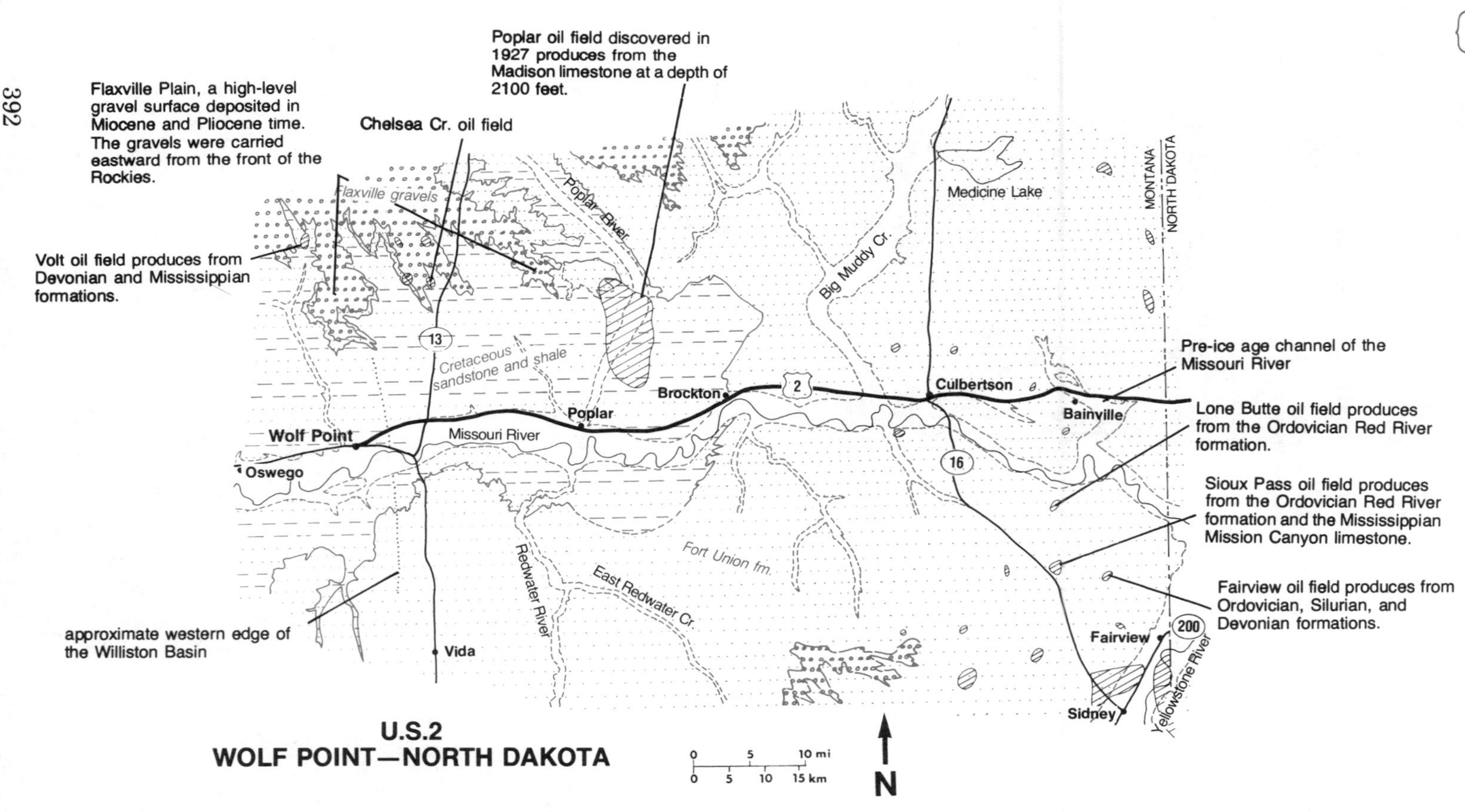

Flaxville Plain, a high-level gravel surface deposited in Miocene and Pliocene time. The gravels were carried eastward from the front of the Rockies.
Chelsea Cr. oil field
Poplar oil field discovered in 1927 produces from the Madison limestone at a depth of 2100 feet.
Volt oil field produces from Devonian and Mississippian formations.
Flaxville gravels
Poplar River
Big Muddy Cr.
Medicine Lake
MONTANA
NORTH DAKOTA
13
Cretaceous sandstone and shale
2
Brockton
Poplar
Culbertson
Bainville
Wolf Point
Missouri River
Oswego
16
Pre-ice age channel of the Missouri River
Lone Butte oil field produces from the Ordovician Red River formation.
Sioux Pass oil field produces from the Ordovician Red River formation and the Mississippian Mission Canyon limestone.
Fairview oil field produces from Ordovician, Silurian, and Devonian formations.
Fort Union fm.
Redwater River
East Redwater Cr.
approximate western edge of the Williston Basin
Vida
Fairview
200
Sidney
Yellowstone River
0 5 10 mi
0 5 10 15 km
N

U.S.2
WOLF POINT—NORTH DAKOTA

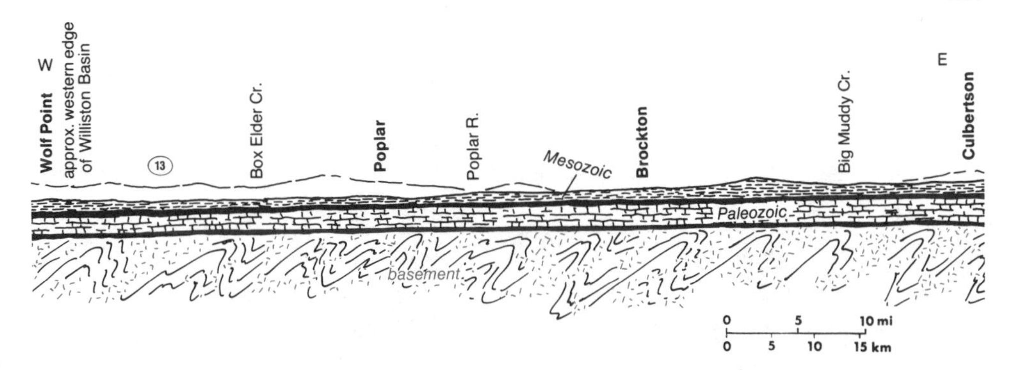

Section along the line of US 2 between Wolf Point and Culbertson. The edge of the Williston Basin is not apparent in this section drawn to true vertical scale.

feet deep. At the end of 1983, a combined total of 72 wells were pumping oil from those fields, which had produced between them about 45 million barrels of oil that sold for a total of about 180 million dollars. At that time, the average production from their wells was about 15 barrels of oil per day—very few wells produce so much after so many years.

A Coal Field for the Next Century

About seven miles west of Culbertson, the highway crosses the edge of the Fort Kipp coal field; exploratory drilling there has revealed two coal seams that total about 14 feet thick, and lie close enough to the surface that they may someday support a big strip mine. The coal lies beneath about 11 thousand acres and adds up to something more than 300 million tons, a modest reserve by the exaggerated standards of eastern Montana. As elswhere in this region, the coal is in the upper part of the Fort Union formation.

The endless High Plains surface—a thin veneer of Flaxville gravel covers the soft sandstones and mudstones of the Fort Union formation.

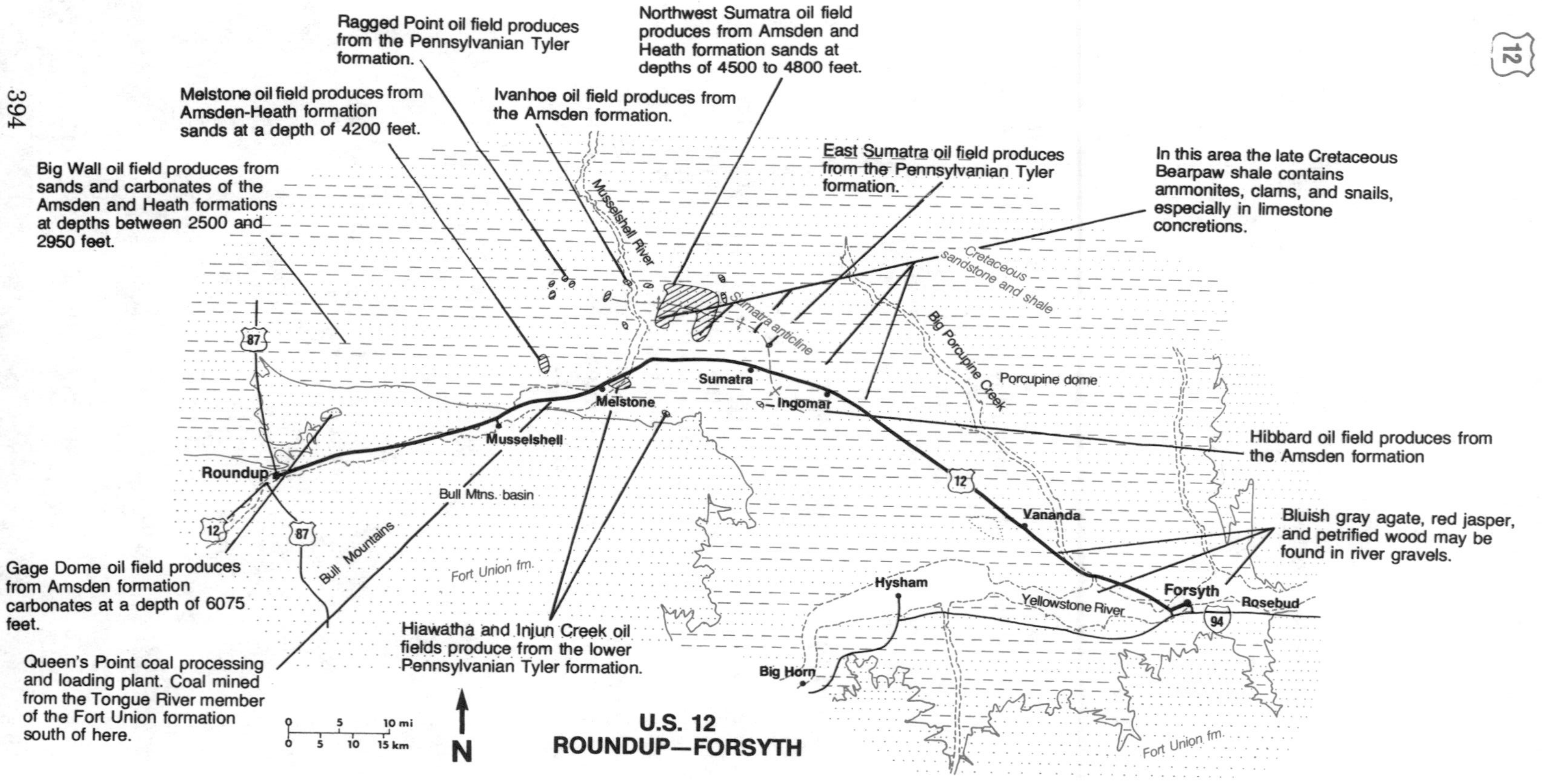

Ragged Point oil field produces from the Pennsylvanian Tyler formation.

Northwest Sumatra oil field produces from Amsden and Heath formation sands at depths of 4500 to 4800 feet.

Melstone oil field produces from Amsden-Heath formation sands at a depth of 4200 feet.

Ivanhoe oil field produces from the Amsden formation.

East Sumatra oil field produces from the Pennsylvanian Tyler formation.

In this area the late Cretaceous Bearpaw shale contains ammonites, clams, and snails, especially in limestone concretions.

Big Wall oil field produces from sands and carbonates of the Amsden and Heath formations at depths between 2500 and 2950 feet.

Musselshell River

Cretaceous sandstone and shale

Sumatra anticline

Big Porcupine Creek

Porcupine dome

87

Sumatra

Melstone

Ingomar

Musselshell

Hibbard oil field produces from the Amsden formation

Roundup

12

Bull Mtns. basin

Vananda

Bluish gray agate, red jasper, and petrified wood may be found in river gravels.

12

87

Bull Mountains

Gage Dome oil field produces from Amsden formation carbonates at a depth of 6075 feet.

Fort Union fm.

Hysham

Forsyth

Rosebud

Yellowstone River

94

Hiawatha and Injun Creek oil fields produce from the lower Pennsylvanian Tyler formation.

Queen's Point coal processing and loading plant. Coal mined from the Tongue River member of the Fort Union formation south of here.

Big Horn

0 5 10 mi

0 5 10 15 km

N

U.S. 12
ROUNDUP—FORSYTH

Fort Union fm.

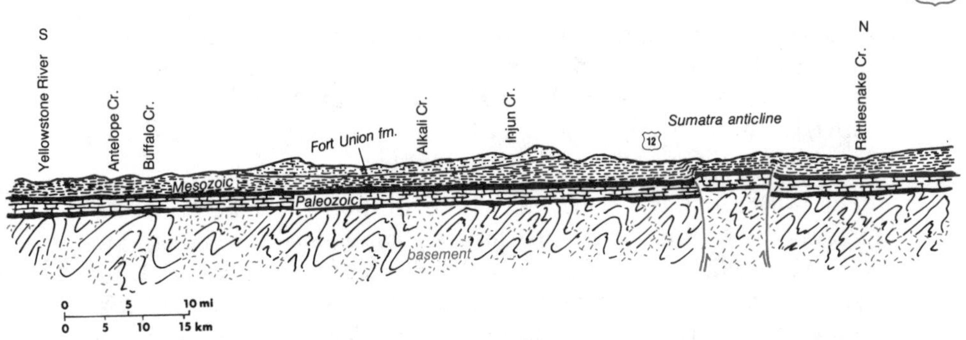

Section across US 12 between Melstone and Sumatra. Like many folds in eastern Montana, the Sumatra anticline drapes over faults in the continental basement.

US 12: Roundup—Forsyth
102 mi./163 km.

US 12 follows the Musselshell River past bluffs of Fort Union sandstone between Roundup and the area a few miles east of Melstone. Between Melstone and Forsyth, the road crosses broad plains eroded in Cretaceous shales and sandstones older than the Fort Union formation. The arch of the Sumatra anticline a few miles north of the road contains a number of oil fields. The course of the river bows northward, apparently because it follows the flank of the fold in the bedrock.

Roundup

Roundup was a busy coal mining town throughout the first half of this century. Several large mines and a number of small ones employed a large labor force, and provided the main economic base for the town. In those days, the highway passed big tipples and sprawling railway yards near the west side of town. Then the coming of diesel and electric locomotives, cheap oil, and natural gas eroded the market for coal. Mining stopped during the 1950s. The tipples are gone now, as are the railway yards and most of the miners. All the mines in the Roundup area were underground, so the scars they left on the landscape were relatively minor.

Oil

The area around Melstone and Sumatra contains a number of small oil fields and one large one, all along the crest of the Sumatra anticline. The oil comes from several sandstone layers within the Tyler formation, river channel sediments deposited during Pennsylvanian time. Exposures of Tyler sandstones in the Big Snowy Mountains tell the story of their formation.

The Mississippian period, about 300 million years ago in round numbers, was a time when shallow sea water flooded most of Montana. The sedimentary formations deposited then in central and eastern Montana include several that contain large amounts of dark organic matter, the raw material of oil. Geologists call such formations source rocks. Then, most of Montana rose above sea level, and the streams of Pennsylvanian time carved an erosional landscape into the Mississippian rocks. The Tyler formation consists largely of sand deposited in those stream channels. It is now full of oil that soaked into it from the Mississippian rocks.

Sandstone makes a good reservoir rock for oil because the pore spaces between the grains are numerous enough to hold quite a bit of oil, large enough to let it seep through the formation. It is fortunate that the Tyler sandstones are there because the pore spaces in the Mississippian source rocks are too small to permit oil to flow through the formation and into a well.

The Sumatra field, the largest in the area, was discovered in 1948. It produced some 40 million barrels from the Tyler sandstone by the end of 1983 when it still had 63 producing wells. Another 30 had gone out of production. The Hiawatha field, about one mile northwest of Melstone, is fairly typical of the small oil fields that produce from Tyler sandstones along the crest of the Sumatra anticline. It was discovered in 1967 at a depth of about 5000 feet, and produced almost one half million barrels of oil by the end of 1983. By then the field was about played out, down to four producing wells.

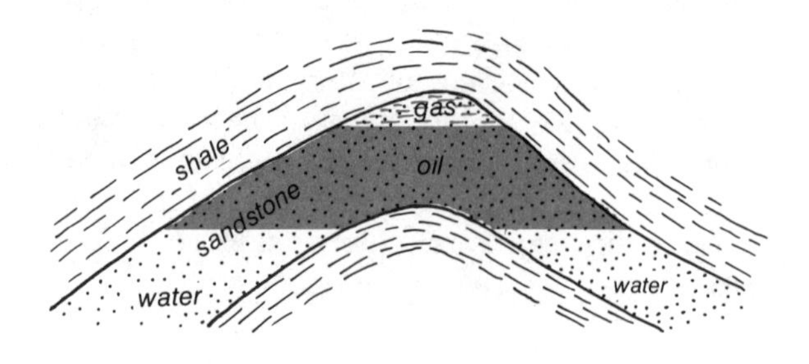

Section through the Sumatra anticline showing how oil accumulates in the crest of the fold.

Porcupine Dome

Between Ingomar and Vananda, the highway skirts the southwestern edge of the Porcupine Dome, a broad uplift in the plains that does not show in the landscape. The Porcupine Dome lies directly southeast of the Cat Creek anticline, and appears to be part of the same structural trend. But unlike the Cat Creek anticline, the Porcupine dome does not produce oil despite the considerable efforts of wildcat drillers. Rocks exposed in the central part of the dome normally lie more than 2000 feet beneath the surface in this part of Montana.

Inside an early drilling rig. —Montana Historical Society photo

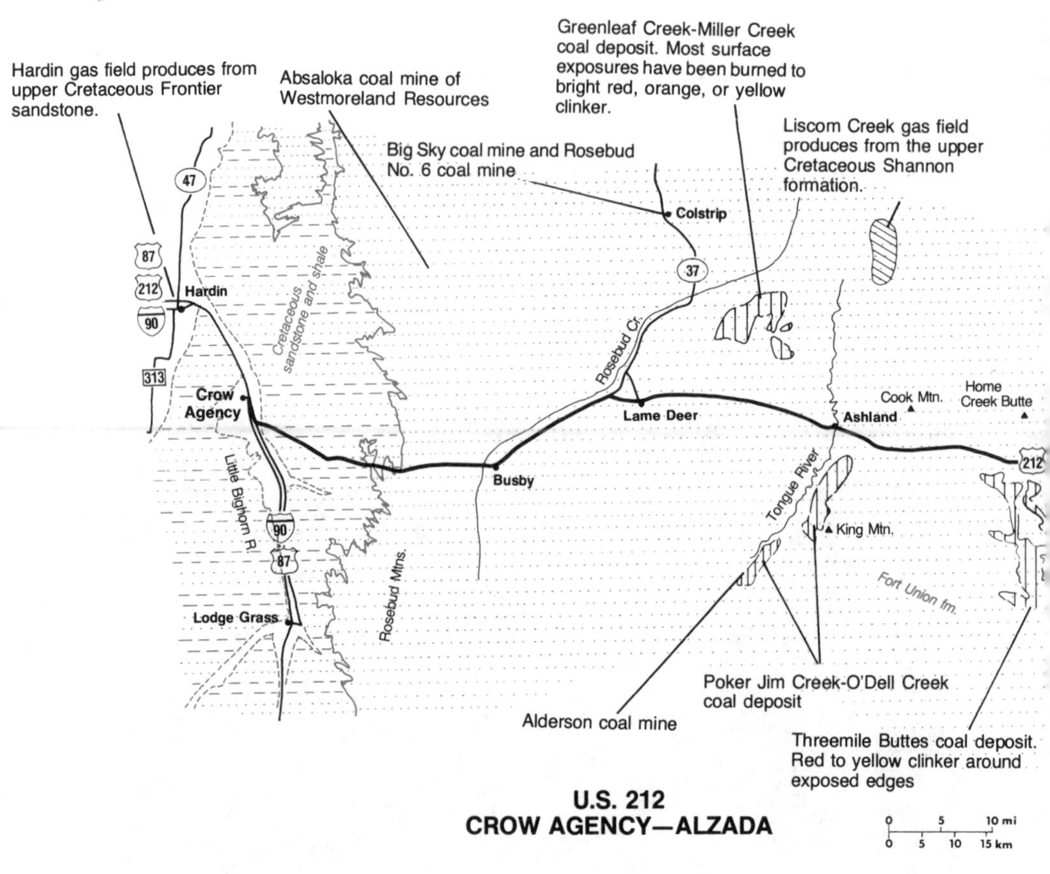

Nearly level plains eroded on soft sedimentary rocks of the Fort Union formation spread across large areas of eastern Montana. —Montana Department of Commerce photo

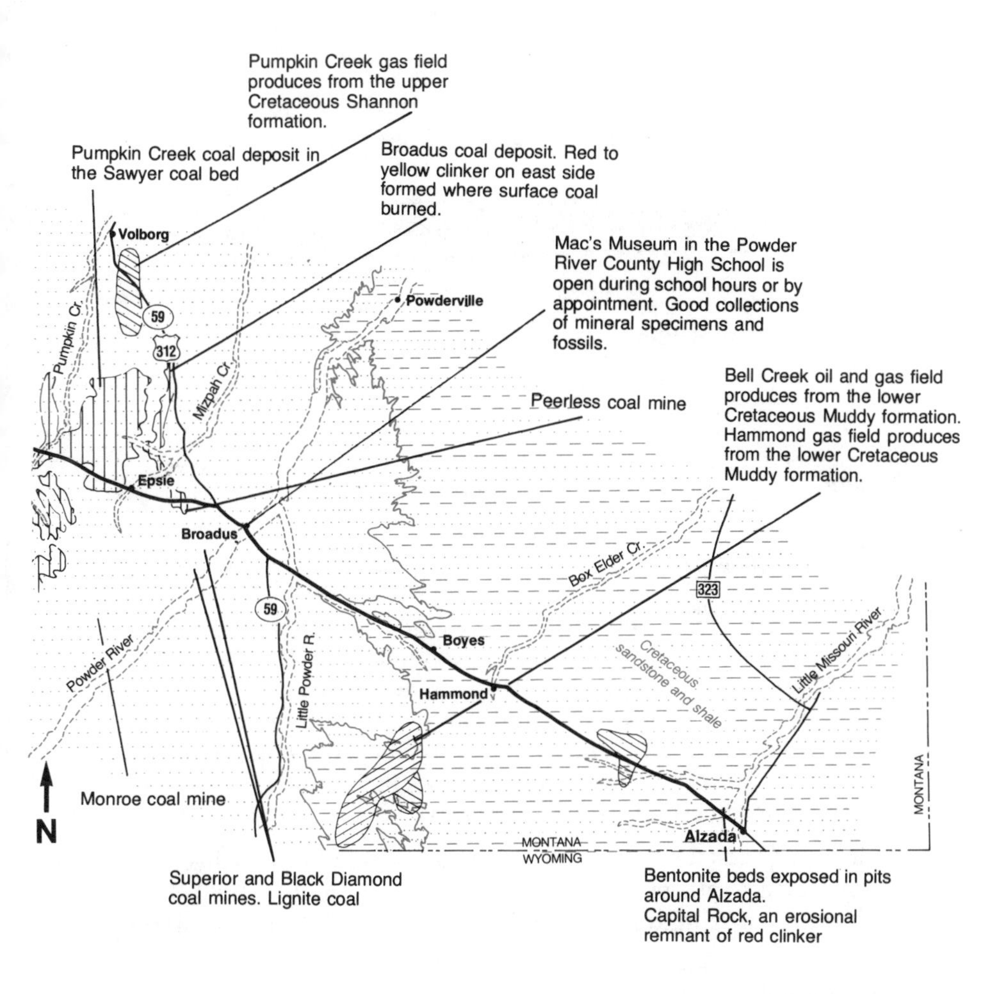
Pumpkin Creek gas field produces from the upper Cretaceous Shannon formation.
Pumpkin Creek coal deposit in the Sawyer coal bed
Broadus coal deposit. Red to yellow clinker on east side formed where surface coal burned.
Mac's Museum in the Powder River County High School is open during school hours or by appointment. Good collections of mineral specimens and fossils.
Bell Creek oil and gas field produces from the lower Cretaceous Muddy formation. Hammond gas field produces from the lower Cretaceous Muddy formation.
Peerless coal mine
Volborg
Powderville
Pumpkin Cr.
Mizpah Cr.
59
312
Epsie
Broadus
Box Elder Cr.
323
59
Boyes
Hammond
Little Powder R.
Powder River
Cretaceous sandstone and shale
Little Missouri River
MONTANA
WYOMING
Alzada
N
Monroe coal mine
Superior and Black Diamond coal mines. Lignite coal
Bentonite beds exposed in pits around Alzada.
Capital Rock, an erosional remnant of red clinker

US 212: Crow Agency—Alzada

164 mi./262 km.

The highway crosses late Cretaceous sedimentary rocks along parts of the route, Fort Union formation along most of it. They look about the same except that the Cretaceous rocks contain widely scattered dinosaur bones, whereas the Fort Union formation does not—those bones are very hard to find. The only rocks resistant enough to weather into hills and make prominent outcrops are the thick beds of Tongue River sandstone in the upper part of the Fort Union formation. That is where most of the coal is too.

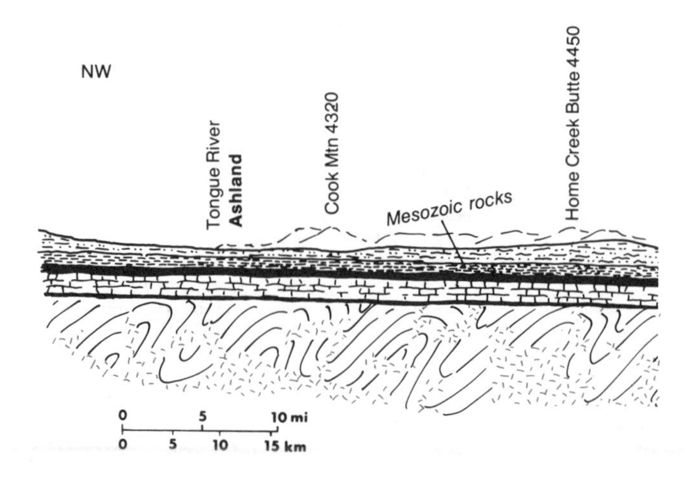

Section along US 212 between Ashland and Broadus. The sedimentary formations lie almost flat in this part of Montana, deeply covering the complex

Powder River Basin

The eastern half of the route passes along the northern end of the Powder River Basin, where the basement rock lies beneath an uncommonly thick accumulation of sedimentary rocks. The Wyoming portion of the Powder River Basin contains a number of highly productive oil fields. Although Montana's much smaller portion contains many fewer fields, it does include one of the biggest in the region. The figures are impressive.

The Bell Creek field, 20 miles southeast of Broadus, was discovered in 1967 at a depth of about 4500 feet. By the end of 1983, it had produced approximately 122 million barrels of oil worth slightly

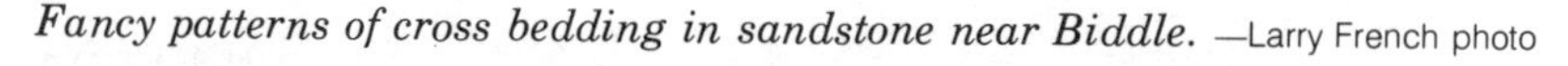

Fancy patterns of cross bedding in sandstone near Biddle. —Larry French photo

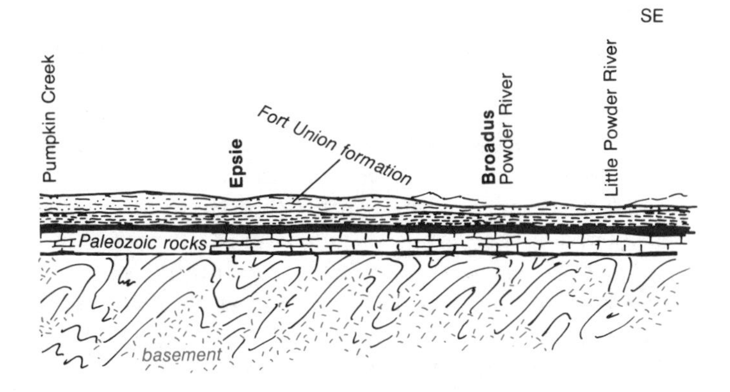

igneous and metamorphic rocks of the Precambrian basement.

more than one billion dollars from a total of 397 wells, of which 210 were still pumping. The same field also produces impressive quantities of natural gas. That was quite a discovery, a giant field by almost any standards.

All that oil and gas comes from the Muddy sandstone, a middle Cretaceous formation perhaps deposited about 90 million years ago as a barrier island along the coast of an inland sea that then covered much of central Montana. The open sea stretched away to the west, and a coastal plain extended east toward the area where the Black Hills later rose. Geologists who have studied the Muddy sandstone in the Bell Creek field compare it to the sand now accumulating along Galveston Island on the coast of Texas. Fine muds deposited in the lagoon behind that barrier island blocked movement of the oil, trapping it in the sandstone.

Painted hill, a knob of Fort Union formation sandstone about seven miles east of Busby. Bands of red clinker along its flanks and crest give the hill its name.
—Montana Bureau of Mines and Geology photo by H. L. James

Coal

West of Broadus, the road passes through areas of rugged hills that support a scatter of trees, the typical landscape of coal country in eastern Montana. This is the giant Pumpkin Creek coal field, which contains about 2.5 billion tons of unmined coal in several major seams—the North American equivalent of the Persian Gulf oil fields. Someday, enormous strip mines will produce that coal. Many other coal seams burned millions of years ago, baking the rock above them to hard clinker that resists erosion, and so maintains the steep hills. Bright red caps on those hills and horizontal bands of red along their sides are the clinker horizons. The Tongue River sandstone, which contains most of the coal, weathers into pinnacles and picturesque outcrops of yellowish boulders.

The coal fields south of Ashland would be a marvel anywhere except in eastern Montana—they contain something well in excess of one billion tons of coal. Most of that vast energy reserve is in the Knoblock seam, which is as much as 60 feet thick in some places. The region contains several other coal fields just as large or larger, but they are not so close to the road.

Sandstone monuments in Medicine Rocks State Park. —U.S. Forest Service photo by K. D. Swan

Montana 7: Wibaux—Ekalaka
90 mi./144 km.

Most of the route between Wibaux and Ekalaka passes through gently rolling prairie landscape, past farms and ranches that grow wheat, sugar beets, and cattle. There is one small area of good bedrock exposure about midway between Wibaux and Baker where a large patch of "gumbo buttes" on both sides of the road provide good badlands exposures of the Fort Union formation. A roadside historical marker points out the wagon ruts of an old trail.

A cluster of large buttes about midway between Baker and Ekalaka provide more good exposures of Fort Union sandstone. Each butte has a prominent red cap, rock baked hard as a coal seam burned out. Had there been no coal seam to slowly smolder and bake the rocks above it, the landscape in the area of the buttes would be as subdued as that in the surrounding prairies.

Just south of Ekalaka, the road passes through a narrow belt of rugged wooded hills, and then starts a long route across a magnificent expanse of high plains landscape. Mile after mile of prairie gently rolls into the infinite horizon. If only there were a herd of buffalo instead of cattle grazing in the distance, this area would look almost as the high plains did when Lewis and Clark saw them. We forget how

7

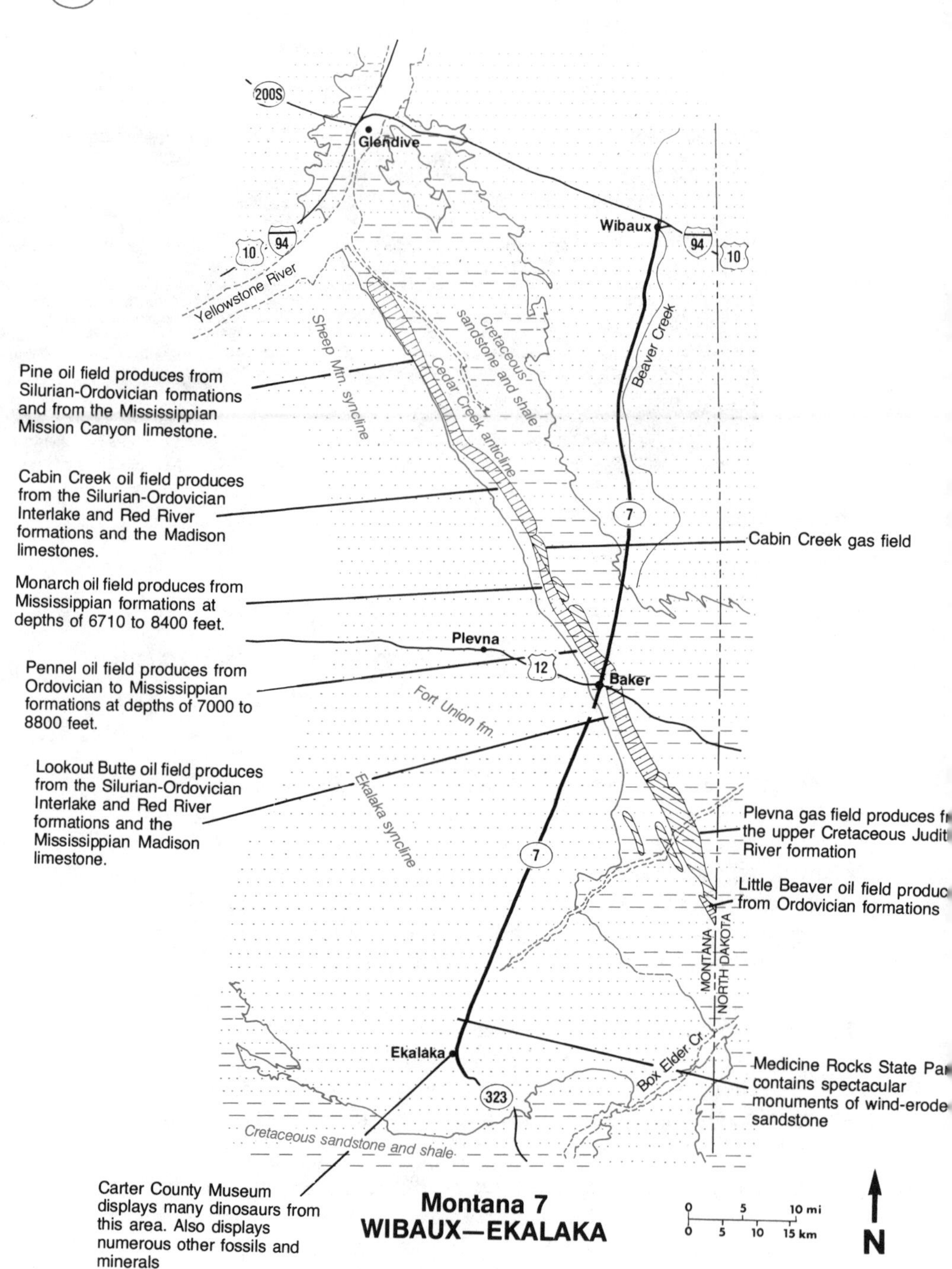

200S
Glendive
Wibaux
94
10
Yellowstone River
Sheep Mtn. syncline
Cretaceous sandstone and shale
Cedar Creek anticline
Beaver Creek
Pine oil field produces from Silurian-Ordovician formations and from the Mississippian Mission Canyon limestone.
Cabin Creek oil field produces from the Silurian-Ordovician Interlake and Red River formations and the Madison limestones.
7
Cabin Creek gas field
Monarch oil field produces from Mississippian formations at depths of 6710 to 8400 feet.
Plevna
12
Baker
Pennel oil field produces from Ordovician to Mississippian formations at depths of 7000 to 8800 feet.
Fort Union fm.
Lookout Butte oil field produces from the Silurian-Ordovician Interlake and Red River formations and the Mississippian Madison limestone.
Ekalaka syncline
Plevna gas field produces f
the upper Cretaceous Judit
River formation
Little Beaver oil field produc
from Ordovician formations
MONTANA
NORTH DAKOTA
Ekalaka
Box Elder Cr.
323
Medicine Rocks State Pa
contains spectacular
monuments of wind-erode
sandstone
Cretaceous sandstone and shale
Carter County Museum displays many dinosaurs from this area. Also displays numerous other fossils and minerals
Montana 7
WIBAUX—EKALAKA
0 5 10 mi
0 5 10 15 km
N

much barbed wire fences alter the appearance of a landscape until we see a large area without them.

The Cedar Creek Anticline

Between Wibaux and Baker, the highway angles across the Cedar Creek anticline, nothing to look at on the ground, but one of the most striking geologic features in this part of Montana. As the name suggests, the Cedar Creek anticline is an arch rather sharply folded into the rocks. Even though it is a strong structure, the Cedar Creek anticline has very little effect on the landscape because none of the various formations it brings to the surface resist erosion well enough to stand up in striking relief. Here and there, you may notice tilted sedimentary layers. They and the older formations the fold brings to the surface enabled early geologists to recognize the Cedar Creek anticline.

Oil and gas move up through the porous rocks until a layer of impermeable shale blocks their movement where it rolls over the arch of an anticline. The first stage in oil and gas exploration is generally to drill on the crests of anticlines exposed at the surface, because those are the easiest to find. In 1912, drilling on the Cedar Creek anticline located the Cedar Creek gas field in Cretaceous sandstones at a depth of 2700 feet. Drilling of more than 300 wells since then has extended the field along virtually the entire crest of the anticline, natural gas beneath an area of about 113,000 acres.

The Cedar Creek anticline also produces oil from more than 200 wells in several fields that extend along nearly the entire length of its crest. Most of the oil comes from the Madison limestone and older Paleozoic formations. The Pennel field north of Baker, for example, produced some 56 million barrels of oil from 1955, the year of its

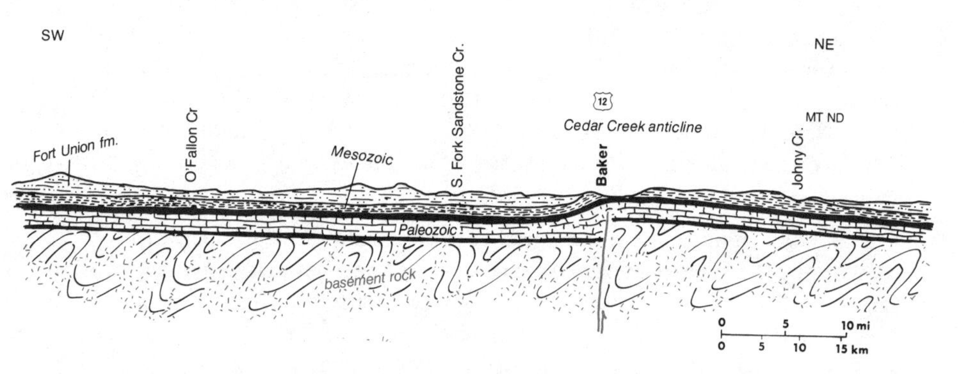

Section across the Cedar Creek anticline near Baker.

Natural arch in sandstone at Medicine Rock state Park.

discovery, until the end of 1983. In 1983, 140 wells, most of them about 7,000 feet deep, produced about three million barrels of oil, still a very productive field. The Cabin Creek field farther north on the crest of the anticline is even more impressive. By the end of 1983, it had produced 82 million barrels of oil from about 80 wells.

Medicine Rocks State Park

About ten miles north of Ekalaka, dozens of bizarre sandstone monuments dot the landscape. Each has steep sides and a surface etched into weird patterns of hollows. Most stand in the centers of small depressions, like miniature castles within their moats. They are remnants of an eroded bed of sandstone.

Look carefully at the vertical faces of the monuments to see thin beds of sandstone lying at angles within the thicker beds, a pattern geologists call crossbedding. The cross beds in the sandstone at Medicine Rocks State Park look like those in modern sand dunes, therefore suggest that this sandstone formed as a dune field. And the sand grains are small and uniform in size, another characteristic typical of dune sands. Those similarities make it seem that the sandstone at Medicine Rocks State Park was originally a sea of sand dunes.

It is easy to detach sand grains from the rock simply by rubbing it with your hand, and there is clear evidence that sand grains are coming off the rock surface naturally at a fast rate. Many visitors to the area deface the rocks by carving their names and the date. But the rock sheds its surface so rapidly that those inscriptions become unreadable within a few years. And there are no crusty growths of lichens on the Medicine Rocks, apparently because their surfaces disintegrate too rapidly to permit those slow growing plants to survive.

The Medicine Rocks appear to be remnants left by wind erosion. There are no small stream channels in the area, as there would be if running water were the agent of erosion. The real clue is the tendency of the rocks to stand in the middle of closed depressions. It would be impossible for any process of erosion that operates on the surface to transport material away from the rocks without first filling those depressions to create a downhill path. However, it is easy to imagine the strong high plains winds howling around those rocks, whipping sand away from their bases to create the depressions, and clawing sand grains away from their surfaces to erode the rocks themselves.

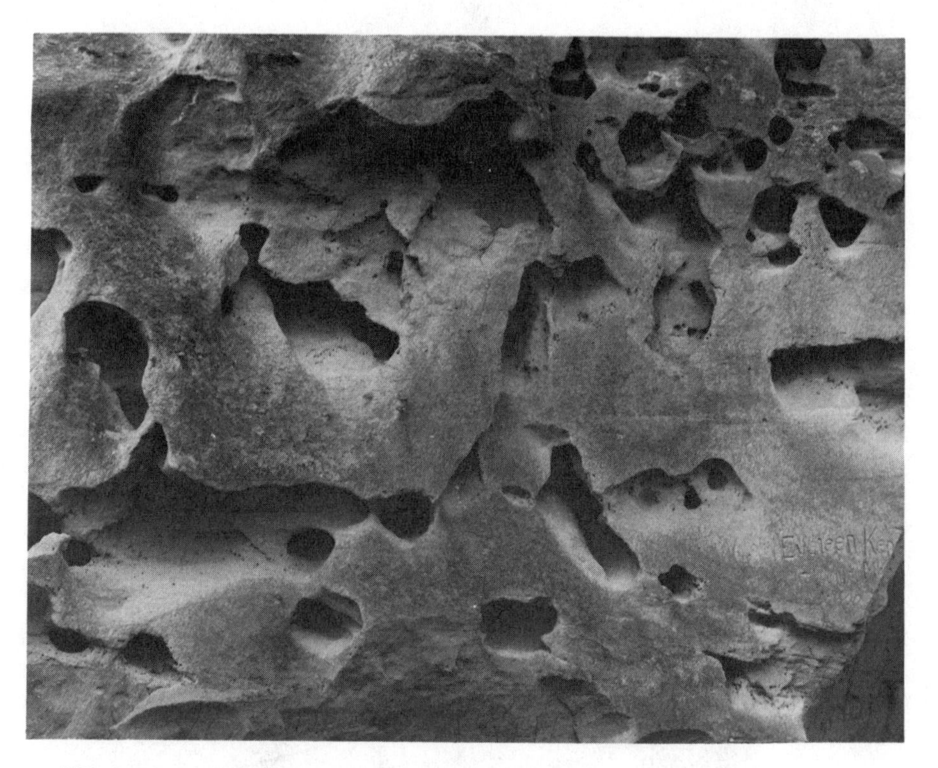

Wind-excavated holes and rapidly weathering grafitti on a sandstone monument, Medicine Rocks State Park.

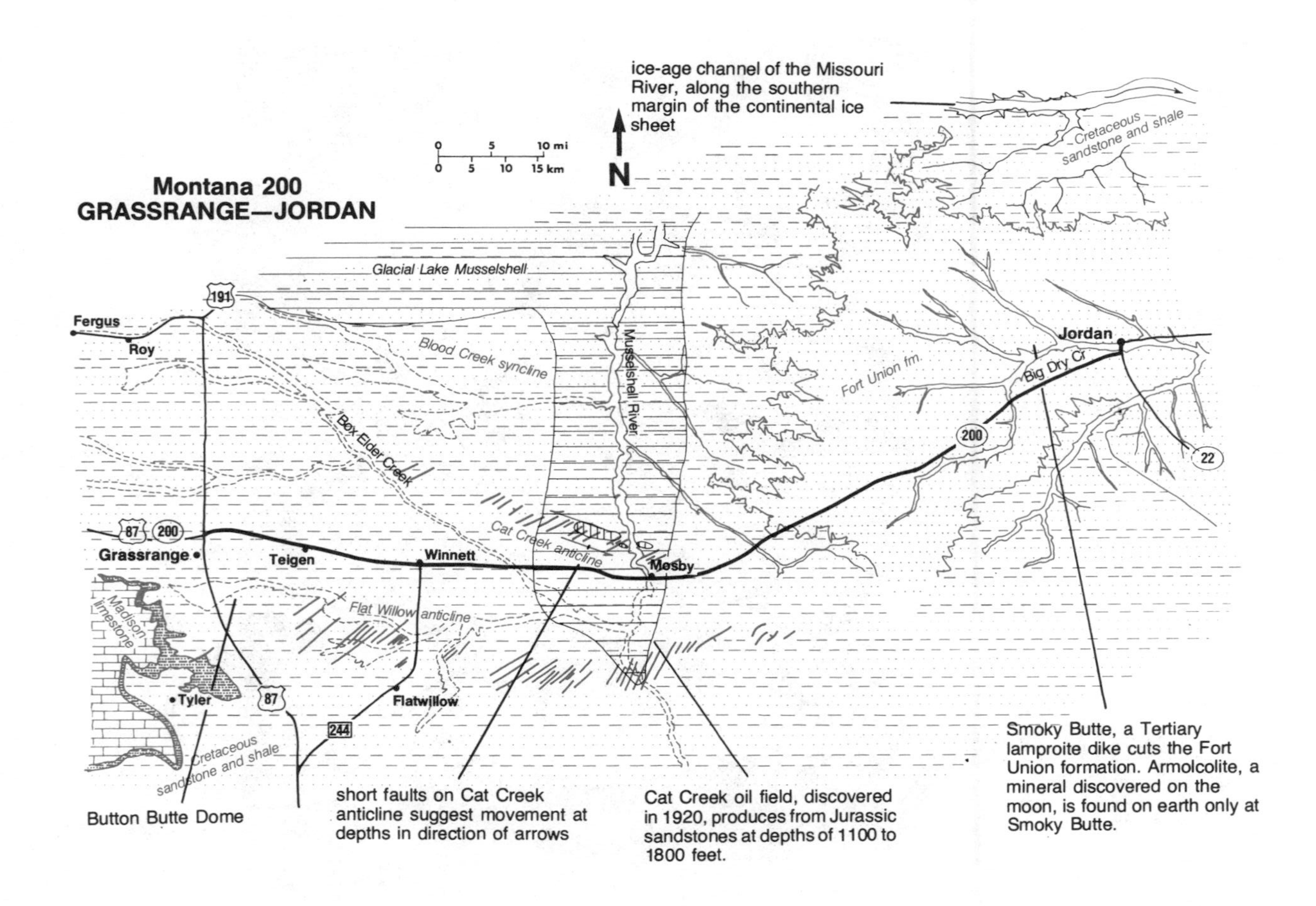

Montana 200
GRASSRANGE—JORDAN
0 5 10 mi
0 5 10 15 km
N
ice-age channel of the Missouri River, along the southern margin of the continental ice sheet
Cretaceous sandstone and shale
Glacial Lake Musselshell
191
Fergus
Roy
Blood Creek syncline
Musselshell River
Fort Union fm.
Jordan
Big Dry Cr
200
22
Box Elder Creek
87
200
Grassrange
Teigen
Winnett
Cat Creek anticline
Mosby
Madison limestone
Flat Willow anticline
Tyler
87
Flatwillow
244
Cretaceous sandstone and shale
Button Butte Dome
short faults on Cat Creek anticline suggest movement at depths in direction of arrows
Cat Creek oil field, discovered in 1920, produces from Jurassic sandstones at depths of 1100 to 1800 feet.
Smoky Butte, a Tertiary lamproite dike cuts the Fort Union formation. Armolcolite, a mineral discovered on the moon, is found on earth only at Smoky Butte.

Drilling in the Cat Creek field about 1920.
—Montana Historical Society photo by Herb Titter.

Montana 200: Grassrange—Jordan
99 mi./158 km.

The western two-thirds of the route crosses Cretaceous formations, some deposited in shallow sea water. They consist mostly of shales, along with some sandstones. The eastern third of the route crosses Fort Union formation, mudstones and some sandstones deposited on land during latest Cretaceous and earliest Tertiary time. None of the bedrock along the way resists erosion very well, so good exposures are mostly in roadcuts, creek banks, and occasional patches of badlands.

Cat Creek Oil Fields

The Judith Mountains, in the distance west of Grassrange, perch near the crest of a series of anticlinal arches, the Cat Creek anticline, that trends generally east and southeast, north of the highway. These folds are part of the southeast trending set of structures that probably formed during Eocene time, about 50 million years ago. They appear at the surface as tilted layers of sedimentary rock that attracted the attention of the early oil geologists, who quickly located the crests of the folds.

A wildcat well discovered oil in the crest of the Cat Creek anticline in 1920, at a depth of 998 feet. That was the second oil discovery in the region. Numerous other wells quickly followed in a frenzied drilling

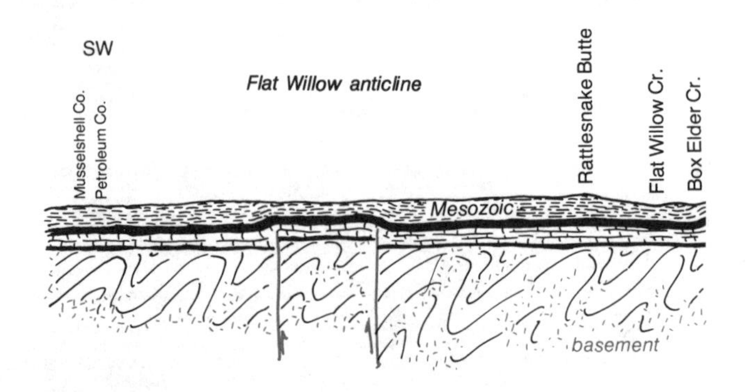

Section across the line of Highway 200 between Winnett and Mosby showing the broad Cat Creek anticline. The folds probably formed as Mesozoic and

boom, and the field has produced oil ever since from three separate pools and several different depths. Although daily production is modest by some standards, the field has produced an immense amount of oil during its long history. It undoubtedly includes the two most profitable square miles in this part of Montana.

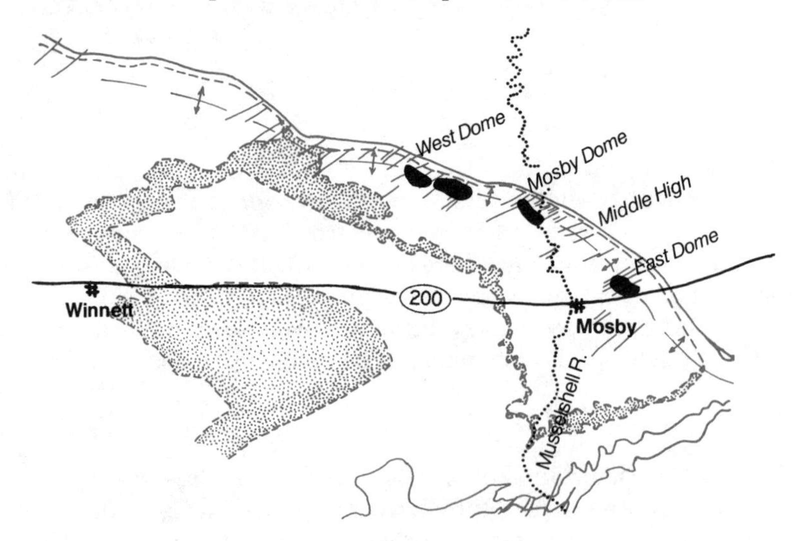

Oil fields in the Cat Creek anticline.

Moon Mineral

Smoky Butte, about eight miles west of Jordan, provided Montana with two of its most interesting geologic discoveries.

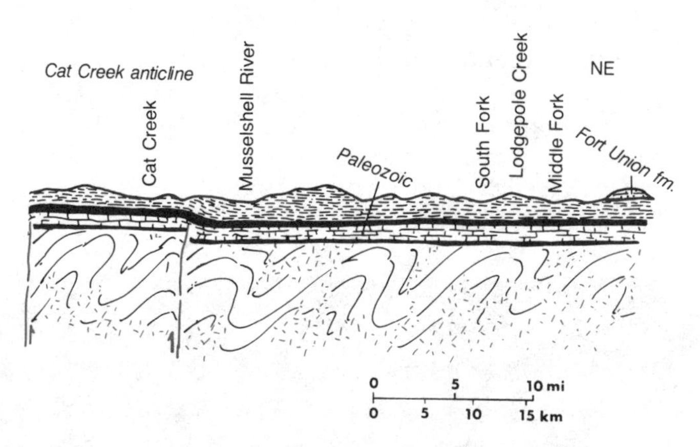

Paleozoic sedimentary formations draped over faulted blocks in the Precambrian basement.

Smoky Butte is a butte because it contains a diatreme, one of those small igneous intrusions that punched up from the earth's mantle about 50 million years ago. The igneous rock of the diatreme and the baked Fort Union formation around it resist erosion well enough to enable Smoky Butte to stand slightly above the surrounding plains. Most of Montana's diatremes are in the central part of the state; this is the easternmost known outpost of that activity. Like those in many diatremes, the rocks in Smoky Butte are abnormally rich in potassium.

A French mineralogist working on samples of the fine grained igneous rock from Smoky Butte found in it a mineral called armalcolite that is otherwise known only in rocks collected on the Moon. No one seems to understand why the rock in Smoky Butte contains so much potassium, or why it contains armalcolite.

The Actual Dust that Killed the Dinosaurs

Rocks exposed in Smoky Butte contain the thin layer of dark sediment that seems to mark the boundary between Cretaceous and Tertiary time everywhere in the world that rocks of that age exist. Dinosaurs thrived until that layer was deposited, and apparently vanished as it was laid down. Geologists generally recognize that layer through its high content of iridium, an element that is abundant in some kinds of meteorites but extremely rare on the earth's surface. At Smoky Butte, the fatal layer contains material far more incriminating than iridium.

The dust layer at Smoky Butte contains mineral grains full of microscopic fractures in a pattern known to occur only in rocks shattered in extremely violent explosions. It also contains an extremely rare mineral called stishovite, a distinctive variety of quartz that likewise forms only in explosion-shocked rocks. The shocked mineral grains and stishovite at Smoky Butte provide the most conclusive evidence so far found to support the theory that the dinosaurs did indeed die under a dark cloud of dust ejected from a big meteorite explosion crater.

The Hell Creek badlands north of Jordan. Rocks in this area have yielded the skeletons of dinosaurs that roamed this area in the last years before their kind were exterminated. —Larry French photo

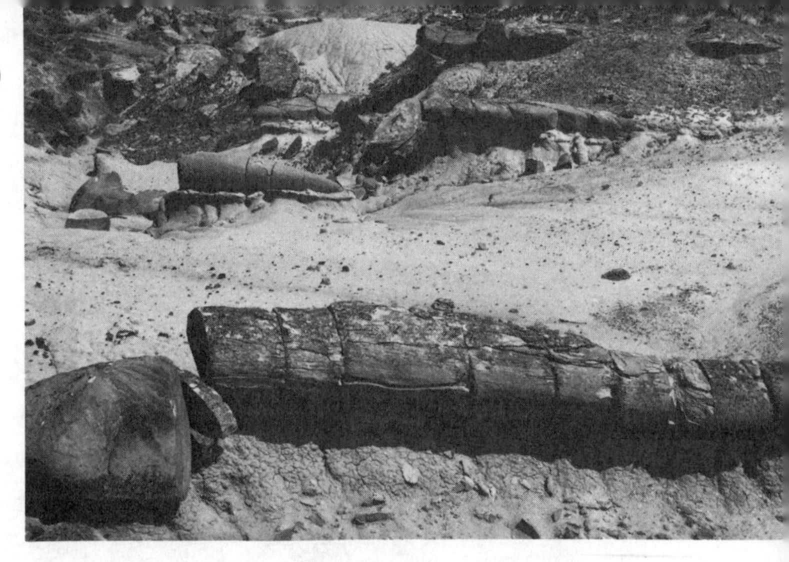

Long stringers of sandstone that was deposited in the channels of small streams look almost like logs as they weather out of the Cretaceous Hell Creek formation in the badlands north of Jordan. Larry French photo

Montana 200: Jordan—Sidney
237 mi./379 km.

East of Jordan, the highway follows Big Dry Creek for about 20 miles, passing through a rough landscape with numerous isolated buttes and patches of badlands topography along the valley walls. The badlands become especially spectacular farther east, along the east bank of Little Dry Creek, where the road passes through a miniature desert. The rocks are exposures of the Fort Union formation baked hard and colorful by burning coal seams. Splashing raindrops and running surface water carved the fantastic shapes as they selectively stripped off the softer material, leaving the more resistant parts of the formation standing up in relief.

Between the area east of Little Dry Creek and Circle, a distance of about 40 miles, the road passes through a softly billowing high plains landscape. Breathtaking views of rolling grasslands and vast geometric expanses of wheat fields and summer fallow open ahead as the road reaches the crest of each broad swell in the landscape. There are no badlands along this stretch of road and no bedrock outcrops except occasional red patches of baked mudstone or sandstone at the crests of widely scattered and isolated buttes.

The landscape between Circle and Sidney heaves up and down in

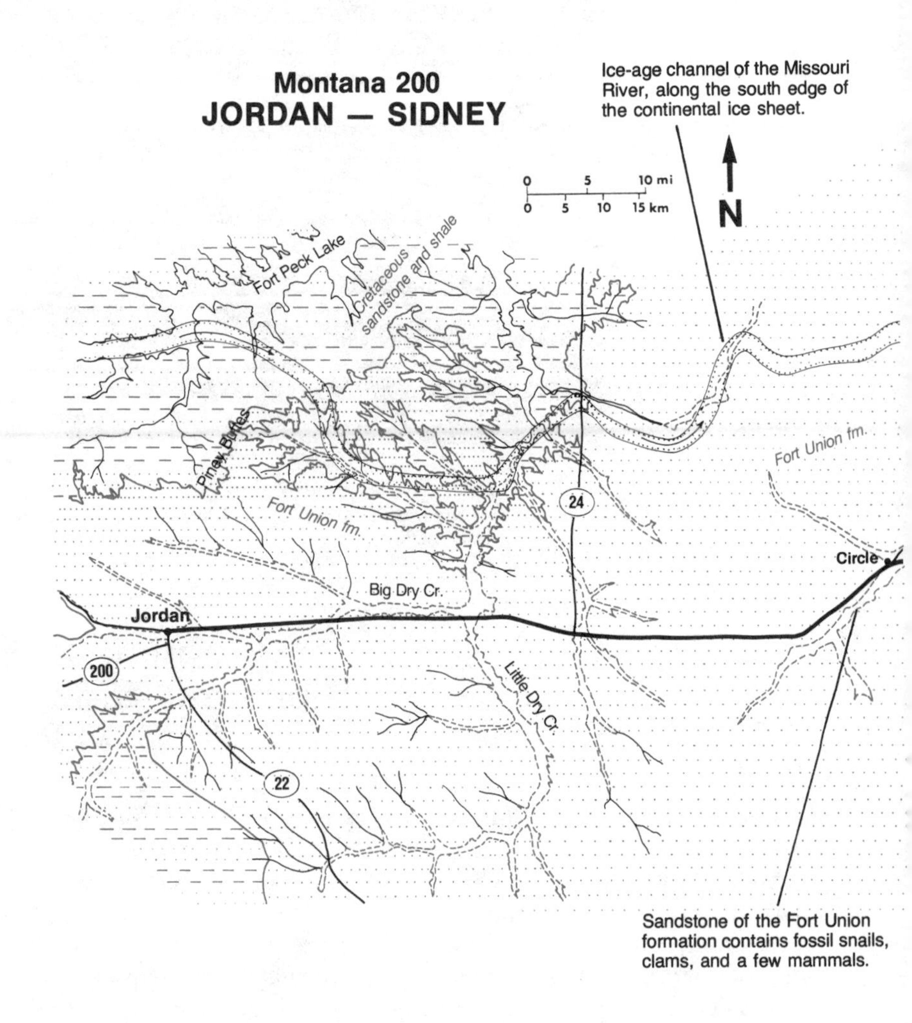

broad swells that carry Montana 200 through stretches of grassland and tightly patterned wheat fields as it rises and falls across the gently billowing landscape. The countryside seems almost intimate as the road passes through the low areas. Then the land expands into immensity as the road reaches the crest of a swell to reveal miles of plains stretching away into distant horizons.

Soft mudstones and sandstones of the Fort Union formation lie beneath this country. Very little of that rock resists erosion well enough to form bedrock outcrops. There is one exception about midway beneath Circle and Sidney, where several isolated monuments of white sandstone stand in a depression at the crest of a

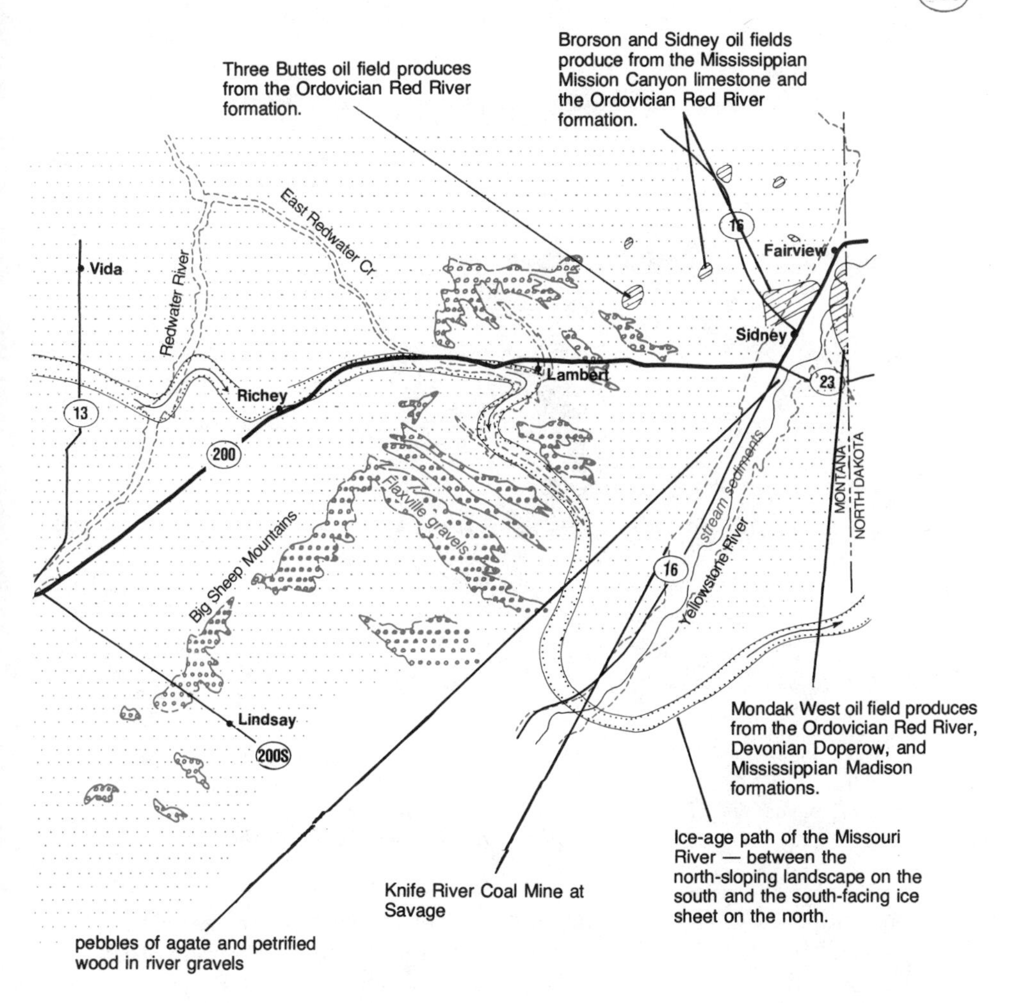
Brorson and Sidney oil fields produce from the Mississippian Mission Canyon limestone and the Ordovician Red River formation.
Three Buttes oil field produces from the Ordovician Red River formation.
East Redwater Cr.
Vida
Redwater River
Fairview
16
Sidney
Richey
Lambert
23
13
200
Flaxville gravels
Big Sheep Mountains
stream sediments
Yellowstone River
MONTANA
NORTH DAKOTA
16
Lindsay
200S
Mondak West oil field produces from the Ordovician Red River, Devonian Doperow, and Mississippian Madison formations.
Ice-age path of the Missouri River — between the north-sloping landscape on the south and the south-facing ice sheet on the north.
Knife River Coal Mine at Savage
pebbles of agate and petrified wood in river gravels

ridge. The sandstone is old dune sand now slightly hardened into weak rock, and the wind is snatching it away grain by grain to excavate the depression and carve the monuments of rock.

Glaciation

During the earlier of the two great ice ages that involved eastern Montana, ice spreading from central Canada reached as far west as Lambert. Glacial deposits mantle the ground between that area and Sidney. Watch for occasional erratic boulders littering the fields. Some of those boulders are granite and gneiss, hard crystalline

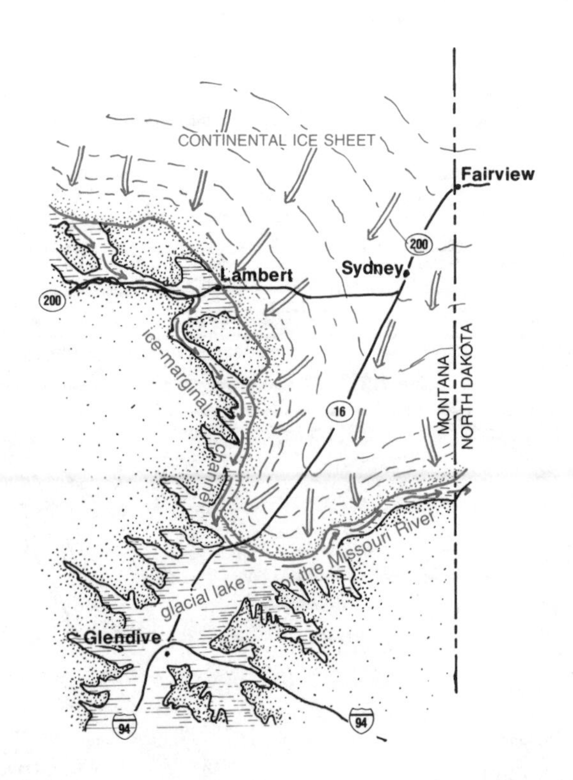

The Bull Lake Glacier deflected the Missouri River to a course between the edge of the ice and higher ground beyond. Meanwhile, the Yellowstone River backed up to form a large glacial lake that extended well south of Glendive.

basement rocks that lie thousands of feet below the surface in this part of Montana. The glacier brought them all the way from central Canada, the nearest place where they are exposed at the surface.

Along most of the route between Richey and Lambert, the highway follows an abandoned valley, the ice-age channel of the Missouri River. During the Bull Lake ice age of 70,000 to 130,000 years ago, the continental glacier pushed the river south, trapping it between the ice and higher ground beyond. So the river followed a course close to the edge of the glacier.

Oil

Several miles west of Sidney, the road crosses a broad and nearly flat river terrace, a remnant of an old floodplain of the Yellowstone River left high as the stream eroded its channel to a lower level. This area is well within the Williston Basin, so it is no surprise to see an oil field, in this case the Sidney field. At the end of 1983, the 26 wells that

dot the terrace had produced a total of more than two million barrels of oil with a market value of almost 43 million dollars.

Canvas walls protect the derrick floor of this oil drilling rig from the cold wind.

Glossary

Agglomerate: A rock composed largely of coarse fragments of volcanic debris.

Alkalic: Igneous rocks that contain an uncommonly large amount of sodium or potassium, or both.

Andesite: A volcanic rock intermediate between rhyolite and basalt in composition, color, and eruptive behavior. Andesite is the common rock of large volcanoes.

Anticline: An arch folded into layered rock.

Augite: A common silicate mineral that contains iron and magnesium, and generally forms rather stumpy crystals. Augite is black, and typically occurs in dark igneous or metamorphic rocks.

Basalt: A common volcanic igneous rock. Basalt is black, and consists mostly of the minerals plagioclase and augite.

Batholith: Any large mass of granite or similar rock, technically one that covers an area greater than about 40 square miles.

Biotite: The black variety of mica common in granites, and light-colored gneisses and schists.

Blue-green Algae: An extremely primitive green plant that appeared on earth during Precambrian time and survives essentially unchanged.

Bull Lake: An ice age that reached its maximum at an unknown time probably between 70,000 and 130,000 years ago.

Clinker: Rock baked hard above a burning coal seam. Most clinker is either red or yellow, and has a bubbly texture.

Crust: The earth's outer skin composed of granite and similar rocks in the continents, or basalt in the oceans.

Diabase: An igneous rock that resembles basalt except that the crystals are large enough to reveal individual crystals of black augite and pale plagioclase feldspar without using a microscope.

Diatreme: A small igneous intrusion generally shaped like a dike or vertical cylinder composed of peridotite or similar dark rock derived directly from the earth's mantle.

Dike: A body of igneous rock that formed as magma filled an open fracture.

Dolomite: A sedimentary carbonate rock composed of the calcium-magnesium carbonate mineral dolomite.

Drumlin: A streamlined hill composed of till laid down beneath a moving glacier.

Fault: A fracture in the earth's crust along which the opposite sides slipped past each other.

Feldspar: A group of abundant aluminum silicate minerals. The potassium feldspars, orthoclase, are typically white or pink. Plagioclase feldspars, which contain sodium and calcium, are typically white or greenish white. Most feldspars occur in blocky crystals.

Fossil: Any trace of a plant or animal preserved in rock.

Gabbro: A coarse-grained igneous rock composed principally of plagioclase feldspar and augite, a coarse-grained equivalent of basalt.

Gneiss: The word refers to quite a variety of coarse-grained metamorphic rocks that resemble each other in having a banded or streaky apearance. Gneisses form through recrystallization at high temperature of various kinds of sedimentary or igneous rocks.

Granite: A coarse-grained igneous rock composed mostly of feldspar and quartz.

Hornblende: A common silicate mineral that crystallizes into glossy black needles in light-colored igneous and metamorphic rocks such as granite, schist, and gneiss.

Igneous: Rocks that crystallized from molten magma.

Laccolith: An igneous intrusion that formed as a blister of magma injected between layers of sedimentary rock.

Lava: Melted rock, magma, that has erupted onto the earth's surface.

Limestone: A carbonate sedimentary rock composed of the calcium carbonate mineral calcite.

Lineation: An internal directional structure within certain rocks that gives them directionality in the same way that a handful of spaghetti noodles has directionality.

Lithosphere: The rigid outer rind of the earth, approximately the outer 60 miles of the mantle.

Magma: Melted rock. Technically, magma becomes lava if it pours out on the surface.

Marble: A coarse-grained metamorphic rock formed through recrystallization of one of the carbonate sedimentary rocks, limestone or dolomite.

Metamorphism: Recrystallization of a rock at high temperature to produce a new rock that commonly differs considerably from the original.

Mica: A group of silicate minerals that tend to split into thin flakes. The common varieties are black biotite and colorless muscovite. Both typically occur in generally light-colored igneous and metamorphic rocks such as granite, gneiss, or schist.

Moraine: A deposit of till.

Muscovite: The colorless variety of mica common in granites and light-colored gneisses and schists.

Oil Seep: A spring that produces oil.

Oil Shale: A fine-grained sedimentary rock that will yield oil if it is roasted. Oil shale is typically dark brown or black, has a waxy appearance and feel, and burns with a yellow, smoky flame.

Ore: Any kind of rock that can be profitably mined.

Orthoclase: A common feldspar mineral that contains potassium. Orthoclase is generally white, beige, or pink, and typically occurs in light-colored igneous and metamorphic rocks.

Outwash: Sediment deposited from glacial meltwater. Like other stream sediments, outwash typically consists of distinct layers of well-sorted sand, gravel, and clay.

Overthrust Fault: A nearly horizontal fracture surface along which the rocks above slid over those below.

Peat: An accumulation of partially decayed plant material. Buried peat eventually turns into coal.

Peridotite: An igneous rock composed essentially of augite and a green silicate mineral called olivine.

Pinedale: The most recent ice age. The Pinedale glaciers reached their maximum about 15,000 years ago and finally melted something less than 10,000 years ago.

Placer: A concentration of heavy minerals within a deposit of stream or beach sediment.

Plagioclase: A group of extremely common feldspar minerals that contain sodium and potassium in varying proportion. Plagioclase is typically white or greenish white, and occurs in a wide variety of igneous and metamorphic rocks.

Plate: A segment of the lithosphere.

Quartz: A very common mineral composed of silicon dioxide. Quartz comes in many varieties of which the commonest is clear and colorless, like glass.

Resurgent Caldera: A large volcano that repeatedly opens an enormous crater as it erupts large volumes of rhyolite, then fills the crater with more rhyolite. The Yellowstone volcano is an example.

Rhyolite: A pale volcanic rock that contains quartz. Rhyolite has the same composition as granite.

Sandstone: A deposit of sand hardened into rock.

Schist: The term refers to a variety of metamorphic rocks that contain enough mica to make them flaky, or enough needle-shaped minerals to make them splintery. Schists form through recrystallization at high temperature of various kinds of sedimentary rocks.

Sedimentary Rocks: Deposited material such as sand, clay, mud, or gravel hardened into rock.

Shonkinite: A dark igneous rock composed mostly of augite and potassium feldspar and essentially similar in composition to basalt except in having a much larger potassium content.

Sill: A layer of igneous rock sandwiched between layers of sedimentary rocks.

Shale: A flaky rock composed largely of clay.

Skarn: A kind of metamorphic rock that forms where granitic magma reacts with limestone. Skarns typically consist of large crystals of garnet and a variety of other minerals.

Syenite: An igneous rock composed mostly of feldspar.

Syncline: A trough folded into layered rocks.

Talc: An extremely soft magnesium silicate mineral that occurs in metamorphic rocks recrystallized at relatively low temperature. Talc has many uses in the ceramic, paper, paint, cosmetics, and other industries.

Thrust Fault: A gently dipping fracture along which the rocks above the fracture surface moved up over those beneath.

Till: Sediment dumped directly from glacial ice. Till typically consists of material of all sizes and shapes indiscriminately mixed and deposited without internal sedimentary layering.

Vein: A zone of distinctive rock developed along a fracture. Many veins contain valuable minerals.

If You Want to Know More. . . .

It is difficult to suggest further reading about the geology of Montana because so much of the literature is available only in large libraries. The best collections within the state are in the libraries of the University of Montana in Missoula, Montana State University in Bozeman, and Montana Tech in Butte.

Another good place to start looking for more information is the Montana Bureau of Mines and Geology. Its offices are on the campus of Montana Tech. The bureau issues numerous modestly priced publications of its own, and also sells publications on Montana geology issued by various other organizations. All are available by mail order or through direct sale at the bureau offices. Write to the Montana Bureau of Mines and Geology, Butte, Montana 59701, for a current list.

Index